Material Aspects of Ferrofluids

Editors

R.P. Pant
CSIR-National Physical Laboratory
New Delhi, India
And
Academy of Scientific and Innovative Research (AcSIR)
Ghaziabad-201002, India

Vidya Nand Singh
CSIR-National Physical Laboratory
New Delhi, India

Komal Jain
CSIR-National Physical Laboratory
New Delhi, India

Arvind Gautam
CSIR-National Physical Laboratory
New Delhi, India

CRC Press
Taylor & Francis Group
Boca Raton London New York

CRC Press is an imprint of the
Taylor & Francis Group, an **informa** business

A SCIENCE PUBLISHERS BOOK

First edition published 2023
by CRC Press
6000 Broken Sound Parkway NW, Suite 300, Boca Raton, FL 33487-2742

and by CRC Press
4 Park Square, Milton Park, Abingdon, Oxon, OX14 4RN

CRC Press is an imprint of Taylor & Francis Group, LLC

Library of Congress Cataloging-in-Publication Data (applied for)

ISBN: 978-1-032-22798-6 (hbk)
ISBN: 978-1-032-22802-0 (pbk)
ISBN: 978-1-003-27424-7 (ebk)

DOI: 10.1201/9781003274247

Typeset in Times New Roman
by Radiant Productions

Foreword

Right from its inception, magnetic fluids also now known as ferrofluids have created a deep interest amongst researchers working in different fields. A ferrofluid is composed of a large number of magnetic nanoparticles, each either coated with a layer of surfactant or charged stabilized, dispersed in a suitable liquid. Its physical as well as fluid mechanical properties can be controlled by an external magnetic field. These characteristics have led to several novel applications in different branches of engineering and technology. From 1978, every 3 yr international conferences are organized in different countries and its proceedings are published. A more recent development is among applications in bioengineering and biomedicines. Several books are also published. The last one, was in 2013. So, after nearly a decade the present book is published and is devoted to Material Aspects of Ferrofluids. It covers 11 large chapters on synthesis, different physical properties like rheological, optical, thermal and biocompatible and some applications of ferrofluid. A majority of the authors have more than two decades of experience in their respective fields. I hope that the book will be useful for professionals as well as research students.

R.V. Mehta

Preface

Nanoengineering is manipulating the countable number of atoms and molecules, making them efficient, innovative materials for contemporary technology. In nanomagnetism, the finite size has attracted the attention of scientists, whether for quantum dot or 2D or nano size particles for spin dynamics in a different orientation. The development of the nano-object having zero defects is the ultimate goal in engineering to make materials for high precession technology. Ferrofluids are manufactured materials with magnetic and liquid characteristics and generate a field gradient. It is a liquid suspension of nanomagnetic particles that act like a liquid magnet and moves up, down or around the obstacles with the influence of an external magnetic field. The study of ferrofluids encompasses basic science like physics, chemistry, material science and mathematics, as well as applied science like engineering and biomedical. The physical properties of these materials are based on the lattice structure, shape, size, density of magnetic particles, functionalization, dispersing medium, coating, etc. The research on ferrofluids was initiated nearly six decades ago and has received interest in a wide range of scientific and technological aspects. Some of the fascinating details of these fluids are detailed in this book. The book comprises 11 chapters which cover a blend of basic research on the technological applications of ferrofluid. The chapters cover a relatively large spectrum of current research activity of magnetic fluid from physicochemical, thermal, rheological biomedical and engineering applications.

Chapter 1 introduces ferrofluids' basic science, physical characteristics and various synthesis methods. It gives an insight into the structural of magnetic properties and stability factors in stable fluid synthesis. The magnetically induced optical properties of ferrofluids materials are promising for designing the devices like optical delay, isolators, polarization modulators, sensors, etc. Therefore a concise account of the optics of magnetic fluids is described in Chapter 2. Similarly, Chapter 3 extends the magneto-optical phenomenon in non-linear effects in the presence of magnetic and electric fields pertaining to shape, size, particle density and hyperpolarizability in ferrofluids. The variation in thermal properties of ferrofluid and its magnetic field dependency on microstructure evolution are discussed in Chapter 4. Chapter 5 deals with charge stabilized ionic ferrofluids, their preparation, colloidal structure and evolution of thermo-diffusion in micro millimeters by Rayleigh scattering for stabilization. Chapter 6 discusses the superior tribological properties of ferrofluids, faming as nano lubricants and gives a detailed experimental account of various parameters affecting these properties. In continuation to the flow properties of ferrofluid, Chapter 7 describes the effective controlled magnetic force ultrafine surface finishing process where the distribution of abrasives can also be controlled. Chapter 8 deals with the ferrofluid-based composites and provides a brief idea of the synthesis, characteristics and quality testing of these composites. Chapter 9 discusses using ferrofluids as potential candidates for bio-medicine, focusing primarily on cancer treatment using the magnetic hyperthermia technique. While Chapter 10 discusses the use of ferrofluids for energy harvesting based on the self-levitation, thermo-magnetic effects and MFs motions such as vibration, sloshing and sensing like temperature, magnetic field and inclination/tilt and thermal applications (cooling and thermal energy storage and conversion). It details various ferrofluid devices and their components and their realization. Lastly, Chapter 11 discusses the use of ferrofluid and their composites system as bioinspired magnetic nanoparticles, textiles, fibers, transformer oil, liquid crystal, etc.

In a book of this type, it is tempting to describe the various developments in their historical sequence. For the reader's convenience, however, we preferred maintaining continuity of thought and describing each aspect of the problem in a separate chapter. We hope this book will interest ferrofluid scientists, young researchers, innovators, active researchers and engineers working in nanomagnetic fluids.

Prof. R.V. Mehta is an eminent Indian physicist and founder of ferrofluid research in India and very kindly agreed to our request to write a chapter. We sincerely thank Prof R.V. Mehta for his interest and encouragement in this book, and especially for writing a forward to this book.

We are immensely thankful to all the authors for their contributions, comments and stimulating discussions. We cannot forget the publisher's kind, patient assistance in preparing the printed version, for which we are sincerely grateful.

R.P. Pant

Contents

1

Design and Development of Ferrofluids Materials

Prashant Kumar,[1,2,3,*] *H. Khanduri,*[1] *Arjun Singh,*[2,5]
Komal Jain,[1] *Saurabh Pathak,*[4] *Arvind Gautam,*[1]
Lan Wang[3] *and R.P. Pant*[1,2,*]

1.1 Introduction

Magnetic materials are an essential part of modern technology. One such important material is ferrofluid (FF). The colloidal suspensions of ultrafine to nano-sized magnetic particles are termed ferrofluids or magnetic fluids. In 1963, S. Papell at NASA first developed a ferrofluid to solve the fuel pumping problem in space rocket engines [1–3]. To achieve this, he prepared a colloidal suspension of ultrafine magnetic particles and mixed it in fuel to control a fuel injection in a rocket using a magnetic field. It opened a new area in material science and technology viz where control of FFs using a magnetic field was established [3–5].

FFs move as a whole up-down or around the obstacle in response to an external magnetic field. This fluid shows a non-Newtonian nature under the influence of a magnetic field and becomes semi-fluidic with increased viscosity of many folds. These colloids have a large particle number density of $\sim 10^{23}/m^{-3}$ [6–8]. Since the particles are magnetic, there is always a possibility of aggregation and sedimentation

[1] Indian Reference Materials Division, CSIR-National Physical Laboratory, Dr. K.S. Krishnan Marg, New Delhi-110012.
[2] Academy of Scientific and Innovative Research (AcSIR), Ghaziabad-201002, India.
[3] School of Science, RMIT University, Melbourne, VIC 3000, Australia.
[4] National Creative Research Initiative Center for Spin Dynamics and SW Devices, Department of Materials Science and Engineering, Seoul National University, Seoul 151-744, South Korea.
[5] Department of physics, Indian Institute of Technology, Jammu-181221.
* Corresponding authors: prashantkhichi92@gmail.com, rppant@nplindia.org

of particles due to Vander Waal's magnetic force of attraction. To overcome this problem, each of the particles is coated with a layer of a suitable surfactant or charge to stabilize. Thus, it comprises three components: solid magnetic nanoparticles, carrier medium and surfactant. The choice of these components wholly depends on the point of use [8–10].

FFs are essential materials due to the control of properties and behavior under the influence of an external magnetic field. They are widely used in many practical and industrial applications such as Magnetic Resonance Imaging (MRI), magnetic sealing, aerospace, data storage, loud speakers, etc. [10–13]. FFs have unique sensational and actuation characteristics, making them suitable for applications like pressure, temperature, vibrational sensors, etc. They also show magneto-optical applications based on dichroism and birefringence. Biomedicals is another area where FFs and MNPs are showing their tremendous potential. Due to their small size, can quickly penetrate through blood and tissues [13–15]. They can be used for targeted drug delivery at specific sites by providing a suitable coating layer. Other applications are magnetic hyperthermia-based cancer treatment to kill cancerous cells under an external a.c. field [16–19].

The properties and applications of FFs are entirely different from their companion Magnetorheological Fluid (MR fluid). MR fluids contain single/multi-domain micron-sized magnetic particles (Fig. 1.1). These particles have high magnetization and retain some amount of coercivity and retentivity. Being a suspension of large particles, MR fluids are unstable and separate into different components very quickly. On the other hand, FFs have less magnetization, negligible coercivity and retentivity showing superparamagnetic nature [5, 20–23].

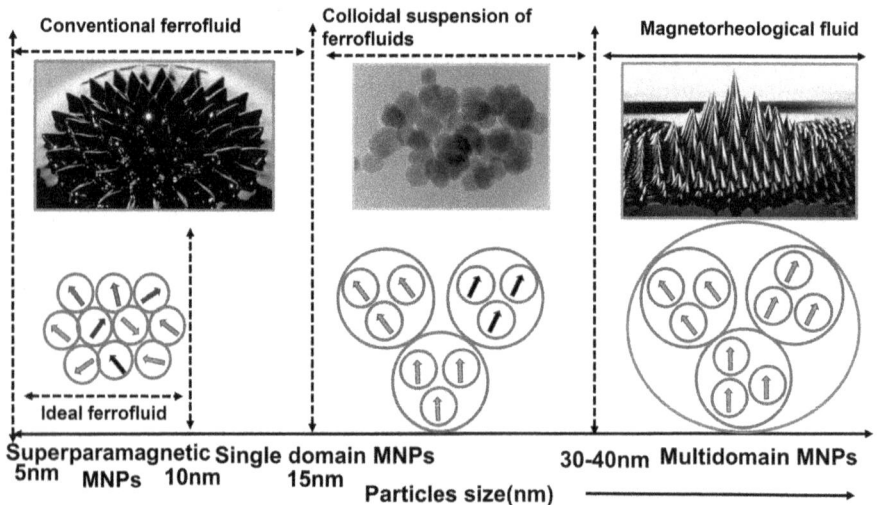

Figure 1.1. Schematic representation of the formation of ideal ferrofluid (size ranges from 5–10 nm) with a single domain superparamagnetic behavior, conventional ferrofluid (size 15–30 nm) with single domain magnetic particles, whereas the formation of magnetorheological fluid (MRF) with multi domain magnetic structure with large size particles generally greater than 30 nm.

1.2 Ferrites

Magnetic particles in ferrofluids are mainly made up of ferrites. These are ionic compounds of certain iron oxides and other metal ions. The magnetic properties of ferrites depend on the metal ions they contain. Several ferrites are naturally occurring minerals in rocks with impure states [24–26].

The classification of ferrites is of two kinds: the first one is based on magnetic properties such as coercivity, retentivity, etc. Based on magnetic properties: Ferrites are of two types, soft ferrites and hard ferrites and the second classification is based on the type of crystal structure of ferrites materials. Based on structural parameters, ferrites are of four types: Spinel ferrite (AB_2O_4), hexagonal ferrites ($AFe_{12}O_{19}$), garnets ($A_3Fe_5O_{12}$) and orthoferrites ($AFeO_3$) [24, 26–28].

1.2.1 Based on magnetic properties

1.2.1.1 Soft ferrites

They are kinds of ferrimagnetic compounds that can be easily magnetized and demagnetized due to small coercivity, high permeability and low magneto crystalline anisotropy. They have a wide range of temperature stability. Figure 1.2 (b) represents the magnetization curve showing a small hysteresis loop. These lead to small energy losses in such kinds of materials. They are synthesized by heating and then cooled slowly [28–30].

Soft ferrites are ceramic insulators or semiconductors with cubic crystal structure MFe_2O_4, where M is transition metal ions, such as Co, Mg, Ni, Mn, Zn, etc. Due to low energy losses, soft ferrites are used as cores in electromagnets and transformers, computer memories, ferrofluids, etc. The main advantage of soft ferrite material is that it can operate over wide ranges of frequencies and possess low eddy current losses. Fe_3O_4, $MnFe_2O_4$, $ZnFe_2O_4$, $NiFe_2O_2$ and $CuFe_2O_4$ are some examples of soft ferrite [31–34].

1.2.1.2 Hard ferrites

They are ferrimagnetic compounds that retain permanent magnetism and require a strong magnetic field to magnetize and demagnetize. The broad magnetization curve shown in Fig. 1.2 (a) possesses large coercivity and retentivity. These materials show higher energy losses than soft materials due to the larger hysteresis curve. They are synthesized by heating and sudden cooling. These hard ferrites play a dominant role in the permanent magnet market due to the low price per unit available. The most important hard ferrites, are barium ferrite ($BaO.6Fe_2O_3$), strontium ferrite ($SrO.6Fe_2O_3$) and M-type ferrites [31, 34, 35].

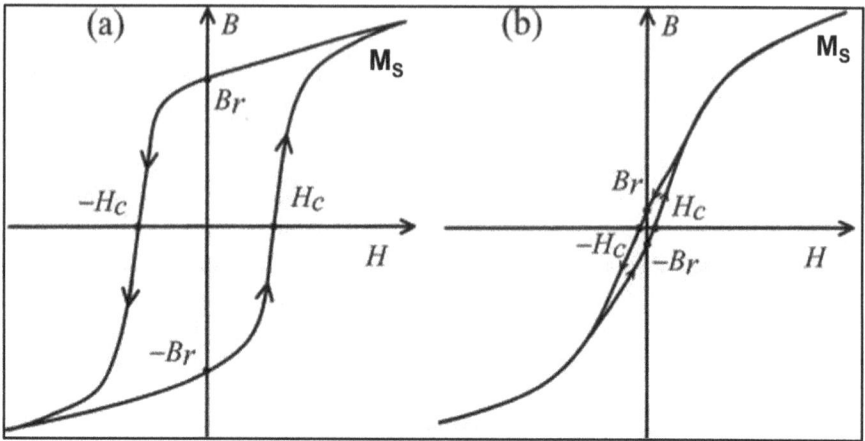

Figure 1.2. Schematic representation of hard and soft ferrite (a) shows the magnetization curve of hard ferrite with large coercivity and retentivity. They offer significant energy losses due to hysteresis and large magneto crystalline anisotropy. (b) shows the magnetization curve of soft ferrite with small coercivity and retentivity. They offer fewer energy losses due to small hysteresis and show low magneto crystalline anisotropy.

1.2.2 Based on the crystal structure of ferrites

Based on the crystal structure, the entire family of ferrites can be categorized into four types, i.e., spinels, garnets, hexaferrite and orthoferrites. Every ferrite has its own unique identity and properties useful for various technological applications. The molar ratio and description of different ferrite compositions are given in Table 1.1.

The spinel structure is named after the mineral spinel ($MgAl_2O_4$ or $MgO.Al_2O_3$), having the same structure which crystallizes in the cubic system. A spinel ferrite is made of a complex oxide crystal structure with a face-centered cubic structure of oxygen and metal ions located at interstitial sites with a general unit formula as MFe_2O_4. M is a divalent metallic cation such as a transition or post-transition metallic cation (M = Mn, Mg, Co, Ni, Zn). The crystal of the spinel has two lattice sites, namely tetrahedral (A) and octahedral (B), occupied by the cations of different valences [36–38].

Garnet ferrites possess the structure of the silicate mineral garnet. These ferrites crystallize in the dodecahedral or 12-sided structure with large trivalent rare-earth ions having significant magnetic moments. The general formula is $R_3Fe_5O_{12}$, where R = Sm, Y, Eu, Gd and Tb are rare earth elements with all metal ions in a trivalent state only [32, 39].

Table 1.1. Different types of ferrites, i.e., spinel, garnet, hexagonal and ortho ferrites.

Sr no.	Types	Molar ratio	Description
1	Spinel	$FeO.Fe_2O_3$–MO	MO is a transition metal oxide
2	Garnet	$R^{3+}Fe_5O_{12}$	R = Er, Sm, Y, Eu, Gd, and Tb
3	Hexagonal	$A^{2+}Fe_{12}O_{19}$	A^{2+} = Ba, Ca, and Sr
4	Ortho	$RFeO_3$	R is yttrium or a rare earth ion.

The important types of hexagonal ferrites are M-, Z-, Y-, W-, X- and U-type ferrites. Most of these ferrites fall under ferrimagnetic, and few are antiferromagnetic. These ferrites get their name from the hexagonal magnetoplumbite structure. It has hexagonal symmetry with a major preferred axis called the c-axis and a minor axis called the a-axis. The best-known compounds in this class are the following:

- M-type ferrites, such as $BaFe_{12}O_{19}$, Z-type ferrites $(Ba_3Me_2Fe_{24}O_{41})$, Y-type ferrites $(Ba_2Me_2Fe_{12}O_{22})$, W-type ferrites $(BaMe_2Fe_{16}O_{27})$, X-type ferrites $(Ba_2Me_2Fe_{28}O_{46})$, U-type ferrites $(Ba_4Me_2Fe_{36}O_{60})$, such as $Ba_4Co_2Fe_{36}O_{60}$ or Co_2U [12, 17, 40, 41].

Last, in the category, the ortho ferrites given by the formula $RFeO_3$, where R is yttrium or a rare earth ion. These ferrites have distorted perovskite structures. Here Fe ions are situated at the center of the octahedral formed by linking with six near oxygen anions and R cations occupying the octahedral interstitial spaces. Most studies on ferrofluids are based on spinel ferrites. Therefore, spinel ferrite-based ferrofluids have been discussed in detail in this chapter.

1.2.2.1 Spinel ferrites

A spinel ferrite has an oxide crystal structure with a Face-Centered Cubic (FCC) core with a unit formula of MFe_2O_4 (M = Mn, Mg, Co, Ni, Zn). Depending on the chemical nature, charge state and stabilization energy, the metal cations can be arranged in two different crystallographic sites: tetrahedral and octahedral. There are 64 tetrahedral and 32 octahedral interstitial sites made by oxide FCC structure. Out of these, eight tetrahedral sites are occupied by M^{2+} ions and 16 octahedral sites by Fe^{3+} are shown in Fig. 1.3 [42–44]. Based on the interstitial sites occupied by the cations, it can be

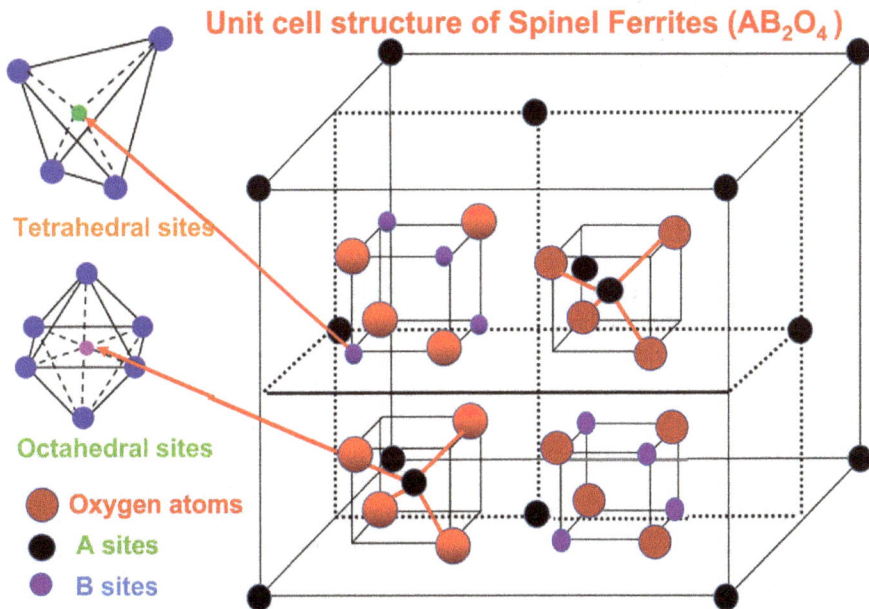

Figure 1.3. Cubic spinel structure of nano ferrites and cations occupancy of magnetic ions in each site, i.e., tetrahedral(A) and octahedral(B) sites.

classified into three groups: normal spinel ferrites, inverse spinel ferrites and mixed spinel ferrites. The site occupation depends on the electrostatic contribution to the lattice energy, the cation radii, cation charges and crystal field effects. In normal/ standard spinel ferrite structures, M^{2+} cations are located at tetrahedral sites and Fe^{3+} cations are found at octahedral sites, for example, in $ZnFe_2O_4$ [36, 44, 45]. While in inverse spinel ferrites, Fe^{3+} cations are distributed equally between tetrahedral and octahedral sites and M^{2+} cations occupy octahedral sites only, for example, $NiFe_2O_4$, $CoFe_2O_4$, Fe_3O_4, etc. Finally, mixed spinel structure, M^{2+} and Fe^{3+} cations, randomly occupy both sites, as in $MgFe_2O_4$ and $MnFe_2O_4$ [28, 30, 32].

1.2.3 *Physiochemical properties of Spinel ferrites*

The ferrites properties depend on the preparation methods, chemical structure and compositions. The variations in the properties can be achieved by doping a few impurities as per the applications in consideration. The following are a few physical properties of the spinel ferrites [28, 32, 46, 47]:

(1) Generally, black or gray as the absorption occurs mainly in the visible range.

(2) A high dielectric constant of the order 10^3 at lower frequencies and falling to about 10 to 20 at microwave frequencies due to the close-packed structure of oxygen ions.

(3) Difficult to measure the melting point due to losing oxygen at high temperatures.

(4) Non-conductors of electricity act as semiconductors under the influence of an electric field.

(5) Very hard and brittle owing to their crystal structure and strong ionic bonds.

(6) Show good magnetic properties such as;

> ➢ Magnetic anisotropy: Most of the ferrites have an easy axis of magnetization.
> ➢ Hysteresis and permeability: Ferrites have either thin hysteresis loops or square loops. The ferrites with a thin loop are classified as soft ferrites and used in devices such as transformers and inductors. While ferrites with square loops are called hard ferrites and are used in memory and switching devices.

1.3 Ferrofluids

Ferrofluids are a new class of magnetic materials that combine fluid and magnetic properties comparable to those of a solid magnet. The combination of fluidity and the capability of interacting with magnetic fields are achieved in magnetic nanofluids because of their composition. Magnetic nanofluids constitute single domain particles suitably dispersed in a carrier liquid medium (two-phase) with the help of a surfactant (three-component). The two phases of this system are the solid magnetic phase (called a dispersed phase), which is colloidally dispersed in a magnetically passive liquid phase (continuous phase) [1, 48, 49]. Figure 1.4 shows the various components of ferrofluid. The three components required to prepare the magnetic nanofluids are:

1. Magnetic phase/nanoparticles
2. Carrier liquid
3. Surfactant (dispersant)

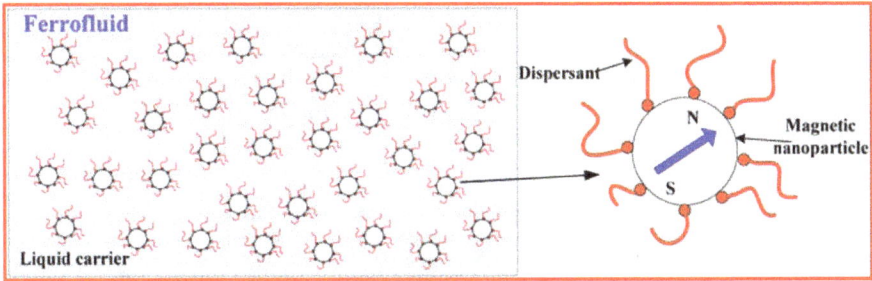

Figure 1.4. Shows various constituents of a magnetic fluid, i.e., carrier liquid, dispersant and magnetic nanoparticles.

1.3.1 Magnetic nanoparticles

The size of the magnetic nanoparticles to be dispersed should be the single domain which generally ranges from 10–50 nm [21, 35]. The stability criterion restricts the size. The magnetic nanoparticles should be small enough to be suitably dispersed in a carrier liquid but not too small. Since the size is less than 10–20 nm, their magnetic properties tend to disappear. Therefore, it is advantageous to develop the ability to prepare particles with a high level of magnetizability, having controlled size distribution while still retaining uniformly other important properties such as composition, crystallinity and morphology [50–53].

1.3.2 Carrier liquid

Carrier liquid plays a dominant role in deciding the suitability of magnetic nanofluid for different applications, as it governs the overall physical properties of a magnetic nanofluid. A carrier liquid should be chosen to cater to the need of the specific applications. Some common carrier liquids are water, hydrocarbon, ester, mineral, silicon oil, fluorocarbon, etc. [2, 7, 11]. The properties of carrier liquids depend on the ultimate use. In general, the carrier liquid must have the following characteristics.

➢ It should be magnetically passive and chemically stable.

➢ Its vapor pressure, viscosity and other physical properties must be chosen according to the type of application.

➢ It should be compatible with the surfactant.

Low evaporation rate, low viscosity and chemical inertness are desirable properties for the carrier liquid [10, 54, 55].

1.3.3 Surfactant or dispersant

The requirement for a surfactant is rather specific because it must prevent magnetic nanoparticles from aggregating. It should have an anchor group with a high affinity for the surface of the particles and a flexible tail of a proper length, which is soluble in the carrier liquid. The surfactant, therefore, has to be tailored to the particle surface as well as to the carrier liquid. The surfactant should also be chemically stable in the environment in which it is used. For sterically stabilized magnetic nanofluid, steric repulsion between particles is usually achieved by coating the surface of the nanoparticles with a long-chain monomolecular layer of an appropriate surfactant, which accounts for the stability of a magnetic nanofluid [8, 10, 56].

Molecules of a simple surfactant consist of a linear chain of non-polar hydrocarbon atoms at one end and an anchor polar group of atoms at the other. The anchor group is called the polar head. The chain is called a non-polar tail. The surfactant is so chosen that its molecules interact with the magnetic nanoparticles through a bond of functional groups to form a tightly bonded monomolecular layer around the particles. The long-chain fragment of surfactant molecules must resemble that of the carrier liquid, so that the thermal motion is not inhibited. Oleic acid is the most commonly used dispersing agent for hydrocarbon-based, with relatively low molecular weight in magnetic nanofluid. Oleic acid is used as the surfactant, so studying the mode of adsorption of surfactant becomes mandatory [2, 11, 46, 57–59].

1.3.4 Mode of adsorption of surfactant

For preparing the magnetic nanofluid having heptane (less viscous and volatile hydrocarbon) as the carrier liquid, oleic acid is used as a surfactant. Oleic acid contains a polar head and a non-polar tail.

Non-polar tail   **Polar head**

$$CH_3(CH_2)_7\ CH=CH(CH_2)_7\ \underset{\displaystyle \underset{O}{\|}}{C}\text{-OH}$$

Oleic acid structure-Tail—Non-polar end (hydrocarbon chain) and Polar head-carboxylic group

The polar head reacts with the surface-bound -OH- ions and gets adsorbed on the nanoparticle's surface and the non-polar chain provides a hydrophobic sheath to individual nanoparticles. Expecting the anchor part, the adsorbed molecules (tails) perform thermal movements. When a second particle approaches closely, the chains have to bend aside, and their motion is restricted severely. Thus, long chain molecules come close together when magnetic nanoparticles are coated with a layer or shell of surfactant. A repulsive force (referred to as steric force) appears, thereby preventing the aggregation of nanoparticles and avoiding sedimentation [1, 8, 10, 11, 60].

1.3.5 Selection of different components of ferrofluid

The magnetic properties of nanofluids are determined by the weight content of solids materials, which can be 25%. The choice of magnetic materials is quite broad. Still, producing colloidal suspensions with sufficient stability requires good compatibility of MNPs with a surfactant and carrier liquid, which significantly reduces the range. An essential component of the magnetic fluid is a carrier liquid, which can be polar or nonpolar. The function of the carrier liquid is to provide an environment where the particles of magnetic material are "suspended". The choice of the liquid depends significantly on the MNPs application [4, 36, 37, 61]. To date, a broad spectrum of carrier liquids is used: water, ethanol, pentanol, glycols, perfluoropolyether, synthetic complex esters, transformer oils, freons, styrene, methyl ethyl ketone, naphtha, various synthetic hydrocarbons and organic solvents, such as heptane, benzene, toluene, mineral and organic-silicone oils, vegetable oils (sunflower, rapeseed, castor oil) and silicone oils. The boiling point, the vapor pressure at increased temperatures and the freezing point are the necessary parameters for choosing such a liquid. The carrier liquid must not react chemically with the magnetic phase and materials used in the device [14, 31, 34, 62].

The final component of MNPs is a stabilizer of magnetic nanoparticles in the carrier liquid. In a volume, the MNPs are subjected to intensive Brownian motion and can collide. The substance-stabilizer is required to prevent the coagulation of magnetic nanoparticles with a collision at the expense of magnetic forces or Van der Waals forces and prevent this collision in general. Stabilizers also prevent oxidation of the surface of magnetic nanoparticles. Usually, stabilizers are surfactants or polymers, adsorbed physically or chemically on the nanoparticle surface [9, 41, 63–65]. Oleic acid is the most common surfactant used to stabilize the magnetic nanoparticles, obtained by conventional chemical methods based on rapid neutralization of di- and trivalent iron salts by an excess of an aqueous solution of alkali. Wine and citric acids, derivatives of fatty acids, sodium oleate, dodecyl amine, sodium carboxymethylcellulose and other chemicals are also used for these purposes. Functional properties of MFs define the use of certain surfactants. Thus, the chemical substance is used to stabilize nanoparticles in magnetic fluids for biological and medical purposes [16, 18, 66].

1.3.6 *The motion of magnetic nanoparticles in a ferrofluid*

In a gradient field (∇H), the whole fluid responds as a homogeneous magnetic liquid, which moves to the region of the highest flux. It means that magnetic nanofluids can be precisely positioned and controlled by an external magnetic field (H) [2, 3, 67]. The forces holding the magnetic fluid are proportional to the external field's gradient and the fluid's magnetization value. It means that the retention force of a magnetic nanofluid can be adjusted by changing either the magnetization of the fluid or the magnetic field in the region. Magnetic colloids contain a single magnetic domain and have a permanent magnetic moment proportional to their volume. Although magnetic colloids are ferromagnetic on the molecular scale, they resemble a paramagnet on the colloidal scale, with a significant difference that the magnetic moments of magnetic colloids are much larger than the magnetic moments in paramagnets with typical values of order 10^{-19} Am^2 for magnetic colloids compared to 10^{-23} Am^2 for paramagnets. Figure 1.5 schematic shows the structure of coated magnetic nanoparticles and oleic acid structure.

For this reason, magnetic particles constituting magnetic nanofluids are sometimes called SP. To be SP, the dipole moment of each particle must be free to rotate on the time scale of experiments [20, 22, 68]. Two modes of rotation are

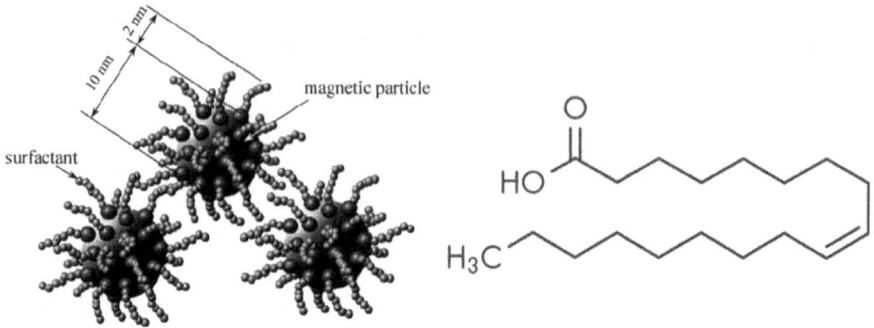

Figure 1.5. Schematic view of coated magnetic nanofluid particles and oleic acid structure [69].

operative in magnetic colloids. One is the Brownian rotation and the other is the Néel rotation, which will be described later in this chapter.

The stability of the magnetic nanofluid depends on the type of different interactions that exist between magnetic nanoparticles. Various attractive and repulsive forces exist between the nanoparticles in the dispersed phase of magnetic nanoparticles in a magnetic nanofluid.

1.4 Various interactions in Ferrofluid

In ultrafine-magnetic nano-particle systems, different kinds of interparticle interactions exist and the interaction strength varies with the concentration of the volume. The different types of characteristics of ferrofluid materials due to these interactions have been described here.

1.4.1 Magnetic interparticle interactions (attraction) or Dipole-Dipole interaction

The size of the particles usually encountered in a magnetic fluid is such that they act as superparamagnetic and single domains. Thus, they are magnetized fully in one direction, which can fluctuate thermally. The magnetic attraction energy for two equal spheres with their dipoles along the line is given as follows [7, 17, 49]:

$$E_M = -\frac{\pi M^2 \, d^3}{72\mu_o \, x(1+h)^3} \tag{1.1}$$

where $h = x/d$; x is the center-to-center distance, d is the diameter of the particles and M is magnetization in an external field. In the absence of a magnetic field, the dipole rotates thermally. The limiting value of $E_m < k_B T$ is

$$(E_M)_{H=0} = -\frac{E_M^2}{6k_B T} \tag{1.2}$$

It is a long-range potential and varies slowly with distance. It can be easily seen that for small particles, e.g., 100Å for Fe_3O_4, randomizing Brownian energy is always predominant over magnetic attraction. Hence, thermal agitation prevents agglomeration due to this attraction.

In the single /multi-domain size magnetic nanoparticles system, each nanoparticle behaves like a tiny magnet and interacts with each neighboring dipole. As a result, dipole-dipole interaction energy becomes dominant when nanoparticles are near. This interaction is attractive and long-range compared to van der Waal's interaction as E_{dd} is proportional to $1/R^3$ [9, 70].

Therefore, overcoming this attractive interaction by a repulsive potential is necessary to keep nanoparticles in a stable colloidal state. In a magnetic nanofluid, two different mechanisms are used to provide the repulsive potential viz; (1) Steric repulsion. (2) Electro-static repulsion.

1.4.2 Vander Waal's force (attraction)

This type of Vander Waals force appears between the neutral particles because of the fluctuating electric dipoles. For two equal spheres, van der Waal's attraction energy (E_v) is calculated by Hamaker and is given by the Equation [1, 19, 70].

$$E_V = -\frac{A}{6}\left[\frac{2}{S^2-4}+\frac{2}{S^2}+ln\left(\frac{S^2-4}{S^2}\right)\right] \tag{1.3}$$

where, $S=\left(\dfrac{2R}{D}+2\right)$ and Hamaker constant is $=\dfrac{3\pi NP^2}{4}hv_o$, here, N' is the number of atoms per m³, and the value of Hamaker constant A is taken as 10^{-19} nt.m. R and D are the distance between and diameter of magnetic nanoparticles, P is the polarizability and hv_o is ionization energy. It is an attractive short-range interaction irrespective of size and is proportional to $1/R^6$. Therm agitation does not prevent aggregation since particles, irrespective of size, have a considerable negative potential energy when in close contact.

1.4.3 Steric repulsion

Steric and entropic repulsion is encountered with particles that have long flexible molecules attached to their surface. The molecules can be simple linear chains with a polar group at one end, e.g., oleic acid or long polymer with many polar groups along the chain, so adsorption occurs along the loop. Except for the anchor part. The adsorbed molecules perform thermal movements. When a second particle approaches close by , the position of the chains can take is restricted. A rigorous calculation of the repulsion energy for adsorbed molecules is a difficult task and therefore, a simple estimate for the entropic effect for the short chains, which are often used in magnetic fluids, is carried out and given by the Equation [8, 12, 19] (4).

$$E_a =\frac{2}{3}\pi Nk_B T\left(\delta-\frac{x}{2}\right)^2\frac{(1.5d+2\delta+\frac{x}{2})}{\delta} \tag{1.4}$$

where N is the number of molecules anchored per unit area; δ is the thickness of the stabilizing layer and x is the distance between two surfaces.

1.4.4 Electro-static repulsion (Ionic magnetic nanofluid)

In the case of ionic magnetic nanofluids, the stability is maintained by an ionic surfactant, such as, TMAH, with water as a carrier medium. In this case, magnetic nanoparticles are provided an electric charge; therefore, a Coulomb repulsive potential keeps these nanoparticles away from each other. If this repulsive potential exceeds the attractive one, the magnetic nanofluid will remain stable. Such magnetic fluids are known as ionic magnetic nanofluids. In such magnetic nanofluid, the magnetic nanoparticles usually acquire electrical charges due to [7, 49, 71]:

 (i) Ions leaving the surface of the nanoparticles.

 (ii) Adsorption of ions formed by the liquid, e.g., OH⁻ or H⁺ in an aqueous medium.

(iii) Adsorption of ions formed by a substance when dissolved in liquid, e.g., phosphate, fatty acid, etc.

The electric repulsion between charged magnetic nanoparticles is the coulomb repulsion, which is controlled by the screening action of the surrounding counter ions in a liquid. In the colloid world, the counter charge layer is called the 'double layer', as shown in Fig. 1.6.

The coulomb repulsion is effective only when the magnetic nanoparticles are close; hence, the double layer gets overlapped. This result in a repulsive force and its strength depends on [5, 11]:

 (i) The size of the magnetic nanoparticles,

 (ii) The concentration of ions in liquid

(iii) The surface potential (usually 10–100 mV).

Specifically, TMAH as an ionic surfactant coats the magnetic particles with hydroxide anions, which attract tetramethylammonium cations, forming a diffuse shell around each particle and creating repulsion between particles. Both surfactant and ionic magnetic fluids are stable and the overall stability can be calculated theoretically while assuming certain physical parameters. The net interaction curve is an algebraic sum of Van der Waals attractive energy, attractive magnetic energy and steric repulsion energy, whose decisive roles are shown in the figure below.

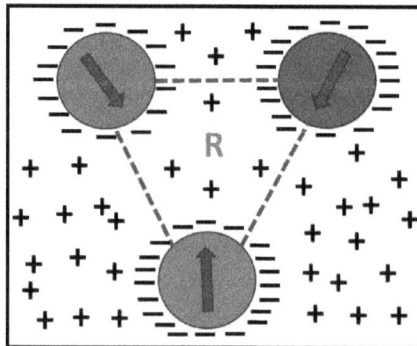

Figure 1.6. Ionic magnetic fluid where each magnetic particle has to be the same surface charge so the electrostatic repulsion occurs. The presence of the countercharge leads the system to be neutral.

1.4.5 Net potential energy

As described above, the resultant interparticle forces between two or more particles will decide the stability of magnetic nanofluids' colloidal suspension. For particles of size 10 nm and surfactant coating thickness of 2 nm, a variation of interaction energy is calculated, and the same is shown in Fig. 1.7. The dark line shows the net potential energy curve (ET) resultant. In order to prevent agglomeration $E_T > 25$ kT [9, 19, 67]. The effect of the particle size on the net potential energy curve is shown in Fig. 1.7, for surfactant thickness (δ) ~ 2 nm, and the number density of particles is 10^{18}/m^3. The following conclusion is drawn from Fig. 1.7:

1. For ~ 5 nm, particle size pertains only to repulsion, even at a short distance.

2. For ~ 10 nm, particle size has the minimum potential energy at about 4 nm; therefore, the thermal energy keeps them in a stable dispersed state.

3. For ~ 15 nm, particle size possesses minimum potential energy, so deep and the slope is so steep that all the particles end up in aggregates regardless of the magnetic field.

Thus, the stable magnetic nanofluid possesses both fluidic and magnetic behavior. To explain the magnetohydrodynamic behavior of these magnetic nanofluids, Rosenzweig gave the hydrodynamic equation known as ferrohydrodynamic. Next the essential aspects of this ferrohydrodynamic [12, 24] are briefly described.

1.5 General aspects for stability in ferrofluids

The stability of these magnetic nanofluids or colloids is a crucial issue that can be considered in a physicochemical view. The critical energy terms per particle in this regard can be written as follows: Magnetic energy = μ_oMHV, Thermal energy = k_BT, Gravitational energy = $\Delta\rho$VgL, where k_B is Boltzmann's constant, T is temperature, μ_o is the permeability of free space, V is the particle volume and L is the elevation in a gravitational field [15, 72]. Each of these factors is briefly described below.

1.5.1 Stability in a magnetic field gradient

When the magnetic field is applied, MNPs are attracted to the higher-intensity regions of a magnetic field. At the same time, thermal motion counteracts the force of the field and provides statistical motions that allow particles to move all the portions of fluid. The magnetic energy μ_oMHV represents the reversible work in removing a magnetized particle from a point in the fluid, where the field is H, to a point in the fluid that is outside the field:

$$w = -\int_{H}^{0}\left(\mu_o \, M \frac{dH}{dS} \, V\right) dS \approx \mu_o \, MH \qquad (1.5)$$

When some part of the fluid is located in a field-free region, then stability against segregation is favored by a high ratio of the thermal energy to the magnetic energy [9, 69]:

$$\frac{K_B T}{\mu_o MHV} \geq 1 \tag{1.6}$$

By assuming a spherical-sized particle of diameter d, the above expression for maximum particle size becomes

$$d \leq \left(\frac{KT}{\mu_o MHV}\right)^{\frac{1}{3}} \tag{1.7}$$

Under normal conditions, the value of d is ~ 10 nm for stable dispersion in the field gradient.

1.5.2 *Stability against gravitational sedimentation*

The relative influence of gravity on magnetism can be understood by taking the ratio of respective energies,

$$\frac{\Delta \rho VgL}{\mu_o MHV} = \frac{gravitational\ energy}{magnetic\ energy} \tag{1.8}$$

Under normal conditions, the ratio is near 0.047, indicating its lesser contribution in making the fluid unstable than the gradient effects produced by a magnetic field [7, 72].

1.5.3 *Stability against agglomerative behavior*

A typical magnetic fluid consists of 10^{16} particles/mL, where particle collisions are frequent. The energy required to separate a dipole-dipole pair of particles with a diameter, d with surface-to-surface distance, scan be written as [2, 67, 73].

$$E_{dd} = \frac{1}{12} \mu_o M^2 V \tag{1.9}$$

But thermal energy is available to disrupt these agglomerates with the effectiveness of disruption governed by the ratio.

$$\frac{24kT}{\mu_o M^2 V} = \frac{thermal\ energy}{dipole\text{-}dipole\ contact\ energy} \tag{1.10}$$

The ratio needs to be greater than unity for the stability of dispersion which limits the particle size given by [4, 8].

$$d \leq \left(\frac{144kT}{\mu_o M^2 \pi}\right)^{\frac{1}{3}} \tag{1.11}$$

For magnetite particles at room temperature, the value of d is close to 10 nm.

1.5.4 *Surface adsorption and steric stabilization*

Apart from the above physical factors, attractive Van der Waals interactions also prevail. To prevent the particles from coming close to each other, they are coated with long chain molecules. The polar groups of adsorbed species associated with the particle surface, either physically/chemically with tails, are chosen similarly to the surrounding fluid matrix. The chemical compatibility of the surfactant and carrier medium is a basic requirement while dispersing particles of ferrofluid. The overall stability of ferrofluids can be calculated theoretically while assuming specific physical parameters. The net interaction curve is an algebraic sum of Van der Waals attractive energy, attractive magnetic energy and steric repulsion energy [69], whose decisive roles are shown in Fig. 1.7.

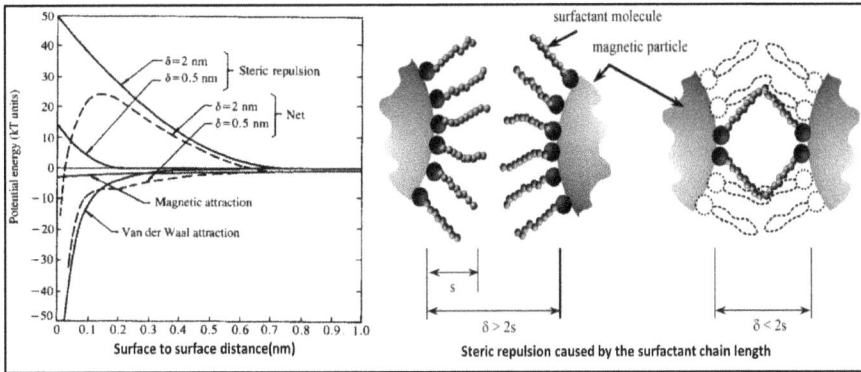

Figure 1.7. Potential energy versus surface-to-surface separation of sterically protected colloidal magnetite particles (d = 10 nm, N = 10^{18} molecules/m^3, δ is length of tail) [9,74].

1.6 Synthesis methods of mixed-ferrites based ferrofluid

Various synthesis techniques are adopted for synthesizing mixed ferrites-based magnetic nanofluid, i.e., top-down and bottom-up. The process and synthesis method of mixed ferrites mainly consist of two parts, the first is the preparation of the MNPs in the desired size, and the second is the stabilization of the dispersed MNPs in the suitable carrier medium. As for the synthesis of MNPs, various synthesis methods are reported in literature. These MNPs are widely used for NFs preparation and in multiple applications such as magnetic data storage and devices, optical sensor and actuators, catalysis, magnetic fluid hyperthermia, targeted drug delivery, etc. There are several methods for the synthesis of MNPs. In some studies, MNPs are categorized by the phase where the MNPs exist: solid, liquid and gas phase synthesis. In the solid phase, these MNPs are obtained through annealing and combustion in the atmosphere and vacuum. In the liquid phase, the MNPs are synthesized by chemical coprecipitation, hydrothermal, chemical reduction microemulsion, thermal decomposition, ball milling, microwave and solution-thermal methods. In the gas phase, they are produced by various methods such as spraying, pyrolysis, chemical gas-phase deposition, arc discharge and laser pyrolysis [14, 18, 31, 71].

The methods for preparing the MNPs can also be divided into two categories. The first category includes the methods for obtaining these MNPs by grinding the solid materials. The second one includes the opposite methods based on the MNPs assembly of atoms, ions and molecules. These approaches to produce MNPs are called the "bottom-up" and "top-down" approaches. Compared with the methods of production of MNPs by the principle of grinding, i.e., ball milling, the concept of "bottom-up" assembly has a larger number of opportunities to control the shape, size, structure and physical and chemical properties of MNPs. All the above methods apply simultaneous synthesis and dispersion of MNPs in a suitable carrier liquid, excluding such processing stages as drying, storage, transportation and dispersion of magnetic nanoparticles. This process minimizes agglomeration of MNPs and generates uniformly dispersed particles in the carrier medium. Its disadvantages are the presence of impurities, for instance, due to unreacted reagent residues and the products of material abrasion in the ball mills at grinding the compact material [22, 37].

For this reason, the bottom-up approach is simpler and more precise in synthesizing small nanoparticles less than 100 nm. The top-down approach is preferred for synthesizing thin films and nanoparticles whose size is larger than 100 nm. These two methodologies are shown in Fig. 1.8. There are several preparation methods of mixed ferrites-based MNPs such as chemical co-precipitation, hydrothermal, solvothermal, sol-gel, solid-state reaction, microwave, microemulsion, thermal decomposition, combustion and ball-milling have been applied for the synthesis of MNPs [43, 66, 75].

Usually, the first method for synthesizing MNPs and stable magnetic nanofluid was grinding the bulk materials in the ball milling technique, with the addition of a suitable surfactant, i.e., oleic acid and heptane. This prepared MNPs was stable for 1 yr.

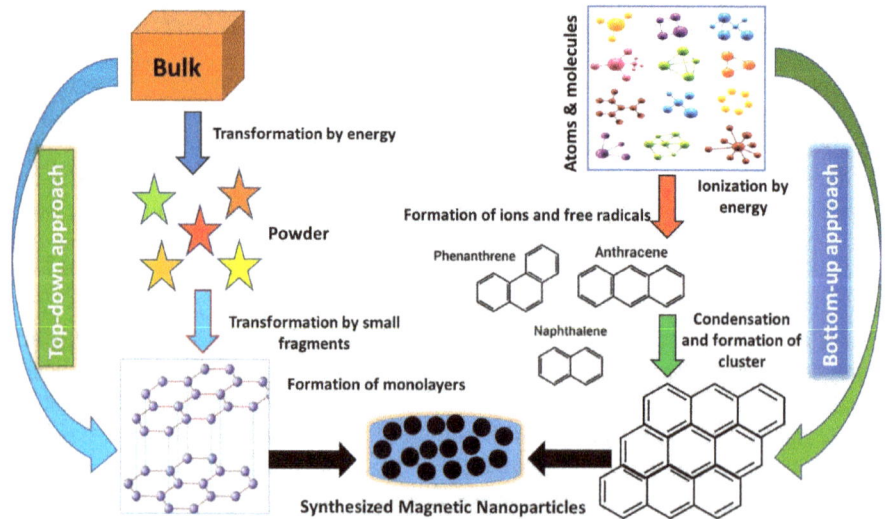

Figure 1.8. Synthesis mechanism of magnetic ferrofluid by top-down and bottom-up approach method.

The main disadvantage of this method is the long time of the MNPs production: from several days to several months, resulting in low productivity. Other disadvantages are as follows: contamination of the magnetic fluid by the product of ball abrasion and a wide spread of particles by size. However, this method was used for a long time and improved, so, despite its shortcomings, it is still in use. This method of MNPs synthesis is the one-step physical "top-down" method, carried out in the solid phase.

1.6.1 Top-down approach

In the top-down approach, the starting material is a bulk material of the same material to be synthesized and broken into smaller fragments or particles when an energy source is applied. The energy applied can be mechanical, chemical, thermal or could be another form of energy such as laser irradiation. In Pulsed Laser Deposition, the energy is absorbed by the material and transformed into chemical and/or thermal energy to break (inter) molecular bonds of the bulk material. This approach usually results in smaller flakes or particles with a wide size distribution, which is considered a disadvantage of the top-down method. It is possible to overcome this and synthesize nanoparticles of a precise size using a focused ion beam or lithography, but it requires expensive equipment. In the top-down technique, MNPs can be prepared by breaking down many materials. This method has many limitations, such as high synthesis temperature required, crystal defects, distribution of ions, inhomogeneous products and surface structure deficiency and the existence of layers in the materials [39, 75, 76].

The most common methods used in this approach are as follows.

- Laser pyrolysis (chemical gas-phase deposition)
- High energy ball milling (ceramics method)

1.6.1.1 Laser pyrolysis (Chemical gas-phase deposition)

In this chemical method, the vapor of metal in the gas phase can be obtained. In this chemical gas-phase deposition technique, the carrier gas flow delivers the precursors to the reaction chamber, where the vacuum conditions and high temperature (> 800–900°C) are maintained. The reactions occur in the high-temperature chamber, and the reaction products interact to form clusters or nanoparticles. The growth and aggregation of MNPs are suppressed by the rapid expansion of the two-phase flow at the outlet of the reaction chamber. Subsequent thermal treatment of particles in different gas media allows modification of the synthesized nanoparticles' composition, morphology and the crystal structure. In particular, the chemical gas-phase deposition method synthesizes the particles of iron oxide in the reaction of ferric trichloride with water at temperatures of 800–1000°C. The method of laser pyrolysis consists in heating the moving mixture of gases by continuous CO_2-laser for initiating and maintaining the chemical reaction until the achievement of the critical concentration in the reaction zone; homogeneous nucleation occurs there, and nanoparticles are formed. Nanoparticles formed during the reaction are entrained by the gas flow and trapped by the filter at the outlet [17, 32, 77].

1.6.1.2 High-energy ball milling (Ceramic method)

Ball milling is a solid-state synthesis mechanical technique commonly used to grind powders into fine particles. It is a user-friendly, environmentally friendly and a cost-effective process that demands mass-scale production in industries. As the name suggests, the ball milling technique consists of balls and a mill drum. Based on specific applications, there are different types of ball mills. However, most ball mills consist of a hollow cylindrical chamber rotating along their axis, which gets filled with free balls. Generally, these balls are made up of ceramics, steel, stainless steel, silicon carbide, tungsten carbide balls, etc., and provide a vast amount of mechanical energy to the material, which is expected to be crushed into a fine powder. The crushing is done by the impact and attrition between the balls and the powder. This process is done for several hours to get a uniform fine powder. Ball milling is an easier method that usually works on a top-down approach for synthesizing nanoparticles in a powdered form. The process starts with powdered materials of the desired composition having larger particle sizes (usually in microns). This powder is placed in a container in the presence of two or more heavy metals (usually, the ratio of 2:1 between the amount of material and mass of balls is preferred), which are free to move inside this vessel. The balls are generally made of hardened steel or tungsten carbide, and the container is very tight. The balls are then made to rotate around a predefined axis to apply a force on the material so it can be powdered into smaller particle sizes. There is a possibility of the addition of impurity, due to the balls and material of the container, which is reduced by providing an inert gas atmosphere. The efficiency is also reduced if the container is more than half filled by powdered material. As this process is mechanical force-based, heat (up to 1000°C) is generated, which can be controlled through cryocooling to the desired temperature. Too low a temperature can also lead to the formation of amorphous nanoparticles. Controlling the rotation speed of balls and the time of the milling process, the particle size can be tuned [18, 28, 64, 78].

1.6.2 Bottom-up approach

In materials science, nanomaterials are synthesized by interacting with atoms or molecular species through a set of chemical reactions provided by the techniques. The precursor is typically a liquid or gas that is ionized, dissociated, sublimated or evaporated and then condensed to form either an amorphous or/and crystalline nanoparticle. This approach produces nanoparticles with fewer defects, homogenous chemical composition, less contamination and particles with a narrow size distribution. In the bottom-up method, MNPs can be made by building up the atoms. The main advantages of this method are obtaining the homogeneous and uniform MNPs without any presence of impurity with the distribution of ions from their lattice sites that give the narrow size distribution of MNPs [57, 59, 79, 80].

The most common methods used in this approach are as follows.

- Chemical coprecipitation
- Hydrothermal
- Solvothermal

- Sol-gel
- Reverse Micelle
- Microwave-assisted

1.6.2.1 Chemical co-precipitation method

The co-precipitation method is the most common, prevalent method used for preparing MNPs due to the high yield capacity and simplicity in producing ultrafine MNPs production in a reasonable amount and cost in a relatively straightforward way. In this regard, co-precipitation is the most convenient, economical and less time-consuming, has high mass production and is frequently employed among all other methods to achieve uniform-sized MNPs. This technique can efficiently distribute large amounts of MNPs in an aqueous media. Using this technique leads to co-precipitation of the precursor materials in a solution at high or room temperature with the base addition under an inert atmosphere. The possibility of scaling up and achieving large quantities of MNPs are essential benefits of the co-precipitation route. As a result, the industry uses co-precipitation as the leading chemical technique to produce thousands of nanograms of nano-sized MNPs. However, the critical challenges of the co-precipitation method are the morphological control, crystalline quality and distribution of MNPs particle size [35, 42, 59].

The method is also commonly used to synthesize biocompatible MNPs, which have applications *in vivo* biomedical fields like hyperthermia and targeted drug delivery. This technique uses aqueous solutions of divalent and trivalent transition metal salts that are uniformly mixed in 1:2 mole ratios with continuous and vigorous stirring in an alkaline medium. The coprecipitation method needs careful monitoring of pH to achieve high-quality MNPs. The common disadvantage of this technique is the relatively low crystallinity of synthesized SFs; hence a subsequent thermal treatment is necessary to achieve better crystallinity. This method is preferred to be the suitable route for synthesizing water-dispersible MNPs. Various SFs have been synthesized by the coprecipitation method, including $CoFe_2O_4$, $ZnFe_2O_4$, $MnFe_2O_4$, Fe_3O_4, $NiFe_2O_4$, $MgFe_2O_4$ and rare earth ions doped MFe_2O_4 [56, 62, 81, 82].

1.6.2.2 Hydrothermal method

This method synthesizes MNPs at a large scale and with high yields. Hydrothermal reactions are carried out in the aqueous phase with high-pressure reactors or autoclaves in which the pressure is mostly higher than 125 atm and the temperature rise to 100–200°C. Water acts as a reactant in these conditions that results in acceleration of the hydrolysis reaction kinetics with the increase in the precursor materials solubility showing considerably more mobility due to the lower viscosity of water and also leads to the increase in the uniformity, crystallinity and faster Ostwald ripening of the production (magnetic nanoparticles). The surfactant is often used in the solution process to control the shape, size, crystal structure and morphology. This method synthesizes various MNPs with high crystallinity and water solubility. Generally, this technique produces water-dispersible MNPs with high magnetic response and crystallinity. However, sometimes the particle size of the MNPs is too large for special applications and causes problems due to colloidal stability.

Moreover, the reaction time tends to become longer. Therefore, it is not easy to control the aspect ratio and shape of the synthesized magnetic nanoparticles. The main advantages of this method are the production of ultra-fine and narrow particle size distribution, reasonable nucleation control, good morphology at high reaction rates and different reaction temperatures and pressure levels, high yield and simplicity [39, 46, 47, 83].

1.6.2.3 Micro-emulsion or Micelle method

A microemulsion or micelle is a colloidal suspension that is thermodynamically stable. This method is used to synthesize the MNPs of spinel ferrites. It is divided into water reverse micelles dispersed in oil and oil micelles dispersed in water depending on the ratio of water and oil, as well as the hydrophobic-hydrophilic balance of surfactant. In this technique, two immiscible liquids exist in a single phase by helping surfactants. Water/oil is one of the microemulsion solvents through which the immiscible solvent's dispersion prepares the solution. The surfactant stabilizes the solution (e.g., dodecyl sulfate). Generally, the aqueous phase, which consists of the magnetic nanoparticle's precursor inside, is dispersed as 1–100 nm nanodroplets. Surfactant molecules surround aqueous droplets, forming the so-called "micelles" as nanoreactors. Subsequently, magnetic nanoparticles are formed inside the micelles, which confine the particles and restrict the nucleation of particles, their growth and aggregation. Finally, for precipitation of the MNPs, a second emulsion is added to the solution. The advantage of this procedure is the easy MNPs size control by modulating the particles to a small nanometer size. However, the shape and size distribution by the microemulsion method is in a wide range. If one introduces a polar solvent such as water into the critical micelle solution, reverse micelles are formed—by homogeneously dispersed solvent molecules surrounding by the polar head groups. The small droplets of encapsulated polar solvents in reverse micelles act as reactors (the sites for reactions). A large number of reverse micelles containing reagents and precipitating agents diffuse into each other, leading to nucleation and particle growth. This method is very attractive in terms of control of the function of particle size distribution during the preparation of the magnetic nanofluid. The type and amount of surfactant control the particle size. The disadvantages of this method are low productivity, the presence of surfactants in magnetic fluid, the need for an organic solvent and some MNPs synthesized by this method $Zn_{0.2}Co_{1.8}Fe_2O_4$, $ZnFe_2O_4$, $NiFe_2O_4$, $CoFe_2O_4$ [14, 24, 43, 76].

1.6.2.4 Sol-gel method

The sol-gel synthesis method is widely used for the preparation of MNPs, in which the metal alkoxide solution undergoes hydrolysis and condensation polymerization reactions to produce gels and any volatile impurities are removed by heat treatment after the synthesis reaction. The advantages of this method include low-cost, simple, carried out at a relatively low temperature without special equipment and achievable narrow particle size distribution. However, particular care should be taken to control better reaction parameters such as the stirring rate, the concentration of sol and annealing temperature. The major limitation of this method is the lack of purity of the final product; thus, thermal treatment is required after the synthesis to achieve high purity. Nevertheless, the homogeneity, composition control, particle size and particle distribution can be achieved well by the sol-gel method. Nanocomposites based on MNPs dispersed in silica gel matrix have been synthesized by this method, and some MNPs synthesized by this method are Fe_3O_4, $CoFe_2O_4$, $MnFe_2O_4$, $NiFe_2O_4$, $CuFe_2O_4$ and $ZnFe_2O_4$ [18, 39, 77, 84].

1.6.2.5 Solvothermal method

Solvothermal (or hydrothermal if the water is solvent) is one of the most eco-friendly and promising synthesis methods. It employs the usability of aqueous or nonaqueous solvents to better control particle size distribution and morphology of MNPs. The shape, size and morphology of MNPs were altered by adjusting the experimental reaction conditions, such as temperature, time, solvent, precursor materials and surfactants. Due to its simplicity in the procedure, commercial production of MNPs is feasible with enhanced physical and chemical properties. A wide range of MNPs have been synthesized, and some examples are MWCNT@ $CoFe_2O_4$, rGO@ $MnFe_2O_4$, Ni–Zn ferrite and metal-doped $MgFe_2O_4$ [18, 31, 48, 77].

1.6.2.6 Microwave-assisted synthesis method

Microwave-assisted synthesis works based on aligning dipoles of the material in an external field via the excitation produced by microwave electromagnetic radiations and are usually executed in combination with a known synthesis strategy. This method is quite advantageous as the synthesis process can be tuned to yield the desired size and shape of magnetic iron oxide nanoparticles. The alignment or orientation of molecules by the external electrical field may produce internal heat, which reduces the processing time and energy required. This is especially due to the heating homogeneity of microwaves. The reaction time can be reduced to a great extent by adopting a microwave-assisted synthesis process. Moreover, this method controls the magnetic properties of ferrite nanoparticles through agitation in experimental measures. Some MNPs synthesized by this method are $MgFe_2O_4$, $MnFe_2O_4$, Ni Fe_2O_4, $CoFe_2O_4$, $ZnFe_2O_4$, Fe_3O_4, Ni Fe_2O_4 and $MgFe_2O_4$ [32, 45, 52, 58, 75, 76].

1.6.2.7 Other methods

Besides the above mentioned methods, some other developed methods for synthesizing spinel ferrite nanoparticles include sonochemical, electrochemical, biological, chemical vapor deposition, sputtering, spry-pyrolysis, ion irradiation and nano casting. Each technique has its significance in targeting one or more specific scientific needs.

Different types of ferrofluids synthesized by different methods and using different coating agents, i.e., surfactant, carrier medium and particle size, are listed in Table 1.2.

Table 1.2. Reported synthesis of different types of ferrofluid in different mediums and surfactants.

Sr. no	Magnetic nanoparticles	Carrier medium	Surfactant	Particles size(nm)	References
1.	Fe_3O_4	Water	Oleic acid and Sodium Oleate	7.5	[55]
2.	Fe_3O_4	Kerosene	Oleic acid	15	[8]
3.	$CoFe_2O_4$	Water	Oleic acid and Sodium Oleate	14	[38]
4.	$CoFe_2O_4$	EG (ethylene glycol)	--	12	[61]
5.	Fe_3O_4	Water and Kerosene	Oleic acid and Sodium Oleate	10	[4]
6.	$Mn_xFe_{3-x}O_4$	Kerosene	Oleic acid	8	[85]
7.	$Co_xFe_{3-x}O_4$	Kerosene	Oleic acid	12	[59]
8.	Fe_3O_4, $Fe_{0.8}Mn_{0.2}Fe_2O_4$, $Fe_{0.8}Ni_{0.2}Fe_2O_4$	Kerosene	Oleic acid	8	[36]
9.	Fe_3O_4	Silicon and fluorine oil	Polydimethylsiloxane and perfluoropolyether	12	[86]
10.	Fe_3O_4 and $CoFe_2O_4$	Water	Oleic acid and Sodium Oleate	10 & 15	[68]
11.	$CoFe_2O_4$	Kerosene	Oleic acid	8	[79]
12.	$Co_{1-x}Zn_xFe_3O_4$	Heptane	Oleic acid	12	[87]
13.	α- Fe_2O_3	Glycerol	-----	5	[48]
14.	Fe_3O_4	Kerosene	Oleic acid	15	[8]
15.	Fe_2O_3	Water	Sodium dodecylbenzene sulfonate (SDBS)	5–25	[88]
16.	$Co_{1-x}Zn_xFe_3O_4$	Kerosene	Oleic acid	8-10	[37]
17.	$CoFe_2O_4$	Kerosene	Oleic acid	10	[57]
18.	Fe_3O_4	Kerosene	Oleic acid	15	[56]

1.7 Properties of ferrofluids

The magnetic nanofluid acts like a homogenous magnetizable medium. The particles are sufficiently small so that the ferrofluid retains its liquid characteristics even in a magnetic field and substantial magnetic forces can be exerted to induce a fluid motion. The magnetic liquid can be considered an ultrafine particle system with interparticle spacings large enough to approximate the particles as non-interacting. Ferrofluids are sensitive to the magnetic field, but do not retain a permanent magnetization leaving zero remanences and zero coercivity after removal of the magnetic field across them. Hence, ferrofluids are magnetically soft and this unusual behavior of ferrofluids comes from the reduced size of magnetic particles. By studying the physical properties of a magnetic nanofluid, one can decide its suitability in a particular application. The external magnetic field can modify many physical properties such as the magnetic, rheological, density and dielectric constant of the magnetic fluid [2, 7, 69]. Some important properties, such as hydrodynamics and magnetic and rheological properties of magnetic nanofluid, are described below.

1.7.1 *Hydrodynamic properties of magnetic ferrofluids*

Hydrodynamics deals with the laws of motion of liquids and gases and their interaction with solids. Hydrodynamics may be called magneto-hydrodynamics, electro-hydrodynamics or hydrodynamics, depending on the body's forces. In hydrodynamics, the body force is gravity, while in magneto-hydrodynamics, the body force is **j x B**, i.e., Lorentz force. Assuming that magnetization induced in the fluid is always parallel to the direction of the applied field, the body force in magnetic fluids becomes Fm = V/4π **M H** [3, 7, 9, 74]. Thus the equation of motion of a magnetic fluid becomes P dq/dt = F_p + Fg + Fv + Fm,n where d/dt is the substantial derivative and q is the velocity of the fluid element, Fp, the force due to the pressure gradient, Fg the gravitational force and Fv is the viscous force, Fm is magnetic force. For the laminar flow of an incompressible fluid (F = 0) in a steady state, the modified Bernoulli's Equation is [69],

$$p + \frac{1}{2}pq^2 + pgh'' - \frac{1}{4}\pi \int_0^H MdH = constant \qquad (1.12)$$

where p and q are the fluid's pressure and velocity, g is the acceleration due to gravity, ρ is the density of the fluid, h″ is an elevation in the gravitational field and H is the applied magnetic field.

The above equation is useful for investigating velocity and pressure distribution in a gradient magnetic field. Comparing this with the familiar Bernoulli's Equation, it can be seen that the magnetic fluid introduces an additional term in the relation. The combination of this term with the remaining labels gives rise to certain novel and useful phenomena like:

 (i) The fluid can be suspended in space,
 (ii) Its specific gravity can be varied by changing the external magnetic field,
 (iii) The stable levitation of non-magnetic as well as magnetized objects can be attained,
 (iv) The generation of fluid motion can be achieved by thermal or magnetic means,
 (v) The ability to flow and conduct magnetic flux,
 (vi) The formation of stable liquid spikes in the perpendicular magnetic field.

1.7.2 *Magnetic properties*

1.7.2.1 General magnetic properties

For a magnetic nanofluid, magnetization, i.e., the magnetic moment per unit volume, is the most important property. Magnetic nanofluids usually exhibit superparamagnetic behavior. The typical nanoparticle size in a nanofluid should be about 10–12 nm, which is smaller than the magnetic domain particles size. Due to the smaller size, magnetic nanofluid contains only single domain particle. Hence each particle is permanently magnetized. The nanoparticles experience appreciable thermal translation that maintains them in suspension against the force of gravity.

In contrast, in the absence of a field, thermal reorientation of nanoparticles results in zero remanences and coercivity, i.e., nanofluids are magnetically soft. The magnetic moments of nanofluids are much larger than the magnetic moment of a paramagnet; for this reason, nanofluids are called superparamagnetic. Therefore, they have a permanent magnetic moment proportional to their volume. How do these magnetic properties of nanofluids determine their suitability for a particular application? The properties of magnetic nanofluids can be studied by various techniques and classified as static and dynamic magnetic properties [3, 67].

1.7.2.2 Static magnetic properties

Equilibrium magnetization in a steady applied magnetic field is the static magnetic property of magnetic fluids. Under normal conditions, a fluid consists of millions of tiny magnets moving randomly in a carrier medium due to thermal agitation. Thus, the fluid resembles a paramagnetic gas. Under the influence of an external magnetic field, particles attempt to align along the direction of the field, while the Brownian motion attempts to destroy the alignment. The degree of alignment increases with the field strength. When the area is switched off, the fluid attains its original state, i.e., no hysteresis. Since these materials' magnetic moment is enormous compared to paramagnetic, these fluids are said to exhibit superparamagnetic behavior.

Therefore, as was seen earlier, the magnetic nanofluid is a soft magnetic material. Since the thermal energy (kT) is greater than the magnetic energy, the particle performs the Brownian motion [10, 57, 89]. This random motion of the MNPs in the fluid resemble paramagnetic gas. When such a magnetic fluid is exposed to an external magnetic field, the particles initially oriented randomly in the suspension tend to align in the direction of the applied magnetic field. Due to significant magnetic moments (resultant) compared to the superparamagnetic MNPs, the fluid resembles a superparamagnetic material. How this varies with the external magnetic field and temperature determine its stability for a particular application. To describe the magnetization of magnetic nanofluid several theoretical models have been used. The simplest is the Langevin theory, where each magnetic MNPs is considered to be independent; therefore, the magnetization M as a function of applied field H is given by [20, 35, 75]:

$$M = \varphi_m M_d L(\alpha) \tag{1.13}$$

where M_d is the domain of magnetization of MNPs, φ_m is the volume fraction of the dispersed MNPs and $L(\alpha)$ is the Langevin function, given as;

$$L(\alpha) = Coth(\alpha) - 1/\alpha \tag{1.14}$$

Here the Langevin parameters, $\alpha = \dfrac{\mu H}{kT}$, where H is the applied magnetic field and μ is the magnetic moments of the single/domains particles $\mu = \mu_o M_d V$, where

V is the volume of MNPs and is the permeability of free space [20]. Here α can be described as;

$$\alpha = \frac{\mu_o M_d VH}{kT} \qquad (1.15)$$

Using Equation (11), the particle size and the domain magnetization of the particles may be obtained by fitting the experimental data of M/M_s versus H.

1.7.2.3 Dynamic magnetic properties

When an alternating magnetic field is applied to a magnetic nanofluid, two kinds of the orientation of the particles takes place, viz: Brownian and Neel. In the Brownian rotation, the magnetic moment of the particles can rotate by mechanical rotation for the carrier liquid. Here the magnetization is frozen with particles and changes its direction only if the particles rotate mechanically. In the Neel rotation, the magnetic moment within the grain attempts to align along the direction of the field without mechanical rotation.

1.7.3 Relaxation mechanism in ferrofluid

The system of single-domain ferromagnetic MNPs immersed in a solid or liquid phase exhibits an interesting relaxation phenomenon, a subject matter of many experimental and theoretical analyses. In MNPs systems, the thermal energy promotes the fluctuation of the magnetic moments. Therefore, to understand the role that temperature plays in the magnetic behavior of the particles, it is necessary to investigate the dynamics of the particle moments as a function of Temperature (T). The high- and low- T limit cases can be figured out quickly. At very high T, the thermal energy is much larger than the anisotropy energy barrier ($k_B T \gg EA$), k_B the Boltzmann constant. So, the magnetic anisotropy influences the orientation of the magnetic moments of the particles, which fluctuate freely with temperature. In this case, a paramagnetic-like behavior is observed, and the particles are in the SPM state. On the contrary, at very low T the particle moment remains confined along the anisotropy direction (local energy minimum) because thermal fluctuations cannot switch its orientation out. When this happens, the particles are said to be in the blocked state ($k_B T \ll EA$) [7, 47, 73]. The rotation mechanism of ferrimagnetic MNPs in such a matrix is the result of the competition of three orientation mechanisms related to

  (i) The external magnetic field

 (ii) The axis of easy magnetization

(iii) The Brownian motion (which acts only when the MNPs are suspended in a fluid matrix)

Mainly, there are two distinct mechanisms by which the magnetization of magnetic fluid may relax after an externally applied magnetic field is removed. They are:

  (i) The Brownian Mechanism—In this mechanism, the rotational motion of the MNPs within the carrier liquid take place along with its magnetic moments

(μ), which are locked in the direction of an easy axis of magnetization. These kinds of MNPs are called hard magnetic materials. The time associated with the rotational diffusion is the Brownian relaxation time (τ_B) having a relation [70];

$$\tau_B = \frac{3V_H \eta_O}{kT} \qquad (1.16)$$

where V_H is the hydrodynamic volume of MNPs and is the viscosity of the carrier liquid.

(ii) The Neel mechanism—In these mechanisms, only the rotation of magnetic moments (μ) within the MNPs takes place and the presence of such a relaxation is permissible in only magnetically soft MNPs. Therefore the magnetic moments may reverse their direction (within the MNPs) by overcoming the energy barrier, which for uniaxial anisotropy, is given by KV, where K is the anisotropy constant of MNPs. The probability of such transition is exp $(\frac{KV}{kT})$, where KV is the anisotropy energy and kT is the thermal energy. This reversal time is referred to as the Neel relaxation time (τ_N) and given by the relation [33, 40, 90];

$$\tau_N = \tau_O \, exp \left(\frac{KV}{kT} \right) \qquad (1.17)$$

where, τ_o is the characteristic time often quoted as approximate value 10^{-9}s.

The size of MNPs implies the existence of a distribution of relation time with both relaxation mechanisms contributing to magnetization phenomenon . They do so with an adequate relaxation time τ_{eff}, where for a particular MNPs;

$$\tau_{eff} = \frac{\tau_N \tau_B}{(\tau_N + \tau_B)} \qquad (1.18)$$

The mechanism with the shortest relaxation time is dominant. Therefore, when $\tau_N \ll \tau_B$, relaxation occurs by the Neel mechanism and the materials are said to possess the "intrinsic superparamagnetic". Whereas $\tau_N \gg \tau_B$, determining the Brownian mechanism and the materials exhibit the "extrinsic superparamagnetic" [20].

1.7.4 Occurrence of spin-glass in magnetic ferrofluids

A spin glass is a disordered magnet with frustrated interactions, augmented by stochastic positions of the spins, where conflicting interactions, namely both ferromagnetic and antiferromagnetic bonds, are randomly distributed with comparable frequency. The term glass refers to the magnetic disorder of spins akin to positional disorder in conventional glass materials. Therefore the key parameters to have a spin glass state are randomness and frustration of spins. Many applications mainly depend on the use of densely packed magnetic nanoparticles. It is thus essential to know how interparticle interaction affects the physical properties of magnetic nanoparticle systems. In particular, it is crucial to understand how interparticle interactions change the thermal stability of the magnetic recording media due to the current effort to shrink the volume of storage devices. It has been suggested that dense nanoparticle samples may exhibit glassy dynamics due to dipolar interparticle interaction; the randomness induces disorder and frustration in the particle positions and anisotropy axis orientations [80, 91, 92]. In frozen ferrofluids, the strength of the dipolar interaction between the single-domain nanoparticles can be continuously varied by changing the particle concentration. With increasing particle concentration, the magnetic behavior may evolve from superparamagnetic to spin-glass-like. The particles in a frozen ferrofluid with an electrically non-conductive medium can be approximated as giant macro spin having strengthened dipolar interactions. Due to random orientations, the dipolar interactions will be both positive and negative. Ferrofluids frozen under zero field are Heisenberg systems with random uniaxial anisotropy, while an Ising system is characterized by parallel uniaxial anisotropy. Another type of spin-glass ordering/disordering has been proposed to take place at the surface of individual nanoparticles. Meiklejohn and Bean have shown that uniaxial exchange anisotropy can exist in a field-cooled fine magnetic particle with a single-domain ferromagnetic core and antiferromagnetic surface. This explanation assumes core-shell structure in each nanoparticle and sub-sequential changes in various magnetically dependent parameters, viz., magnetization, magnetic coercivity, angular coercivity, etc. This type of spin glass transition has been observed in double surfactant water-based ferrofluid containing magnetite nanoparticles [80, 92].

1.7.5 Rheological (viscoelastic) properties of ferrofluid

Viscoelastic properties play an essential role in many scientific and technological applications, and one must determine how an external magnetic field modifies this property. The variation of shear stress on the shear rate also decides whether the fluid is Newtonian and Non-Newtonian. Due to the presence of large numbers of MNPs in the fluids, increases the rate of energy dissipating during the viscous flow. The viscosity of magnetic nanofluid is greater than that of carrier liquid. Under the influence of an external magnetic field, the fluid is subjected to shear deformation, and its viscosity is modified. In the case of magnetic fluids, it is essential to determine how an external magnetic field modifies the rheological properties of magnetic fluids. Rheological properties become more complicated when the fluid is a mixture of several liquids or contains dispersed solid particles. A magnetic fluid usually exhibits Newtonian behavior without a magnetic field where shear stress (F) is linearly dependent on shear rate (c).

In contrast, its viscosity is modified in the presence of a field. The hydrodynamical model developed by Shliomis can explain the field-induced viscosity. This model assumes a fluid particle to possess an intrinsic angular momentum. Here the effect of a homogenous magnetic field on the viscosity of magnetic nanofluid, whose solid MNPs possess the magnetic moment, is considered.

In the absence of a magnetic field, they can be approximated as a non-interacting suspension of particles. For such systems, Einstein gave the colloidal viscosity, η_o

in terms of the carrier fluid viscosity, ηc and the volume fraction of everything suspended in the fluid [69, 74],

$$\eta_o = \eta_c \left(1 + \frac{5}{2} \right) \check{\phi}$$

(1.19)

The linear relation is not accurate for any volume fraction above $\check{\phi} = 0.1$. By assuming a higher order of some critical volume fraction $\check{\phi}$, the viscosity should diverge, i.e., the colloid becomes rigid. Using the volume fraction for closed packed spheres for this critical volume fraction, $\check{\phi} = 0.74$, the relation for the viscosity can be described by Equation 17. It approximates the viscosity fairly well up until $\check{\phi} = 0.3$ [9, 49, 55, 57].

$$\eta_o = \eta_c \left(1 - \frac{5}{2}\phi + \left(\frac{5}{2}\phi_c - 1 \right) \left(\frac{\check{\phi}}{\check{\phi}} \right)^2 \right)^{-1}$$

(1.20)

In the presence of dc—a field, the viscosity is usually greater under the influence of a static field. Still, it can be seen that an alternating magnetic field can decrease the viscosity [89]. In an extreme case, the viscosity can become harmful—the magnetic field contributes to particle rotation that hinders it.

In the case of magnetically hard ferrofluid particles, their moments are fixed for the crystal structure and they relax to a magnetic field via physical rotation of the particles. By ignoring any interaction between the particles, i.e., assuming the suspension is dilute, the particles moving in a flow with vorticity will rotate in the same direction as the vorticity of the flow. The applied field will torque the particles toward lining up their magnetic moments with the field. If the magnetic field is in the same direction as the vorticity, the magnetic moments will line up with the vorticity, and the particles will rotate about their magnetic moments. There is no additional torque from the magnetic field, so the viscosity is left the same as with no field. However, if the applied field is perpendicular to the vorticity, the magnetic torque again will attempt to line the particles up with the field.

In contrast, though, the torque from the shear flow will try to misalign the particles with the field. These competing torques will create an anisotropic viscosity change dependent on the strength of the field. At a large enough applied field, the viscosity will be enough to prevent the rotation of the particles [4, 13, 56] entirely.

The rate of change of the magnetization of particles when rotated relative to the field is written as

$$\frac{d\vec{M}}{dt} = \frac{1}{I}\vec{S} \times \vec{M} - \frac{1}{\tau_B} (\vec{M} - M_O \widetilde{\vec{H}})$$

(1.21)

First, assuming the stationary flow, $dS/dt = 0$ and then $dM/dt = 0$. Then an expansion in M for small deviations around the equilibrium value linearizes the equations, allowing the calculation of the force per unit volume. The viscosity term has an additional portion, referred to as the rotational viscosity, which is field-dependent. Therefore, the rotational viscosity for a field perpendicular to the vorticity of the flow is written as [1, 9, 67],

$$\eta = \frac{3}{2} \Phi \eta_o \left[\frac{\alpha - \tanh(\alpha)}{\alpha + \tanh(\alpha)} \right] \tag{1.22}$$

where θ is the angle between local vorticity and H, Φ is the volume of the dispersed phase $\alpha = \dfrac{\mu H}{k_B T}$ M is the particle's magnetic moment and H is the applied field. Examining this expression in the high field limit, H $\rightarrow \infty$, the fraction involved goes to 1 and the maximum viscosity change is [74]:

$$\frac{\eta_{r,max}}{\eta_o} \rightarrow \frac{3}{2} \phi \tag{1.23}$$

is the volume fraction of the colloid particles. In the low field limit, the ξ dependence can be approximated as $\xi^2/6$, giving the low field behavior as

$$\frac{\eta_r}{\eta_o} = \frac{1}{4} \phi\, \xi^2 = \frac{\mu^2 \phi}{4(k_B T)^2} H^2 \tag{1.24}$$

The theoretical formula agrees well with experimentally observed viscosity in certain magnetic nanofluids, and this equation has been found to fit well with several experimental data.

Application of the homogenous magnetic field on magnetic fluids increases with the viscosity of the magnetic fluid. The effect of an external magnetic field on the viscosity of a dilute suspension of magnetic particles was studied theoretically and experimentally by many workers. Rosensweig has given an expression that may be used to determine the particle size in a magnetic fluid when magnetic fluid is not subjected to a magnetic field [38, 55, 72].

Hence the rheological behavior of colloidal dispersion mainly depends upon the following factors such as:

(i) The concentration of the dispersed medium

(ii) The viscosity of the dispersion medium

(iii) Size and shape of the dispersed phase

(iv) Interaction between the nanoparticles-particles and particles-dispersion medium.

1.8 Perspective applications of ferrofluid

Magnetic nanofluid is a multifunctional medium; even the most unexpected application can be made possible. Numerous applications have been developed and reported, some gaining high commercial value. Magnetic nanofluid is a seal of shafts and roads, as a lubricant, damper and shock absorber, magnetic nanofluid as a heat carrier in loudspeakers, electrical motors and various devices like pressure transducers slope angle data transmitters and accelerometers have been elaborated in various engineering applications [1, 15, 43]. As described earlier magnetic nanofluid acts as an essential working material for several engineering and technology applications where the combination of the liquid state and magnetic properties is essential. An important property of concentrated ferrofluids is that permanent magnets strongly attract them while their liquid character is preserved. The attraction can be strong enough to overcome the force of gravity. Magnetic fluids have found numerous technological applications. Their remarkable magneto-viscous effects are exploited in most cases, which is not present in any other synthetic novel material. Some of their applications are briefly described below.

1. Loudspeaker—This is the most popular and well-established application of the ferrofluid. When the magnetic nanofluid is placed into the air gap between the pole piece and the coil inside the loudspeaker, it helps maintain the voice coil to be concentric with the permanent magnet of the loudspeaker. Ferrofluid dampens extra noises created inside the loudspeaker. It also helps maintain the voice coil's low temperature by conducting heat away from it [3, 7, 45].

2. Biomedical—Almost all applications in medicine exploit the extreme relative size difference between MNPs and living cells. They can be further categorized as follows: magnetic fluid hyperthermia, targeting drug delivery, contrast imaging agent in magnetic resonance imaging (MRI), and magnetic separation of cells. All these applications deal with MNPs in a water-based liquid medium [24, 28, 58].

3. Bearing and sealing—Magnetic fluids form liquid seals around the spinning drive shafts in hard disks. Magnets surround the rotating shaft. A small amount of magnetic fluid will be held in place by its attraction to the magnet in the gap between the magnet and the shaft. The fluid of magnetic particles forms a barrier that prevents debris from entering the interior of the hard drive [15, 67].

4. Printing with magnetic inks—By incorporating nano-sized magnetic particles into the ink, the resulting liquid can generate a magnetic signal. It can subsequently be read by very simple and inexpensive magnetic recording technology. For example, Magnetic Ink Character Recognition (MICR) technology is used by banks to print details on checks to enable automatic processing. When a document that contains this ink needs to be read, it is passed through a machine that magnetizes the ink and then translates the magnetic information into characters [1, 69, 87].

5. Electric power generator—Moving a magnet back and forth through a conductive coil induces an electromotive force that generates an AC in the coil. Ferrofluid bearings levitate the magnets and make their motion easy, thus making electrical power generation devices more efficient. An electric power generator can be designed using a ferrofluid bearing with the principle of an electricity generation law in which the coefficient of friction of the magnet motion is greatly reduced by ferrofluid bearing around permanent magnets [4, 54, 67].

6. Heat transfer—An external magnetic field imposed on a magnetic fluid with varying susceptibility (e.g., because of a temperature gradient) results in

a non uniform magnetic body force, leading to a heat transfer form called thermomagnetic convection. This heat transfer can be useful when conventional convection heat transfer is inadequate, e.g., in miniature microscale devices or under reduced gravity conditions. Special magnetic nanofluids with tunable thermal conductivity to viscosity ratio can be used as multifunctional 'smart materials that can remove heat and arrest vibrations (damper). Such fluids may find applications in microfluidic devices and microelectromechanical systems [15, 88, 93].

7. Optical sensors and devices—Research is underway to create an adaptive optics shape-shifting magnetic mirror from ferrofluid for Earth-based astronomical telescopes. Optical filters are used to select different wavelengths of light. The replacement of filters is cumbersome, especially when the wavelength is changed continuously with tunable-type lasers. Optical filters tunable for different wavelengths by varying the magnetic field can be built using ferrofluid emulsion [23, 54, 83, 90].

8. Spacecraft Propulsion—Magnetic fluids can be made to self-assemble nanometer scale needle-like sharp tips under the influence of a magnetic field. When they reach a critical thinness, the needles begin emitting jets that might be used in the future as a thruster mechanism to propel small satellites such as CubeSats [38, 59].

9. Centrifugal switch—The centrifugal switch cuts off primary windings after the electric motor's rotor to attain a certain speed. The conventional button has large inertia and special care is required for its use in an explosive atmosphere. A ferrofluid-based switch with levers and a reed switch was developed earlier but had a dead zone area of nearly 600 rpm. A switch based on a differential transformer with a ferrofluid core has also been developed in this industry and laboratory, which has almost zero dead zones and very low inertia [9, 49, 69].

10. Dampers and Lubrication—A permanent magnet immersed in a pool of ferrofluid levitates itself from the bottom. This property is used in inertia dampers. A similar damper for the instrument's stepper motors, satellites and dashpot dampers has also been developed. Lubrication oil base or other lubricating fluid base ferrofluid can act as lubricants. The main advantage of ferrofluid lubricants over conventional ones is that an external magnetic field can retain the former at the desired location. In a sealed system, as in food procuring machines, contamination from conventional lubricants can be prevented using ferrofluid lubricants. The development of a low frictional loss, highly reliable engine shaft seal and magnetic oil intake system is based on applying such a lubricant [7, 10, 15, 87].

There are many other applications of magnetic nanofluids in different areas, such as lubrication, actuators and transducers, analytical instrumentation and mechanical engineering, etc.

1.9 Conclusions

This chapter summarizes the introduction of ferrofluid and its properties, ferrites and properties, synthesis methods, properties and plausible application of ferrofluid. Ferrofluids have gained a lot of interest due to their ability to be manipulated on application of an external magnetic field. They can be utilized in different areas such as electronic devices, high-density data storage, biomedical, catalysis, gas sensor, photocatalysis, microwave absorber, magnetic fluids hyperthermia, targeted drug delivery, etc. The specific focus is on the properties, synthesis method (chemical coprecipitation, hydrothermal, solvothermal, sol-gel, microemulsion, spry-pyrolysis and high energy ball milling), crystal structure, magnetic and different applications of MNPs. The influence of the synthesis method on various properties like structural and magnetic of MNPs was also described. The distribution of cations, crystal structure and exchange interaction between the magnetic ions are responsible for the magnetic behavior of MNPs. Finally, it can be said briefly that despite several limitations, high surface area and stable chemical structure, low cost, various morphologies and structures, changeable magnetic, electrical and optical properties are helpful for multiple diverse scientific and technological applications. These applications have made ferrites-based ferrofluids a good market product for commercial benefits.

References

[1] C. Scherer and A. M. Figueiredo Neto. Ferrofluids: Properties and applications. Brazilian J. Phys. 35 (2005).
[2] S. W. Charles. MAGNETIC FLUIDS (Ferrofluids), in North-Holland Delta Series, edited by Dormann, J. L. and D. B. T.-M. P. of F. P. Fiorani (Elsevier, Amsterdam), pp. 267–276 (1992).
[3] K. Raj and R. Moskowitz. Commercial applications of ferrofluids. J. Magn. Magn. Mater. 85: 233 (1990).
[4] S. Pathak, K. Jain, P. Kumar, X. Wang and R. P. Pant. Improved thermal performance of annular fin-shell tube storage system using magnetic fluid. Appl. Energy 239: 1524, (2019).
[5] E. Auzans, D. Zins, E. Blums and R. Massart. Synthesis and properties of Mn-Zn ferrite ferrofluids. J. Mater. Sci. 34: 1253 (1999).
[6] A. P. Philipse, M. P. B. van Bruggen and C. Pathmamanoharan. Magnetic silica dispersions: preparation and stability of surface-modified silica particles with a magnetic core. Langmuir 10: 92 (1994).
[7] B. U. Felderhof, V. V. Sokolov and P. A. Éminov. Ferrofluid dynamics, magnetic relaxation, and irreversible thermodynamics. J. Chem. Phys. 132: 184907 (2010).
[8] W. Yu, H. Xie, L. Chen, and Y. Li. Enhancement of thermal conductivity of kerosene-based Fe_3O_4 Nanofluids Prepared via Phase-Transfer Method, Colloids Surfaces A Physicochem. Eng. Asp. 355: 109 (2010).
[9] R. E. Rosensweig. Heating magnetic fluid with alternating magnetic field. J. Magn. Magn. Mater. 252: 370 (2002).
[10] I. Torres-Díaz and C. Rinaldi. Recent progress in ferrofluids research: Novel applications of magnetically controllable and tunable fluids. Soft Matter 10: 8584 (2014).
[11] N. Fauconnier, A. Bée, J. Roger and J. N. Pons. Synthesis of aqueous magnetic liquids by surface complexation of maghemite nanoparticles. J. Mol. Liq. 83: 233 (1999).
[12] A. H. Latham and M. E. Williams. Controlling transport and chemical functionality of magnetic nanoparticles. Acc. Chem. Res. 41: 411 (2008).
[13] S. Pathak, K. Jain, Noorjahan, V. Kumar and R. P. Pant. Magnetic fluid based high precision temperature sensor. IEEE Sens. J. 17: 2670 (2017).

[14] S. A. Novopashin, M. A. Serebryakova and S. Y. Khmel. Methods of magnetic fluid synthesis (Review), Thermophys. Aeromechanics 22: 397 (2015).

[15] S. Genc and B. Derin. Synthesis and Rheology of Ferrofluids: A Review, Curr. Opin. Chem. Eng. 3: 118 (2014).

[16] P. Tartaj, M. a del P. Morales, S. Veintemillas-Verdaguer, T. Gonz lez-Carre o and C. J. Serna. The preparation of magnetic nanoparticles for applications in biomedicine. J. Phys. D. Appl. Phys. 36: R182 (2003).

[17] D. L. Huber. Synthesis, properties, and applications of iron nanoparticles. Small 1: 482 (2005).

[18] A. H. Lu, E. L. Salabas and F. Schüth. Magnetic nanoparticles: synthesis, protection, functionalization, and application. Angew. Chemie Int. Ed. 46: 1222 (2007).

[19] B. Wright, D. Thomas, H. Hong, L. Groven, J. Puszynski, E. Duke, X. Ye and S. Jin. Magnetic field enhanced thermal conductivity in heat transfer nanofluids containing Ni coated single wall carbon nanotubes. Appl. Phys. Lett. 91: 173116 (2007).

[20] B. D. Cullity. Introduction to Magnetic Materials Second Edition 148: 94 (1974).

[21] M. Sugimoto. The past, present, and future of ferrites. J. Am. Ceram. Soc. 82: 269 (1999).

[22] C. Pereira et al. Superparamagnetic MFe$_2$O$_4$ (M = Fe, Co, Mn) Nanoparticles: Tuning the particle Size and Magnetic properties through a novel one-step coprecipitation route. Chem. Mater. 24: 1496 (2012).

[23] R. Valenzuela. Novel applications of ferrites. Phys. Res. Int. 591839 (2012).

[24] A. Akbarzadeh, M. Samiei and S. Davaran. Magnetic nanoparticles: preparation, physical properties, and applications in biomedicine. Nanoscale Res. Lett. 7: 144 (2012).

[25] C. Pereira et al. Superparamagnetic MFe$_2$O$_4$ (M = Fe, Co, Mn) Nanoparticles: Tuning the particle size and magnetic properties through a novel one-step coprecipitation route. Chem. Mater. 24: 1496 (2012).

[26] A. Kolhatkar, A. Jamison, D. Litvinov, R. Willson and T. Lee. Tuning the magnetic properties of nanoparticles. Int. J. Mol. Sci. 14: 15977 (2013).

[27] P. Pulišová, J. Kováč, A. Voigt and P. Raschman. Structure and magnetic properties of Co and Ni nano-ferrites prepared by a two step direct microemulsions synthesis. J. Magn. Magn. Mater. 341: 93 (2013).

[28] G. Unsoy, U. Gunduz, O. Oprea, D. Ficai, M. Sonmez, M. Radulescu, M. Alexie and A. Ficai. Magnetite: From synthesis to applications. Curr. Top. Med. Chem. 15: 1622 (2015).

[29] D. S. Mathew and R.-S. Juang. An overview of the structure and magnetism of spinel ferrite nanoparticles and their synthesis in microemulsions. Chem. Eng. J. 129: 51 (2007).

[30] W. Wu, Z. Wu, T. Yu, C. Jiang and W.-S. Kim. Recent progress on magnetic iron oxide nanoparticles: synthesis, surface functional strategies and biomedical applications. Sci. Technol. Adv. Mater. 16: 023501 (2015).

[31] H. Shokrollahi. A review of the magnetic properties, synthesis methods and applications of maghemite. J. Magn. Magn. Mater. 426: 74(2017).

[32] A. Ali, H. Zafar, M. Zia, I. ul Haq, A. R. Phull, J. S. Ali and A. Hussain. Synthesis, characterization, applications, and challenges of iron oxide nanoparticles. Nanotechnol. Sci. Appl. 9: 49 (2016).

[33] K. Jain, S. Pathak and R. P. Pant. Enhanced magnetic properties in ordered oriented ferrofibres. RSC Adv. 6: 70943 (2016).

[34] V. F. Cardoso, A. Francesko, C. Ribeiro, M. Bañobre-López, P. Martins and S. Lanceros-Mendez. Advances in magnetic nanoparticles for biomedical applications. Adv. Healthc. Mater. 7: 1700845 (2018).

[35] T. Tatarchuk, M. Bououdina, J. Judith Vijaya and L. John Kennedy. Spinel ferrite nanoparticles: synthesis, crystal structure, properties, and perspective applications. *In*: O. Fesenko and L. Yatsenko (eds.). Springer Proceedings in Physics, Vol. 195 (Springer International Publishing, Cham), pp. 305–325 (2017).

[36] M. Victory, R. P. Pant and S. Phanjoubam. Synthesis and characterization of Oleic acid coated Fe–Mn ferrite based ferrofluid. Mater. Chem. Phys. 240: 122210 (2020).

[37] A. Singh, S. Pathak, P. Kumar, P. Sharma, A. Rathi, G. A. Basheed, K. K. Maurya and R. P. Pant. Tuning the magnetocrystalline anisotropy and spin dynamics in CoxZn1-XFe2O4 ($0 \leq x \leq 1$) Nanoferrites. J. Magn. Magn. Mater. 493: 165737 (2020).

[38] P. B. Kharat, J. S. Kounsalye, M. V. Shisode and K. M. Jadhav. Preparation and thermophysical investigations of CoFe{sub 2}O{sub 4}-Based Nanofluid: A potential heat transfer agent. J. Supercond. Nov. Magn. 32: 341 (2019).

[39] A. Vedrtnam, K. Kalauni, S. Dubey, A. Kumar, A. Vedrtnam, K. Kalauni, S. Dubey and A. Kumar. A comprehensive study on structure, properties, synthesis and characterization of ferrites. AIMS Mater. Sci. 7: 800–835 (2020).

[40] Y. Zhang, L. Sun, Y. Fu, Z. C. Huang, X. J. Bai, Y. Zhai, J. Du and H. R. Zhai. The shape anisotropy in the magnetic field-assisted self-assembly chain-like structure of magnetite. J. Phys. Chem. C 113: 8152 (2009).

[41] P. A. Vinosha, A. Manikandan, A. S. Judith Ceicilia, A. Dinesh, G. Francisco Nirmala, A. C. Preetha, Y. Slimani, M. A. Almessiere, A. Baykal and B. Xavier. Review on recent advances of zinc substituted cobalt ferrite nanoparticles: synthesis characterization and diverse applications. Ceram. Int. 47: 10512 (2021).

[42] P. Kumar, S. Pathak, A. Singh, H. Khanduri, G. A. Basheed, L. Wang and R. P. Pant. Microwave spin resonance investigation on the effect of the post-processing annealing of CoFe2O4 nanoparticles. Nanoscale Adv. 2: 1939 (2020).

[43] F. Sharifianjazi et al. Magnetic CoFe2O4 nanoparticles doped with metal ions: A review. Ceram. Int. 46: 18391 (2020).

[44] P. Kumar, H. Khanduri, S. Pathak, A. Singh, G. A. Basheed and R. P. Pant. Temperature selectivity for single phase hydrothermal synthesis of PEG-400 coated magnetite nanoparticles, Dalt. Trans. 49: 8672 (2020).

[45] B. Shen and S. Sun. Chemical synthesis of magnetic nanoparticles for permanent magnet applications, Chem. - A Eur. J. 26: 6757 (2020).

[46] P. Kumar, S. Pathak, A. Singh, H. Khanduri, Kuldeep, K. Jain, J. Tawale, L. Wang, G. A. Basheed, and R. P. Pant. Enhanced static and dynamic magnetic properties of PEG-400 coated CoFe2−xErxO4 (0.7 ≤ x ≤0) nanoferrites. J. Alloys Compd. 887: 161418 (2021).

[47] P. Kumar, S. Pathak, K. Jain, A. Singh, Kuldeep, G. A. Basheed and R. P. Pant. Low-temperature large-scale hydrothermal synthesis of optically active PEG-200 capped single domain MnFe2O4 nanoparticles. J. Alloys Compd. 904: 163992 (2022).

[48] M. Abareshi, S. H. Sajjadi, S. M. Zebarjad and E. K. Goharshadi. Fabrication, characterization, and measurement of viscosity of α-Fe2O3-Glycerol Nanofluids. J. Mol. Liq. 163: 27 (2011).

[49] M. T. López-López, A. Gómez-Ramírez, L. Rodríguez-Arco, J. D. G. Durán, L. Iskakova and A. Zubarev. Colloids on the frontier of ferrofluids. Rheological Properties, Langmuir 28: 6232 (2012).

[50] W. Wu, Q. He and C. Jiang. Magnetic iron oxide nanoparticles: synthesis and surface functionalization strategies. Nanoscale Res. Lett. 3: 397 (2008).

[51] C. M. B. Henderson, J. M. Charnock and D. A. Plant. Cation occupancies in Mg, Co, Ni, Zn, Al Ferrite Spinels: A Multi-Element EXAFS Study. J. Phys. Condens. Matter 19: 76214 (2007).

[52] A. Ali, H. Zafar, M. Zia, I. ul Haq, A. R. Phull, J. S. Ali and A. Hussain. Synthesis, characterization, applications, and challenges of iron oxide nanoparticles. Nanotechnol. Sci. Appl. 9: 49 (2016).

[53] J. Park et al. One-nanometer-scale size-controlled synthesis of monodisperse magnetic iron oxide nanoparticles. Angew. Chemie - Int. Ed. 44: 2872 (2005).

[54] K. Jain, S. Pathak, P. Kumar, A. Singh and R. P. Pant. Dynamic magneto-optical inversion in magnetic fluid using NanoMOKE. J. Magn. Magn. Mater. 475: 782 (2019).

[55] L. Wang, Y. Wang, X. Yan, X. Wang and B. Feng. Investigation on viscosity of Fe3O4 nanofluid under magnetic field. Int. Commun. Heat Mass Transf. 72: 23 (2016).

[56] G. Noorjahan, A. Basheed, K. Jain, S. Pathak and R. P. Pant. Dipolar interaction and magneto-viscoelasticity in nanomagnetic fluid, J. Nanosci. Nanotechnol. 18: 2746 (2018).

[57] A. Mishra, S. Pathak, P. Kumar, A. Singh, K. Jain, R. Chaturvedi, D. Singh, G. A. Basheed and R. P. Pant. Measurement of static and dynamic magneto-viscoelasticity in facile varying PH synthesized CoFe2O4-based magnetic fluid. IEEE Trans. Magn. 55: 1 (2019).

[58] K. Kuldeep, Jain, P. Kumar, R. P. Pant and G. A. Basheed. Tunning of rheological and magnetic properties of Ni doped magnetite based magnetic nanofluid. Phys. B Condens. Matter 643: 414136 (2022).

[59] P. Kumar, S. Pathak, A. Singh, Kuldeep, H. Khanduri, X. Wang, G. A. Basheed and R. P. Pant. Optimization of cobalt concentration for improved magnetic characteristics and stability of CoxFe3-XO4 mixed ferrite nanomagnetic fluids. Mater. Chem. Phys. 265: 124476 (2021).

[60] R. Betancourt-Galindo, O. Ayala-Valenzuela, L. A. García-Cerda, O. Rodríguez Fernández, J. Matutes-Aquino, G. Ramos and H. Yee-Madeira. Synthesis and magneto-structural study of CoxFe3−xO4 nanoparticles/. J. Magn. Magn. Mater. 294: e33 (2005).

[61] P. B. Kharat, A. R. Chavan, A. V. Humbe and K. M. Jadhav. Evaluation of thermoacoustics parameters of CoFe2O4–Ethylene glycol nanofluid using ultrasonic velocity technique. J. Mater. Sci. Mater. Electron. 30: 1175 (2019).

[62] S. Noorjahan, Pathak, K. Jain and R. P. Pant. Improved magneto-viscoelasticity of cross-linked PVA Hydrogels using magnetic nanoparticles. Colloids Surfaces A Physicochem. Eng. Asp. 539: 273 (2018).

[63] T. Tago, T. Hatsuta, K. Miyajima, M. Kishida, S. Tashiro and K. Wakabayashi. Novel synthesis of silica-coated ferrite nanoparticles prepared using water-in-oil microemulsion. J. Am. Ceram. Soc. 85: 2188 (2002).

[64] T. Hyeon. Chemical synthesis of magnetic nanoparticles. Chem. Commun. 927 (2003).

[65] D. Makovec, A. Košak, A. Žnidaršič and M. Drofenik. The synthesis of spinel–ferrite nanoparticles using precipitation in microemulsions for ferrofluid applications. J. Magn. Magn. Mater. 289: 32 (2005).

[66] J. Park et al. One-nanometer-scale size-controlled synthesis of monodisperse magnetic iron oxide nanoparticles. Angew. Chemie 117: 2932 (2005).

[67] S. Pathak, R. Zhang, K. Bun, K. Zhang, B. Gayen and X. Wang. Development of a novel wind to electrical energy converter of passive ferrofluid levitation through its parameter modelling and optimization. Sustain. Energy Technol. Assessments 48: 101641 (2021).

[68] A. Karimi, S. S. S. Afghahi, H. Shariatmadar and M. Ashjaee. Experimental investigation on thermal conductivity of MFe2O4 (M=Fe and Co) Magnetic nanofluids under influence of magnetic field. Thermochim. Acta 598: 59 (2014).

[69] S. Odenbach. Recent progress in magnetic fluid research. J. Phys. Condens. Matter 16: R1135 (2004).

[70] Y. Bao, A. B. Pakhomov and K. M. Krishnan. Brownian magnetic relaxation of water-based cobalt nanoparticle ferrofluids. J. Appl. Phys. 99: 08H107 (2006).

[71] K. Ueno, H. Tokuda and M. Watanabe. Ionicity in ionic liquids: Correlation with ionic structure and physicochemical properties. Phys. Chem. Chem. Phys. 12: 1649 (2010).

[72] W. Huang and X. Wang. Study on the properties and stability of ionic liquid-based ferrofluids. Colloid Polym. Sci. 290: 1695 (2012).

[73] Y. Bao, A. B. Pakhomov and K. M. Krishnan. Brownian magnetic relaxation of water-based cobalt nanoparticle ferrofluids. J. Appl. Phys. 99: 08H107 (2006).

[74] D. Y. Borin, V. V Korolev, A. G. Ramazanova, S. Odenbach, O. V Balmasova, V. I. Yashkova and D. V. Korolev. Magnetoviscous effect in ferrofluids with different dispersion media. J. Magn. Magn. Mater. 416: 110 (2016).

[75] L. S. Arias, J. P. Pessan, A. P. M. Vieira, T. M. T. De Lima, A. C. B. Delbem and D. R. Monteiro. Iron oxide nanoparticles for biomedical applications: a perspective on synthesis. Drugs, Antimicrobial Activity, and Toxicity, Antibiotics 7 (2018).

[76] A. Ali, T. Shah, R. Ullah, P. Zhou, M. Guo, M. Ovais, Z. Tan and Y. K. Rui. Review on recent progress in magnetic nanoparticles: synthesis, characterization, and diverse applications. Front. Chem. 9: 548 (2021).

[77] T. Dippong, E. A. Levei and O. Cadar. Recent advances in synthesis and applications of MFe2O4 (M = Co, Cu, Mn, Ni, Zn) Nanoparticles, Nanomaterials.

[78] V. M. Chakka, B. Altuncevahir, Z. Q. Jin, Y. Li and J. P. Liu. Magnetic nanoparticles produced by surfactant-assisted ball milling. J. Appl. Phys. 99: 08E912 (2006).

[79] M. Chand, S. Kumar, A. Shankar, R. Porwal and R. P. Pant. The size induced effect on rheological properties of co-ferrite based ferrofluid. J. Non. Cryst. Solids 361: 38 (2013).

[80] S. Pathak, R. Verma, S. Singhal, R. Chaturvedi, P. Kumar, P. Sharma, R. P. Pant and X. Wang. Spin dynamics investigations of multifunctional ambient scalable Fe3O4 surface decorated ZnO magnetic nanocomposite using FMR. Sci. Rep. 11: 3799 (2021).

[81] M. Saghafi, S. A. Hosseini, S. Zangeneh, A. H. Moghanian, V. Salarvand, S. Vahedi and S. Mohajerzadeh. Charge storage properties of mixed ternary transition metal ferrites MZnFe Oxides (M=Al, Mg, Cu, Fe, Ni) Prepared by Hydrothermal Method. SN Appl. Sci. 1: 1303 (2019).

[82] N. Jahan, S. Pathak, K. Jain and R. P. Pant. Enchancment in viscoelastic properties of flake-shaped iron based magnetorheological fluid using ferrofluid. Colloids Surfaces A Physicochem. Eng. Asp. 529: 88 (2017).

[83] P. Kumar, S. Pathak, A. Singh, K. Jain, H. Khanduri, L. Wang, S.-K. Kim and R. P. Pant. Observation of intrinsic fluorescence in cobalt ferrite magnetic nanoparticles by Mn2+ substitution and tuning the spin dynamics by cation distribution. J. Mater. Chem. C (2022).

[84] A. C. H. Barreto, F. J. N. Maia, V. R. Santiago, V. G. P. Ribeiro, J. C. Denardin, G. Mele, L. Carbone, D. Lomonaco, S. E. Mazzetto and P. B. A. Fechine. Novel ferrofluids coated with a renewable material obtained from cashew nut shell liquid. Microfluid. Nanofluidics 12: 677 (2012).

[85] M. Victory and S. Phanjoubam. Synthesis and characterization of Fe-Mn and Fe-Ni nanoparticles by co-precipitation method for ferrofluid applications. Https://Doi.Org/10.1080/10584587.2019.16749 80 204, 112 (2020).

[86] J.-H. Kim, K.-B. Park and K.-S. Kim. Preparation and characterization of silicone and fluorine-oil-based ferrofluids. Compos. Res. 30: 41 (2017).

[87] R. Arulmurugan, G. Vaidyanathan, S. Sendhilnathan and B. Jeyadevan. Co–zn ferrite nanoparticles for ferrofluid preparation: study on magnetic properties. Phys. B Condens. Matter 363: 225 (2005).

[88] B. Wright, D. Thomas, H. Hong, L. Groven, J. Puszynski, E. Duke, X. Ye and S. Jin. Magnetic field enhanced thermal conductivity in heat transfer nanofluids containing Ni coated single wall carbon nanotubes. Appl. Phys. Lett. 91: 173116 (2007).

[89] S. Pathak, R. Verma, P. Kumar, A. Singh, S. Singhal, P. Sharma, K. Jain, R. P. Pant and X. Wang. Facile synthesis, static, and dynamic magnetic characteristics of varying size double-surfactant-coated mesoscopic magnetic nanoparticles dispersed stable aqueous magnetic fluids. Nanomaterials 11: 3009 (2021).

[90] D. Lisjak and A. Mertelj. Anisotropic magnetic nanoparticles: a review of their properties, syntheses and potential applications. Prog. Mater. Sci. 95: 286 (2018).

[91] Y. A. Koksharov, S. P. Gubin, I. D. Kosobudsky, G. Y. Yurkov, D. A. Pankratov, L. A. Ponomarenko, M. G. Mikheev, M. Beltran, Y. Khodorkovsky and A. M. Tishin. Electron paramagnetic resonance spectra near the spin-glass transition in iron oxide nanoparticles. Phys. Rev. B 63: 12407 (2000).

[92] A. Shankar, M. Chand, G. A. Basheed, S. Thakur and R. P. Pant. Low temperature FMR investigations on double surfactant water based ferrofluid. J. Magn. Magn. Mater. 374: 696 (2015).

[93] S. Ruta, R. Chantrell and O. Hovorka. Unified model of hyperthermia via hysteresis heating in systems of interacting magnetic nanoparticles. Sci. Rep. 5: 9090 (2015).

2

A Concise Account of Optics of Magnetic Fluids
From "Baravais Iron" to Ferrofluids

R.V. Mehta

◇◇

2.1 Introduction

The word ferrofluid was first used by R.E. Rosensweig, who established the ferrohydrodynamics-hydrodynamics of magnetizable fluid [1]. Liquid like water, hydrocarbon or an ester containing nano-sized ferro or ferromagnetic particles is called a ferrofluid. To provide stability, each particle is usually coated by a stabilizing layer—called surfactant which is typically a short chain organic molecule with a polar head and hydrocarbon chain. For several years ferrofluid was the trade name of the product developed by Ferrofluidics Corporation U.S.A., founded by Rosensweig and Moskovitz [2]. It has become a synonym for all colloidal suspensions of nano-sized ferro or ferri-magnetic particles either stabilized by a surfactant or charge-stabilized particles. When not exposed to an externally applied magnetic field, these colloids act just as normal fluid. But in the presence of a magnetic field, their physical properties change significantly and exhibit several novel and intriguing phenomena [3]. We shall here focus only on the optical properties of such fluids.

2.2 History

Majorana first studied optical effects in magnetic colloids under an external magnetic field in 1902. He used an aged iron oxide colloid called 'Bravias iron' and observed magnetically induced birefringence and dichroism [4, 5]. He found that (a) the effects are proportional to the square of the applied field and (b) in a birefringent dichroic

Retired Professor, Department of Physics, Maharaja Krishnkumarsinhji Bhavnagar University, India.
Email: rvm@mkbhavuni.edu.in

colloid; the linearly polarized wave was retarded more and absorbed to a greater extent. These effects on the suspended iron particles were created by Schmauss [6]. Later Cotton and Mouton interpreted these due to the ultramicroscopic nature of the particles [7]. They also found the relationship between the fluid's extraordinary and ordinary refractive index: $(n_e\text{-}n_o) = 2\ (n\text{-}n_o)$, where n is the refractive index in zero field fields. Langevin gave the theory of these effects based on his theory of paramagnetism [8]. Heller and Quimfe showed that similar relations apply to other optical parameters like turbidity and absorption [9, 10]. The application of such colloids to study the domain structure of ferromagnetic surfaces was first demonstrated by Bitter. Chain formation of ferrofluids in a parallel field was initially predicted by Heaps [11]. Further work to determine magnetization and other useful parameters from magneto-optical effects of such colloids was investigated by Elmore, Muller and Garrood, respectively [12, 13, 14].

The works mentioned above were carried out during 1900–1940. Almost after 20 yr, one of these authors and his collaborators published a paper on magneto optical effects in magnetite colloids [15]. It may be noted here that this colloid was almost similar to that of the charged stabilized ferrofluid reported by Massart in 1981 [16]. We observed that the freshly prepared colloid exhibited an entirely different effect than the aged one (Figs. 2.1 and 2.2). Here, $Q_{(L,R,K)}$ are parameters proportional to the transmissivity of the fluid in the different field configurations. Q is the ratio of extinction coefficients of the fluid in the field and zero fields. Suffix k represents that the applied field is in the direction parallel to that of the light propagation through the sample. While 'L' and 'R' represent the polarization vector parallel and perpendicular to the propagation direction when the field is in the direction perpendicular (transverse) to the propagation direction. Dipole scattering by a cloud of magnetically oriented particles predicts that $(Q_{II}\text{-}1) = 2(1\text{-}Q_\perp)$ and $Q_\perp = Q_{.K}$ [15, 17]. It was seen that the results given in Fig. 2.2 could be explained based on dipole scattering theory [15, 17]. While for Fig. 2.1, none of these predictions hold. It can be due to aggregated chains of nano-sized particles initially developed due to magnetic dipole attractions between particles. On subsequent aging, the aggregates dissociate into independent particles. This aged colloid was found to be very stable even after several years. It exhibited transmission changes that can be explained based on dipole scattering theory [18]. At the same time, magneto-optical effects in ferromagnetic suspensions and sols were also studied by Bibik and his collaborators and by Skibin et al. [19, 20].

The first paper on the magneto-optical effect on a commercially available ferrofluid was published by Hayes in 1975 [21] and then followed by another paper in 1977 [22]. They followed a similar method as described in Reference 15 and concluded that particles form chains that grow on an increase in the fields in a magnetic field. Size and saturation magnetization was different for this colloid. It is expected as the ferrofluid used surfactant coating to stabilize the ferrofluid, unlike the charge stabilized fluid used in Reference 15. In another paper [22] they used a pulsed magnetic field to determine the magnetic moment and particle size. They concluded that coagulation occurs in water-based fluid on dilution and cannot be prevented by ultrasonic vibration. A commercially available water-based ferrofluid (EMG-7) was also studied by one of us (R.V.M.) [18]. The field direction parallel to

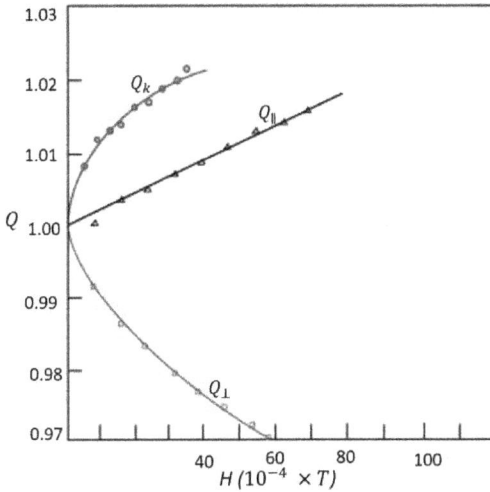

Figure 2.1. Variation of Q with applied field for the freshly prepared fluid. Note the behavior is quite different [15].

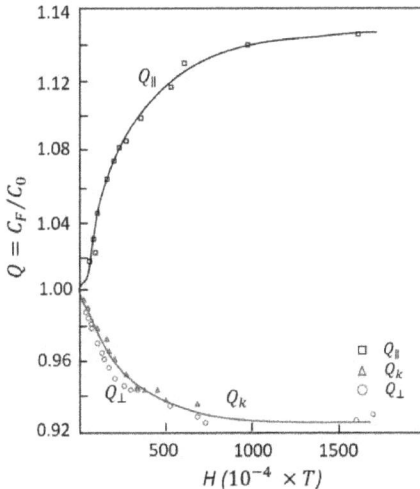

Figure 2.2. Variation of Q with the field the aged fluid.

the direction of propagation, the transmitted light was found to vary with time [18]. The extinction increases with the applied field (Figs. 2.3 and 2.4).

The theoretical curve of $[3L(h)^{-1}]$ vs. log(h) and experimental graph log(QL − Q⊥) versus Log H are matched with fitting parameters (Fig. 2.4), concluding that the system is polydispersed. From the abscissa, a shift in the magnetic moment and, subsequently, domain magnetization of the particle was found to be 6600, which is in good agreement with the domain magnetization (~ 6000) of magnetite. Soon after the advent of science and technology, ferrofluids research in magneto-optical effects increased steadily. Compared to earlier works where the predominantly aqueous medium was used, there is now the choice of

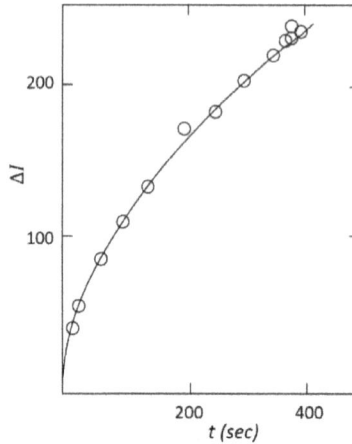

Figure 2.3. Variation of change in intensity with time [18].

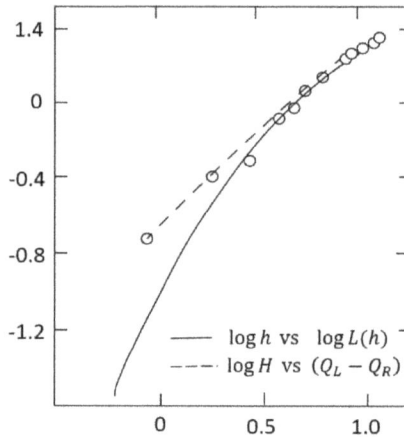

Figure 2.4. Log-Log matching of theoretical and experimental curves [18].

liquid medium like hydrocarbons, esters, etc., where magnetic particles can be dispersed [24–26]. One can use ferrite particles like iron, cobalt, nickel or ferromagnetic particles like cobalt or nickel [27–29]. Various surfactant coatings, like oleic acid, lauric acid, lecithin, etc., can also be used [30–33]. Recently green synthesis is also used for magnetite-based ferrofluid [34]. Each one will have different characteristics. These developments have led to several novel applications of the ferrofluid in photonics and revealed several intriguing effects, leading to research in theoretical developments in optics [35–40]. A detailed discussion of all these aspects is beyond the scope of the present chapter and only salient features of magneto-optics of ferrofluids will be described. We have tried to incorporate specific necessary investigations of several authors. Some other important works of many authors may have been missed, as there are a vast number of papers in this field and even today, this number is growing. This lapse is purely unintentional.

In earlier works, geometrical optics was used to explain magneto-optical effects in ferromagnetic suspension and particles were assumed to form chains in the presence of applied fields [12]. Linear chains in water base fluids are also observed in certain ferrofluids [21, 41–43] and shadow approximation is usually used [40, 43]. Such chains are not desirable in certain applications. Moreover, the proper surfactant can avoid chain formation in a non-polar medium [44]. Since individual particles are nano-sized, geometrical optics is not valid and dipole scattering theory should be more appropriate [45]. A detailed discussion of this theory is discussed elsewhere [44, 45]. Unlike shadow approximation, now the transmission depends on incident light polarization. It was shown that the field-induced extinction coefficients for the field direction transverse to the direction of propagation, the extinction coefficients and birefringence are given by [46].

$$C'_{ext} = -4\pi kN \, Im[(\alpha_e) + (\alpha_e - \alpha_o)L(h)] \tag{2.1}$$

$$(C^r_{ext}) = -4\pi kN \, Im \left[\frac{(\alpha_e) + \alpha_o}{2} - \frac{(\alpha_e) + (\alpha_o)}{2} L(h) \right] \tag{2.2}$$

$$n' - n^r = 2\pi kNRe \left(\frac{(\alpha_e) + (\alpha_o)}{2} \right) L(h) \tag{2.3}$$

$$(C_{ext})^K = (C_{ext})^r) \tag{2.4}$$

In the absence of an applied field, all the extinction coefficients reduce to

$$(C_{ext})^0) = -4\pi kN \, Im \left[\frac{2\alpha_e + \alpha_o}{3} \right] \tag{2.5}$$

It will be convenient to eliminate concentration N from the above expression. For this, we divided Equations 2.1–2.4 by Equation 2.5.

$$(Q_{ext})' = \frac{C'_{ext}}{(C_{ext})^0} = (Q') \, \infty + \frac{3Im((\alpha_e - \alpha_o))}{(\alpha_e) + 2\alpha_o} \tag{2.6}$$

$$(Q_{ext})^r) = (Q_{ext})^K = (Q^r)_\infty - \frac{1}{2}[(Q') - (Q')_\infty] \tag{2.7}$$

where

$$(Q')_\infty = 2 \, (Q^r)_\infty = \frac{3Im((\alpha_e - \alpha_o))}{(\alpha_e) + 2\alpha_o} \tag{2.8}$$

Q_x may be (l, r or k) are calculated by measuring the change in intensity of the transmitted light, i.e.,

$$Q_x = 1 - \frac{\ln\left\{1 + \frac{\Delta I_o}{I_o}\right\}}{\ln\left(\frac{i_o}{i}\right)} \tag{2.9}$$

Dipole scattering is strictly applicable for the size $\ll \lambda/20$, but it can be extended to a larger size in the Rayleigh-Gans region if the particles absorb ferrofluid like particles, which are generally iron cobalt, nickel or ferrites. In several experimental works, the above theory has satisfactorily explained the observed nature of dichroism and birefringence induced by a static magnetic field [15, 17, 18, 21, 47]. If the results fit approximately, then it was inferred that the system is polydispersed and using the log-log matching technique, size distribution was determined [18]. Particle size and magnetization of individual particles measured by the magneto-optical method agreed well with other methods. These expressions are modified in case the system contains particles with a permanent dipole moment and possesses magnetic anisotropy or is composed of a mixture of two kinds of particles or both nano and micron-sized particles. Details are given elsewhere, along with experimental observations [48, 49].

In 1980, Scholten analyzed the origin of large induced optical anisotropy in ferrofluids [50]. He listed several alternatives considered by previous investigators, such as the orientation of single particles having a permanent dipole moment [15], the orientation of superparamagnetic particles with shape anisotropy [21], chain formation [20], etc., and also added another possibility of the orientation of pre-existing small aggregates. Scholten developed a theory based on the last possibility and, after comparing the theoretical predictions with experiments, concluded that the large induced birefringence and dichroism are most likely due to pre-existing small aggregates. Later, Xu and Ridler used this concept of pre-existing small aggregates and developed a theory based on oscillating dipoles formed by two or more small particles [51]. By comparing theory and experiments, it was inferred that good matching could be observed if the sample contains only very small aggregates. It may be remarked here that, in the study, it was assumed that single nano-sized magnetic particles in a well-stabilized ferrofluid were less than 500 Oe. The experimental results above extended the two papers' fields to a few kOe [50, 51]. The water-based ferrofluid obtained from ferrofluidics also exhibited a chain formation when 1 KOe field was applied [52]. It was shown that this chain could be used as spatial filtering. It may be inferred from these findings that at a field larger than 1 KOe, the magneto-static attractive energy (m.H) will be larger than the steric repulsion. It may lead to chain formation. Moreover, the direction of the applied field and that of incident light also play an essential role. When the direction of the field is parallel to the propagation direction, the observed extinction will be independent of the direction of polarization of the light. Only in a transverse configuration will extinction coefficients depend on the direction of the electric vector of the incident light. This fact seems to have been overlooked by some authors [53]. An alternative explanation of the observed effect is possible. It is explained below.

It was shown earlier that when a system contains both nano-sized and micron-sized magnetic particles, the extinction coefficients first decrease with the field to a minimum. After that, it increases and crosses the zero field value when the incident light transverse electric vector is perpendicular to the applied field (Fig. 2.5) [49]. We have already shown that the field-dependent extinction coefficients are almost equal for $(Q_{ext})^r = (Q_{ext})^K$ (Fig. 2.6). From this, we can infer that if measurements are carried out only for longitudinal configuration, one cannot discriminate between whether the

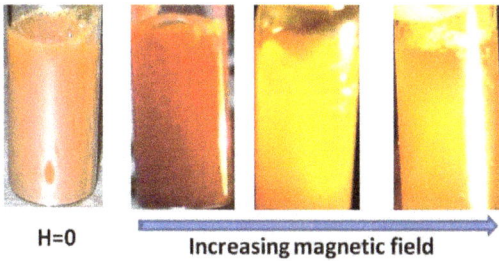

H=0   Increasing magnetic field

Figure 2.5. At *H* = 0, a typical aqueous ferrofluid brownish color is observed. It changes color under a magnetic field, with increasing strength of the field from left to right [87].

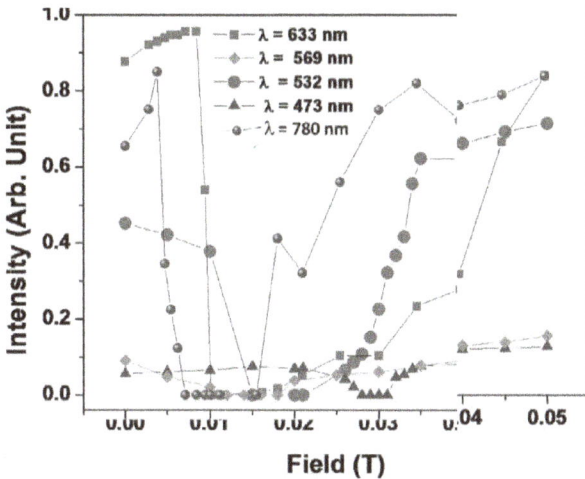

Figure 2.6. Shows the transmitted light's intensity from the bidisperse kerosene-based ferrofluid as a function of an applied field for five different wavelengths. The sample contained three μm diameter magnetite spheres stably dispersed in the ferrofluid. Decreasing the wavelength of the incident light, the stop band of field is obtained at higher applied magnetic field [87].

observed effect is due to either field-dependent aggregation of nano-sized particles (Mie particles) or due to pre-existing small and large particles.

In all the above investigations, a diluted ferrofluid was used and the sample was filled in a spectrophotometric cell. So what was the actual condition in the original sample may not be known. Taketomi first used undiluted ferrofluid to avoid this condition, which is opaque in large path length cells. He prepared a thin film of ferrofluid by sandwiching fluid between two glass cells separated by a spacer with a small hole [54]. The film was placed between an electromagnet with an air core to let the light pass through the ferrofluid with the direction of propagation of light and the field being orthogonal. He observed a very large birefringence and developed a theory to explain this as the field-dependent orientation of the chain formed in the ferrofluid. Detailed investigations on concentration dependence of the effects, types of the ferrofluid, etc., were published in subsequent publications [55, 56]. A similar commercial device is now available called Ferrocell[(c)] [57]. One author and some others recently published a paper comparing magneto-optical effects in the

Ferrocell[(c)] and the fluid having a light path in a 1 mm cell [58]. It was shown that the cell contains a mixture of large and small magnetic particles. Like this, several exciting novels and interesting magneto-optical effects were demonstrated by several authors [59–62]. This device exhibits appealing patterns when exposed to a magnetic field. The light-polarizing study can be used to characterize the magnetic field and control the quality of a compound magnet [58]. Another interesting paper is the quantum field of a magnet using the ferrocell, also shown to perform like a ferrolens [59]. The combination of ferrofluid and dispersion of nonmagnetic particles has also revealed exciting phenomena and is promising for biotechnological and other applications [63]. Among these, carbon nanotube plus ferrofluid, ferrofluid plus liquid crystals and halloysite plus ferrofluid are prominent [64, 63, 45]. So far, the effect of static magnetic field application on ferrofluids' optical properties have been described. Investigations are also carried out in the dynamic magnetic field [65, 66].

2.3 Photonics with ferrofluids

One of the most promising applications of magneto-optical application of ferrofluids is quantum information processing. They can perform all the complex operations without the loss of energy or damage to the mechanical motion. Over and above tunable photonic devices, magnetic sensors, optical tweezers, optical traps, etc., can be designed with a combination of optics and magnetics. Here some of these applications not described in earlier reviews, will be briefly discussed. Other applications were described in earlier reviews [67–70].

2.3.1 Photonic Band Gap (PBG) materials

In a PBG material, electromagnetic waves of certain frequencies are controlled. E. Yablonovitch originally mooted this idea. To verify this concept, he worked out a prototype with a 3-D diamond hole lattice in plexiglass. With this, he demonstrated the capability of PBG material to control the propagation of E.M. waves [72]. Around the same time, S. John proposed the concept of strong localization based on the quantum analog of the Maxwell equation with the Schrodinger wave equation [73]. In such materials, specific frequencies are stopped or allowed to pass. Several applications of these materials were discussed by researchers [74–76]. Among these are novel L.E.D.s, optical fibers, filters, reflectors, low-threshold lasers, etc. Earlier dielectric and metallo-dielectric materials were designed to have PBGs materials. Thermo-optic tuning of liquid crystalline-based PBG material is also efficient in building certain optical devices [77]. There are 1D, 2D or 3D dimensional PBG. The Bragg is the best-known 1D PBG material. Like an electronic band-gap, the PBG is caused by a lattice or a crystal structure. They can be active or passive. Special techniques to prepare such materials are described in books and research papers [78–80].

Magnetic fluids offer another simpler way to design PBGs at certain frequencies [81, 82]. Z. Ge et al. fabricated a highly tunable superparamagnetic nanocrystal and showed that it is possible to tune a stop band over the entire

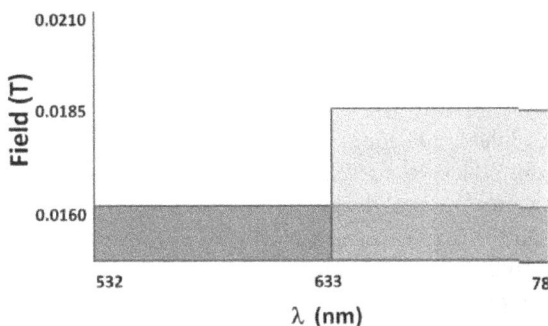

Figure 2.7. Variation of the critical field with wavelength [87].

visible spectrum and the system has a fast response [83]. Ferrofluids are also called superparamagnetic fluids as they exhibit a very large magnetization under the influence of static magnetic fields and have zero remanences and coercivity. It is shown that this fluid also exhibits the PBG effect under applied fields (Fig. 2.7) [84]. In this case, nanomagnetic particles formed linear arrays when a magnetic field is applied. These arrays act as Bragg grating. Bragg multiple scattering in all these cases is responsible for magnetically tunable PBG [4]. Mie resonance is an alternative technique that may induce a photonic band gap [85]. It is also possible to use a bidisperse ferrofluid as a PBG material. In such a fluid, a small number of micron-sized magnetic spheres are dispersed in a ferrofluid comprised of magnetic nanoparticles dispersed in a hydrocarbon liquid [86]. Magnetically tunable PBG effect was demonstrated in such systems [87]. To distinguish between the effects of Bragg scattering and Mie scattering, water-based and kerosene-based ferrofluid was filled in 5 ml glass bottles, and a rare earth magnet generated a gradient magnetic field. On white light illumination, the fluid reflects brilliant colors that depend on the field gradient's magnitude (~ 0.002 T/m) (Fig. 2.5). This effect is similar to earlier works [88, 89]. The kerosene-based ferrofluid did not show such colors. It is known that chain formation of ferrofluid particles can be formed even at very low field strength, while for hydrocarbon media, chain formation is restricted [2]. Hence the brilliant colors observed in the water-based fluid may be attributed to Bragg reflections from such chains. No such colors were observed in the suspension of large magnetic spheres. The mixtures of the two kerosene-based fluids exhibited filed-induced Mie scattering patterns [90].

2.3.2 *Energy and information storage*

Another novel application of the bidisperse ferrofluid is optical energy storage. It was shown that under critical application of the static magnetic field incident light beam with its polarization direction transverse to the direction of the applied field, the light beam is trapped [91]. The experimental assembly for the setup is shown in Fig. 2.8.

It was confirmed that the light beam is not absorbed, suggesting that the trapped beam remains in the system. One can store this energy as long as the field exists. The only loss may be leakage on the potential well [92, 93].

As soon as the field is removed, this stored energy will be released and one can obtain a more powerful beam compared to the original power of the incident beam (Fig. 2.9 and Fig. 2.10). The application of this finding may be useful in bio-optics or other high-power applications.

Another possible application is the transport of information by magnetic means [94]. Figure 2.11 shows a schematic of experimental assembly to study this phenomenon. It was observed that when an object like a thin wire or a fine letter is placed in the above fluid, the image of the object also disappears at the critical field (Fig. 2.12). Under this condition, the electromagnet cell containing the fluid is also transported to other places without any disturbance, and the field is removed. One can observe the image of the object (Fig. 2.13). It suggests this development of information storage for a limited time may be possible with the phenomenon.

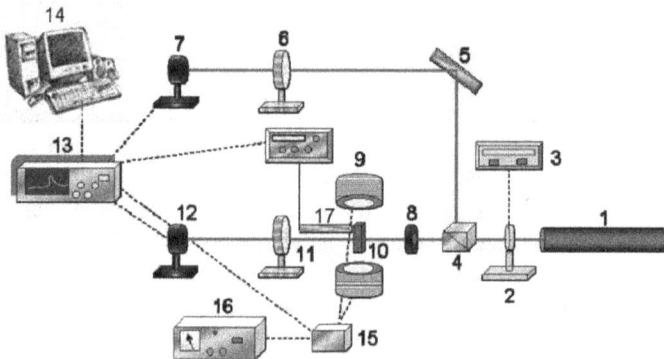

Figure 2.8. Setup used to study the unusual emission of light. 1. He–Ne polarized laser; 2. electromechanical shutter; 3. shutter control; 4. beam splitter; 5. reflector; 6. polaroid; 7. photodetector; 8. iris diaphragm; 9. electromagnet; 10. glass cuvette; 11. polaroid; 12. photodetector; 13. storage oscilloscope; 14. PC; 15. electromagnet power supply; 16. field control device; 17. Hall probe [91].

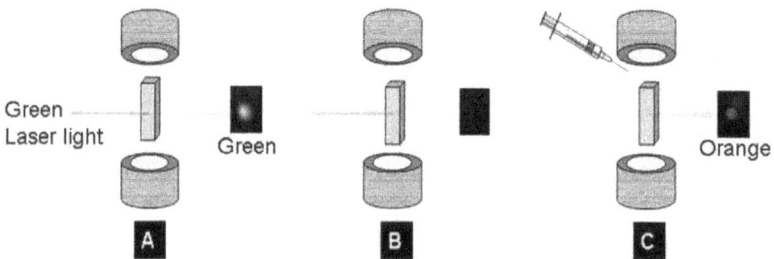

Figure 2.9. Experiment to visualize the presence of light energy having a wavelength of 532 nm even in the absence of incident light. (A) The sample was placed in a zero magnetic field and exposed to the green laser. The C.C.D. image of the transmitted pattern is shown on the right side. (B) At the field H = HC, the sample transmits no light. (C) At the field H = HC, first, the laser light is cut off (the shutter is closed) and Rh B-640 dye is injected into the sample. An orange flash is observed. The appearance of an orange flash even in the absence of the incident green light suggests the presence of light energy corresponding to wavelength 532 nm. This stored energy interacts with dye molecules and emits the orange flash [91].

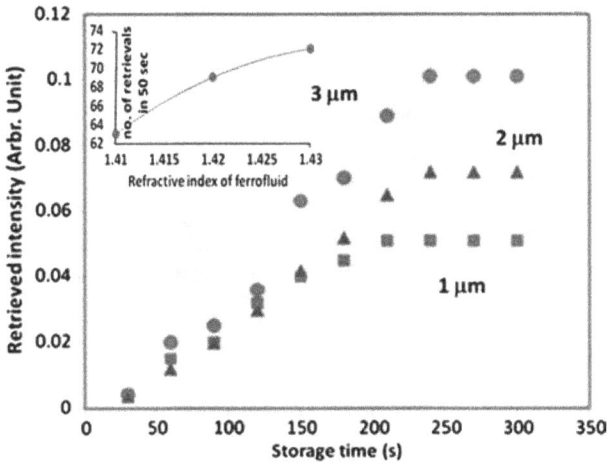

Figure 2.10. Effect of storage time on retrieved intensity for different-sized micrometer-sized magnetizable spheres. Inset shows the number of retrievals in ~ 50 s as a function of the refractive index of the ferrofluid in which the large magnetic particles are dispersed [92].

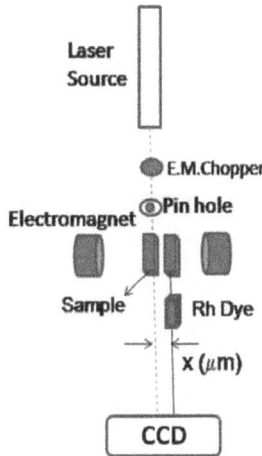

Figure 2.11. Schematics of experimental assembly to observe the trapped light at a distance 'x' cell containing 'RhB-650' dye solution was introduced to confirm the emission of excitation wavelength from the system. For photometric measurements, this cell was removed.

Besides many other optical applications of ferrofluids, like sensing, mapping of magnetic fields, detection of defects, etc., are described in several publications [95–103].

In one such application, magnetic fluid and liquid ethanol are filled in a microstructured optical fiber. This device efficiently senses temperature and the magnetic field [95]. Pathak et al. developed a magnetic fluid-based high precession temperature sensor [96]. Application to visualize defects in ferromagnetic materials was developed by Mahendran et al. [97]. Defects in such materials result in magnetic field gradients.

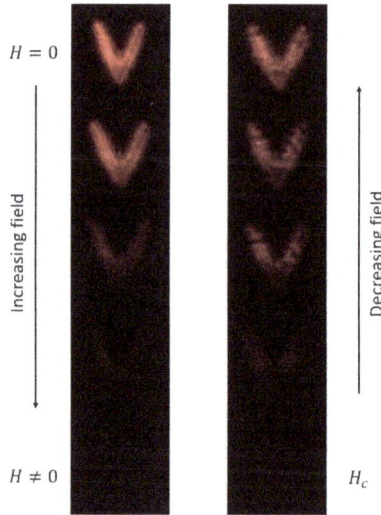

Figure 2.12. C.C.D. images at different fields of a transparent letter 'V' placed in the ferrodispersion. It is observed that the critical field image disappears and reappears again as the field decreases [109].

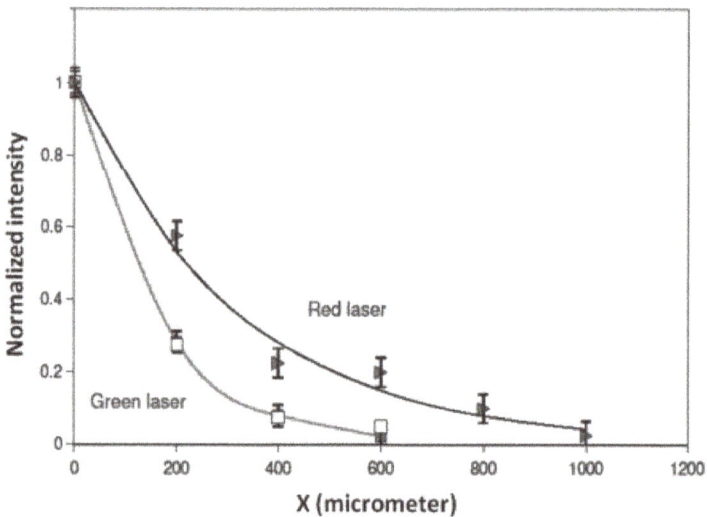

Figure 2.13. Variation of retrieved intensity with distance. © John Wiley and Sons, permission under CCC [94].

The gradients yield an array of ferrofluid droplets that diffracts white light into different colors. Wavelengths of these colors are then correlated with the defect's features. As early as 1965, Muray showed that it is possible to map multipolar magnetic fields. He measured the scattering of polarized light by aligned particles in magnetic fields at different positions and correlated it with magnetic fields [98]. Likewise, Shengli Pu, his associates, and John Philip and his collaborators have published several interesting papers on such applications [99–103].

Another area of such applications is bio-optics. Adoptive liquid crystal lens involving ferrofluid is a recent development [104, 105]. Several applications like optical fiber sensors for label-free D.N.A. detection, treatment of skin cancer with optical capacitor and solar window are promising candidates for the future development in the field of optics of ferrofluids [106–108]. As this is a concise account, it is not possible to cover all aspects. Interested readers are requested to the refer to the original papers

2.4 Summary

In this chapter, magnetically-induced optical properties of magnetic colloids, which mainly consist of single domain magnetic particles, are described from an early history to the present. These colloids are known as ferrofluids, superparamagnetic fluids or magnetic nanofluids. Such fluids have several exciting properties depending on the polarization of incident light and the direction of applied magnetic fields vis a propagation direction of incident light. Tunable optical anisotropy of such fluids is useful in several novel and intriguing applications. Many theoretical and experimental research publications covering academic and applicable areas are available in literature. We could not include all of these as this is a concise account. Even with this brief account, it is observed that the future of magnetically tunable properties of ferrofluids is bright. Soon, we may get a very useful device for health care and home applications.

Acknowledgments

Our thanks are due to Dr. Vishakha Dave for her help in redrawing the figures. Thanks to all my collaborators whose work has been cited in this review. Thanks to the Department of Physics, Maharaja Krishnkumarsinhji Bhavnagar University, Bhavnagar, for providing the facilities.

References

[1] J. L. Neuringer and R. E. Rosensweig. Ferrohydrodynamics. Phys. Fluids 7: 1927–37 (1964).
[2] R. E. Rosensweig. Ferrohydrodynamics. Cambridge Univ. Press, New York (1985).
[3] C. Schere and A. M. Neto. Figueiredo, Braz. J. Phys. 35: 718–27 (2005).
[4] Q. Majorana. Accad. Lincei. Atti. 11: 374 (1902).
[5] Q. Majorana. Compt. Rend. 135: 159 (1902).
[6] A. Schmuss. Ann. Physik. 12: 186–193 (1903).
[7] A. Cotton and H. Mouton. Compt. Rend. 141: 349–351 (1905).
[8] P. Langevin. Radium 7: 249 (1910).
[9] W. Heller and G. Quimfe. Phys. Rev. 61: 382 (1942).
[10] Y. G. Naik and J. N. Desai. Indian J. Pure Appl. Phys. 3: 27–29 (1965).
[11] F. Bitter. Phys. Rev. 38: 1903–1905 (1931).
[12] C. W. Heaps. Phys. Rev. 57: 528–531 (1940).
[13] W. C. Elmore. Phys. Rev. 60: 593–596 (1941).
[14] M. Mueller and M. S. Shamos. Phys. Rev. 62: 631–63 (1942).
[15] M. J. Dave, R. V. Mehta, Y. G. Naik, H. S. Shah and J. N. Desai. Indian J. Pure Appl. Phys. 6: 364–366 (1968).
[16] R. Massart. IEEE Trans. Magnetics, MAG-17, 1247 (1981).

[17] R. V. Mehta, H. S. Shah, J. B. Bhagat and D. M. Bhagat. IEEE Trans. Magn. MAG-16, 1324–1331 (1980).

[18] R. V. Mehta. Magneto-optics of colloids. pp. 377–395. *In*: B. Sedlacek (ed.). Physical Optics of Dynamic Phenomena and Processes in Macromolecular Systems, Walter de Gruyter & Co., Berlin, Germany (1985).

[19] E. E. Bibik, I. S. Lavrov and O. M. Merkushev. Kolloid Zh. 28: 631–634 (1966).

[20] Y. N. Skibin, V. V. Chekanov and Y. L. Raikher. Sov. Phys. JETP 45: 496–499 (1977).

[21] C. F. Hayes. J. Colloid Interface Sci. 52: 239–243 (1975).

[22] C. F. Hayes and S. R. Hwang. J. Colloid Interface Sci. 60: 443–447 (1977).

[23] R. V. Mehta and J. M. Patel. J. Surface Sci. Technology. 5: 49–58 (1989).

[24] R. V. Upadhyay and R. V. Mehta. Praman. J. Phys. 41: 429–442 (1993).

[25] P. C. Scholten. Colloid chemistry of magnetic fluids from Thermomechanics of magnetic fluids, Edited by B. Berkovasky, Hemisphere, Washington D.C. U.S.A. 1–26 (1978).

[26] R. V. Upadhyay, G. M. Sutariya and R. V. Mehta. J. Magn. Magn. Mater. 123: 262 (1993).

[27] R. Patel, K. Parikh, R. V. Upadhyay and R. V. Mehta. Indian. J. Engi. And Mat. Sci. 11: 301–304 (2004).

[28] L. Vekas, D. Bica and O. Martina. Rom. Rep. Phys. 58: 257 (2006).

[29] M. D. Shastry, Y. Babu, P. Goyal, R. Mehta, R. Upadhyay and D. Srinivas. J. Magn. Magn. Matr. 149: 64–66 (1995).

[30] G. M. Sutariya, R. V. Upadhyay and R. V. Mehta. J. Magn. Magn. Matr. 155: 152–155 (1993).

[31] R. V. Upadhyay, K. Parikh and R. V. Mehta. J. Appl. Phys. 88(2): 2799 (2000).

[32] N. Moumen and M. P. Pileni. J. Phys. Chem. 100(5): 1867–1873 (1996).

[33] R. V. Upadhyay, R. V. Mehta!, Kinnari Parekh, D. Srinivas and R. P. Pant. J. Magn. Magn. Matr. 201: 129–132 (1999).

[34] M. A. Moradiya, P. K. Khiriya and P. S. Khare. Asian J. Nanosci. Matr. 4: 263–273 (2021).

[35] L. Luo, S. Pu, S. Dong and J. Tang. Fiber-optic magnetic field sensor using magnetic fluid as the cladding. Sens. Actuators A: Phys. 236: 67–72. https://doi.org/10.1016/j.sna.2015.10.034 (2015).

[36] H. E. Horng et al. Tunable optical switch using magnetic fluids. Appli. Phys. Lett. 85: 5592–5594. https://doi.org/10.1063/1.18335 64 (2004).

[37] H. E. Horng et al. Designing optical-fiber modulators by using magnetic fluids. Opt. Lett. 30: 543–545. https://doi.org/10.1364/ OL.30.000543 (2005).

[38] T. Liu et al. Tunable magneto-optical wavelength filter of long-period fiber grating with magnetic fluids. Appli. Phys. Lett. 91: 121116. https://doi.org/10.1063/1.2787970 (2007).

[39] A. Candiani, W. Margulis, C. Sterner, M. Konstantaki and S. Pissadakis. Phase-shifted Bragg micro structured optical fiber gratings utilizing infiltrated ferrofluids. Opt. Lett. 36: 2548–2550. https:// doi.org/10.1364/OL.36.002548 (2011).

[40] Y. Zhao, X. Liu, R.-Q. Lv, Y.-N. Zhang and Q. Wang. Review on optical fiber sensors based on the refractive index tunability of ferrofluid. J. Light. Technol. 35: 3406–3412 (2017).

[41] W. E. L. Haas and J. H. H. Adams. Diffraction effects in ferrofluids. Appl. Phys. Lett. 27: 571 (1975).

[42] A. Martinet. Birefringence E.T. dichroism lineare des ferrofluids sous champnmagnetique. Rheol. Acta, 13: 260 (1974).

[43] R. V. Mehta, S. P. Vaidya, J. M. Patel and P. M. Vora. Magnetically induced spatial filtering effect magnetic fluids. Appl. Opt. 26: 2297–8. Doi: 10.1364/AO.26.002297 (1987).

[44] X. Bai, S. Pu and L. Wang. Optical relaxation properties of magnetic fluids under externally magnetic fields Opt. Commun. 284: 492 (2011).

[45] Rucha Desai, R. V. Upadhyay and R. V. Mehta. Augmentation of chain formation in magnetic fluids by addition of halloysite nanotubes. J. Phys. D. Appl. Phys. 47: 1, 65501 (2014).

[46] R. V. Mehta. Polarization dependent extinction coefficients of superparamagnetic colloids in transverse and longitudinal configurations of magnetic field. Optical Materials 35: 1436–1442 (2013).

[47] Rajesh Patel, R. V. Upadhyay and R. V. Mehta. Optical properties of magnetic and non-magnetic composites of ferrofluids. J. Magn. Magn. Matr. 300: e-217-e220 (2006).

[48] J. N. Desai, Y. G. Naik, R. V. Mehta and M. J. Dave. Indian. J. Pure and Appl. Phys. 7: 534–538 (1969).

[49] R. V. Mehta, R. V. Upadhyay, Rajesh Patel and Premal Trivedi. Magnetooptical effects in magnetic fluid containing large aggregates. J. Magn. Magn. Matr. 289: 36–38 (2005).

[50] P. C. Scholten. The origin of magnetic birefringence and dichroism in magnetic fluids IEEE Trans. Mag. MAG-16: 221–225 (1980). Doi: 10.1109/TMAG1980.1060595.

[51] M. Xu and P. J. Ridler. Linear dichroism and birefringence effects in magnetic fluids. J. Appl. Phys. 82: 326–332(1997).

[52] R. V. Mehta, S. P. Vaidya, J. M. Patel and P. M. Vora. Magnetically induced, spatial filtering effect with magnetic fluids. Appl. Opt. 26: 2297–8 (1987). Doi:10.1364/AO.26.002297.

[53] John Philip, J. M. Laskar and Baldev Raj. Magnetic field induced extinction of light in a suspension of Fe3O4 nanoparticles. Appl. Phys. Lett. 92: 229119 (2008).

[54] S. Taketomi. Magnetic fluid's anomalous pseudo-Cotton Mouton effects about 10^7 times larger than nitrobenzene. Jpn. J. Appl. Phys. 22: 1137 (1983). Doi.org/10.1143/JJAP.22.1137.

[55] S. Taketomi, M. Ukita, M. Mizukami, H. Miyajima and S. Chikazumi. Magnetooptical effects of Magnetic fluids. J. Phys. Soc. Japan 56: 3362–74 (1987). Doi.org/101143/jPSJ.56.3362.

[56] S. Taketomi, S. Ogawa, H. Miyajima and S. Chikazumi. Magnetic birefringence and dichroism in magnetic fluid IEEE Trans. Magn. Japan 4: 384 (1989).

[57] A. Tufaile, T. A. Vanderelli and A. P. B. Tufaile. Light polarization using Ferrofluids and magnetic fields, pages 7 (2017). Doi.org/10.1155/2017/258717. Advances in Condensed Matter Physics, vol, 2017, Article ID 2583717.

[58] Vishakha Dave, R. V. Mehta and S. P. Bhatnagar. Extinction of light by a Ferrocell and ferrofluid layers: a comparison. Optik-International Journal for Light and Electron Optics 217: 164861 (2020). Doi.org/10.1016/j.jileo.2020.164861.

[59] E. Markoulakis, A. Konstantaras and E. antonidakis. The quantum field of a magnet shown by a nanomagnetic ferrolens. J. Magn. Magn., Mater. 466: 252–259 (2018).

[60] A. Tufaile, M. Snyder, T. A. Vanderelli and A. P. B. Tufaile. Jumping Sundogs, Cat's Eye and ferrofluids. Condensed Matter 5: 45 (2020). Doi:10.3390/condmat5030045.

[61] A. Tufaile, T. A. Vanderelli and A. P. B. Tufaile. Observing the jumping Dogs. J. Appl. Math. Phys. 4: 1977–1988 (2016).

[62] E. Markoulakis, T. Vanderelli and L. Frantzeskakis. Real time display with ferrolens of homogeneous magnetic fields. J. Magn. Magn. Mater. 541(C): 168576 (2021). DOI: 10.1016/j.jmmm.2021.168576.

[63] X. Wang, S. Pu, H. Ji et al. Enhanced magnetic-field induced optical properties of nanostructured magnetic fluids by doping nematic liquid crystals. Nanoscale Res. Lett. 7: 249 (2012). https://doi.org/10.1186/1556-276-7-249.

[64] J.a. Garcia-Merino, C. L. Martinez-Gonzalez, C. R. Torres San Miguel et al. Magneto-conductivity and magnetically-controlled nonlinear optical transmittance in mufti-wall carbon nanotubes. Optic exp. 24: 19552–19557(2016). http://doi.org/10.1364/OE.24.019552.

[65] D. Jamon, F. Donatini, A. Sibini et al. Experimental investigation on the magneto-optic effects of ferrofluids via dynamic measurements. J. Magn. Magn. Matr. 321: 1148–114 (2009). https://doi.org/10.1016/j.jmmm.2008.10.038.

[66] M. Shual, A. Klitnick, Y. Shen et al. Spontaneous liquid crystal and ferromagnetic ordering of colloidal magnetic nanoparticles. Nat. Comm., & 10394 (2016). http://doi.org/10.1038/ncomms10394.

[67] C. Scherer and A. M. Figueiredo Neto. Ferrofluids and applications. Brazilian J. Phys. 35: 718 (2005).

[68] O. Oehisen and I. Sussy. Cervantes-ramirez, approaches on ferrofluid synthesis and applications: Current status and future perspectives, A.C.S. Omega 7: 3134–3150 (2022). https://doi.org/10.1021/acsomega.1c05631.

[69] J. Philip and J. M. Laskar. Optical properties and applications of ferrofluid, A. Review. J. Nanofluids 1(103-20) (2012). DOI:10.1166/jon.2012.1002.

[70] Jing, Dengwei, Sun, le, Jin, Jingyu et al. Magneto-optical transmission in magnetic nanoparticle suspensions for different optical applications: a review. J. appl. Phys. D: Applied Physics 54, id.013001, 17 pp (2021). DOI: 10.1088/1361-6463/abb8fd.

[71] E. Yablonovitch. Inhibited Spontaneous emission in solid-state physics and electronics. Phys. Rev. Lett. 58: 2059(1987). DOI: https://doi.org/10.1103/physRevLett.58.2059.

[72] E. Yablonovitch, T. J. Gmitte and K. M. Leung. Photonic band structure: The face-centred-cubic case employing nonspherical atoms. Phys. Rev. Lett. 67: 2295 (1991). https://doi.org/10.1103/physRevLett.67.2295.

[73] S. John. Strong localization of photons in certain disordered dielectrics Phys. Rev. B 58: 2486–2489 (1987).

[74] G. Guida, A. Lustrac de and A. Priou. Progress in Electromagnetics Research, PIER 41: 1–20 (2003).

[75] R. D. Meade et al. J. of Appl. Phys. 75: 4753–4755 (1994).

[76] O. Painter et al. Science 284: 1819–1821 (1999).

[77] T. T. Larsen, A. Bjarklev, D. S. Hermann and J. Broeng. Optics Express 11: 2589–2596 (2003).

[78] T. Sato, K. Miura, N. Ishino, Y. Othera et al. Photonic Crystals for the Visible Range Fabricated by Auto Cloning Technique and their Application 34: 63–70 (2002).

[79] Shuichi, K. Photonic structures colors in realm of nature. World Scientific, doi.org/10.1142/6496.

[80] Lopez, C. Three-dimensional photonic bandgap materials: semiconductor for light. J. Opt. A: Pure Appl. Opt., 8: RI-R14 (2206).

[81] Z. Ma, Y. Miao and J. Yao. IEEE-Photonics Journal 10: 6803308 (2018).

[82] M. Taghizadeh, F. Bozorgzadeh and M. Ghorbani. Scientific Reports 11: 14325 (2021).

[83] Jianping Ge, Yongxing Hu and Yadong Yin. Angew. Chem. Int. Ed. 46: 7428 (2007).

[84] S. Liu, J. Du, Z. Lin, R. X. Wu and S. T. Chui. Phys. Rev. B78: 15510 (2008).

[85] L. Shi, X. Jiang and C. Li. J. Phys. Condens. Matter 19: 176214 (2007).

[86] R. V. Mehta, R. Patel and R. V. Upadhyay. Phys. Rev. B 74: 195127 (2006).

[87] R. Patel and R. V. Mehta. Eur. Phys. J. Appl. Phys. 52: 30702 (2010). DOI: 10.1051/epjap/2010152.

[88] F. L. Calderon, T. Stora, O. Mondain Monval, P. Poulin and J. Bibette. Phys. Rev. Lett. 72: 2959 (1994).

[89] Jianping Ge, Yongxing Hu and Yadong Yin. Angew. Chem. Int. Ed. 46: 7428 (2007).

[90] H. Bhatt, R. Patel and R. V. Mehta. J. Opt. Soc. Am. A 27,873 (2010).

[91] R. V. Mehta, Rajesh Patel, Bhupendra Chudasama and R. V. Upadhyay. Experimental investigation of magnetically induced unusual emission of light from a ferrodispersion. Opt. Lett. 33: 1987–1989 (2008).

[92] R. Patel and R. V. Mehta. Ferrodispersion: a promising candidate for an optical capacitor. Appl. Opt., 50: G17–G22 (2011).

[93] R. V. Mehta, R. Patel, B. Chudasama, H. B. Desai, S. P. Bhatnagar and R. V. Upadhyay. Magnetically controlled storage and retrieval of light from dispersion of large magnetic spheres in a ferrofluid. Current Sci. 93: 1071–1072 (2007).

[94] Rajesh Patel and R. V. Mehta. Experimental demonstration of magnetic carriage for transport of light trapped in magnetizable Mie Sphere. Adv. Opt. Mat. 1: 703–706 (2013). Doi: doi.org/10.1002/adom.201300123.

[95] E. Wang, P. Chang, J. Li et al. High-sensitivity temperature and magnetic sensor based on magnetic fluid and liquid ethanol filled micro-structured optical fiber Optical fiber Technology 55: 102161 (2020). DOI: 10.1016/,yofte.2020.10216.

[96] S. Pathak, K. Jain, Noorjahan, V. Kumar and R.P. Pant. Magnetic fluid based high recession temperature sensor. IEEE Sensors Journal 17(9): 2670–2675.2675440.

[97] V. Mahendran and John Philip. Nanofluids based optical sensor for rapid visual inspection of defects in ferromagnetic materials. Appl. Phys. Lett. 100: 073104 (2012). Doi: 101063/1.3684969.

[98] J. J. Muray. Scattering of polarized light on magnetically aligned particles in multipole magnetic fields. Appl. Opt. 4: 1011–1016 (1965).

[99] Fan Shi, Xuekun Bai, Feng Wang, Fufei Pang, Shengli Pu and Xianglong Zeng. All-fiber magnetic field sensor based on hollow optical fiber and magnetic fluid. IEEE Sensor Journal 17: 619 (2017).

[100] Min Dai and Shengli Pu. Synthesis and faraday effect of Fe-Al oxide composite ferrofluid. Advances in Materials Physics and Chemistry 5: 344–349 (2015). DOI: 10.4236/ampc.2015.58034.

[101] V. Mahendran and John Philip. Spectral response of magnetic nanofluid to toxic cations: Appl. Phys. Lett. 102: 163109 (2013). Doi: 10.1063/1.4802899.

[102] V. Mahendran. John Philip. An optical technique for fast and ultrasensitive detection of ammonia using magnetic nanofluids. Appl. Phys. Lett. 102: 063107 (2013). Doi: 10.1063/1.4792055.

[103] V. Mahendran and John Philip. Non-enzymatic glucose detection using magnetic nanoemulsions, Appl. Phys. Lett. 105: 123110 (2014). Doi: 10.1063/1.4896522.

[104] J. F. Algorri, D. C. Zografopoulos, V. Urruchi and J. M. Sanchez-Pena. Recent advances in adaptive liquid crystal Lenses. Crystals 9: 272 (2019). https://doi.org./10.3390/crysst9050272.

[105] W. Xiao and S. Hardt. An adaptive liquid microlens driven by a ferrofluid transducer. J. Micromech. Microeng. 20: 055032. https://doi.org/10.1088/0960-1317/20/5/055032.

[106] Hui-Chuan Cheng, S. Xu, Y. Liu et al. Adaptive mechanical-wetting lens actuated by ferrofluids. Opt. Comm. 284: 2118–2121 (2011). https://doi.org/10.1016/j.optcom.2010.12.073.

[107] M. Barozzi et al. Optical fiber sensors for label-free D.N.A. detection. J. light wave Technology 35: 3461–3472 (2017). Doi: 10.1109/JLT.2016.26007024.

[108] K. S. Jayaraman. Magnetic highway to transport stored light. Nature India. Doi: 10.1038/nindia.2013.114.

[109] Vishakha Dave (Late) Rajesh Patel† and Rasbindu V. Mehta. Magnetooptical effect of immersion of nonmagnetic macroscopic objects in the ferrofluid. J. Nanofluids. 5: 1–5 (2016).

3

Behavior of Magnetic Nanoparticles in the Presence of Magnetic and High Intensity Electric Fields
Nonlinear Optical Effects

Antônio Martins Figueiredo Neto[1,*] and *Claudio Scherer*[2]

◇◇

3.1 Introduction

The interaction of magnetic nanoparticles (MNPs) present in a ferrofluid with external magnetic fields is a subject of research extensively studied for many years. Static and variable fields applied in ferrofluids revealed unique phenomena, interesting from these materials' fundamental and technological applications.

On the other hand, the behavior of ferrofluids under the action of electric fields has received less attention. In 2014, Rajnak and co-workers reported observing the clustering of magnetic nanoparticles in a ferrofluid subjected to an ac electric field [1]. The typical electric field strength and frequency range employed were 20 kV/m and from 20 Hz to 2 MHz. More recently, Cherian and co-workers [2] synthesized a complex fluid made of charged superparamagnetic iron oxide nanoparticles stabilized with oleic acid in dodecane, with the addition of bis(2-ethylhexyl) sulfosuccinate sodium salt (aerosol OT), that forms reverse micelles and charges the nanoparticles. When the mixture, present in a holder in the shape of a slab, with planar electrodes, is subjected to a static electric field, a field-induced concentration gradient of nanoparticles is stabilized along the sample. Since the magnetic behavior

---

[1] University of São Paulo, Institute of Physics, São Paulo, Brazil.
[2] Federal University of Rio Grande do Sul, Institute of Physics, Porto Alegre, Brazil.
* Corresponding author: afigueiredo@if.usp.br

of usual ferrofluids depends on the nanoparticle's concentration, the new complex fluid (named *electroferrofluid*) has the advantage of allowing states with voltage-controlled magnetism.

To the best of our knowledge, the first results on the interaction of high-intensity light (frequency of about 500 THz) with magnetic nanoparticles in a ferrofluid were reported in 2007 by Soga and co-workers [3]. Different physical mechanisms may be triggered by high-intensity light reaching MNPs present in ferrofluids, from electronic to thermodiffusion processes, leading to the appearance of MNPs concentration gradients [4]. This triggering depends on the time the sample is illuminated by the light beam, its wavelength and the linear absorption of the sample.

In the first part of this chapter, the experiments that allowed the determination of nonlinear optical parameters of magnetic nanoparticles present in ferrofluids, subjected to high-intensity electric fields, with light in the frequency region of $\sim 10^2$ THz are described. The magnetic moment of each nanoparticle will be used to orient them in the presence of an external magnetic field, $\vec{H}$, allowing the determination of nonlinear optical parameters with the optical electric field parallel and perpendicular to $\vec{H}$.

Later the problem of the interaction of a strong magnetic field with the ferromagnetic particles, giving special attention to the theoretical procedure of performing computer simulation of the rotation of the magnetic moments will be treated. It was shown how to derive the stochastic differential equations for the rotation of magnetic moments and orient the reader to use Stratonovich [5, 6] calculus to perform simulations.

3.2 Nonlinear absorption coefficient, β, refractive index, n_2, free-carrier absorption, σ_{FCA} and the first-order hyperpolarizability, β_{HRS}

When a material is illuminated by a high-intensity optical electrical field, $\vec{E}(\vec{r}, t)$, of the order of $E \sim 10^8$ V/m (e.g., produced by a laser), a polarization, $\vec{P}(\vec{r}, t)$, is induced according to Equation (3.1). Considering a microscopic volume of the material, located in a particular position, the macroscopic polarization depends on the coupling between the field and the electric charge distribution in this volume, and can be written as [7]:

$$\vec{P} = \vec{P}^{(1)} + \vec{P}^{(2)} + \vec{P}^{(3)} + \cdots \qquad (3.1)$$

where the different $\vec{P}^{(n)}$, in the frequency domain, are written as:

$$\vec{P}^{(n)}(\omega) = \varepsilon_0 \int_{-\infty}^{+\infty} \cdots \int_{-\infty}^{+\infty} \chi^{(n)}(-\omega_\sigma; \omega_1, \cdots, \omega_{(n)}) \qquad (3.2)$$
$$\times \vec{E}(\omega_1) \cdots \vec{E}(\omega_n) \delta(\omega - \omega_\sigma) d\omega_1 \cdots d\omega_n,$$

δ and $\chi^{(n)}$ are the Delta function and n^{th} order electric susceptibility, and $\omega_\sigma = \omega_1 + \omega_2 + \cdots \omega_n$. The electric field can be written as a superposition of its monochromatic components as $\vec{E}(t) = \sum_k \Re[\vec{E}_{\omega k} \exp(-i\omega t)]$, with $\omega_k \geq 0$.

The components of the different orders of polarization are written as:

$$P_i^{(1)}(\omega) = \varepsilon_0 \sum_j \chi_{ij}^{(1)}(\omega) E_j(\omega); \tag{3.3}$$

$$P_i^{(2)}(\omega_3) = \varepsilon_0 \sum_{jk} \sum_{nm} \chi_{ijk}^{(2)}(-\omega_3; \omega_n, \omega_m) E_j(\omega_n) E_k(\omega_m); \tag{3.4}$$

$$P_i^{(3)}(\omega_4) = \varepsilon_0 \sum_{jkl} \sum_{nmo} \chi_{ijkl}^{(3)}(-\omega_4; \omega_n, \omega_m, \omega_o) E_j(\omega_n) E_k(\omega_m) E_l(\omega_o), \tag{3.5}$$

where $\omega_3 = \omega_n + \omega_m$; $\omega_3 = \omega_n + \omega_m + \omega_o$.

The optical Kerr effect is characterized by a change in the refractive index of a material which is proportional to the square of the electric field, i.e., the intensity of the light beam, I:

$$\Delta n(I) = n(I) - n_0 = n_2 I, \tag{3.6}$$

where n_0 and n_2 are the linear and nonlinear refractive indices.

In the same way, the Two-Photon Absorption (TPA) may occur if the material is illuminated by a high-intensity light beam:

$$\Delta \alpha(I) = \alpha(I) - \alpha_0 = \beta I, \tag{3.7}$$

where α_0 and β are the linear and nonlinear absorption coefficients. The TPA is a process where the energy gap from the ground state to the excited state is higher than the energy of a single photon ($\hbar\omega$, where $\hbar = h/2\pi$ and h is the Planck's constant) reaching the material. In the condition of the high-intensity beam, two photons may be absorbed by electrons of the material. The cross-section for this process, σ_{TPA}, is related to β, according to Equation 8:

$$\beta(\omega) = \frac{\sigma_{TPA}}{\hbar\omega} N, \tag{3.8}$$

where N is the numerical density of absorbers.

The nonlinear electric susceptibility is related to these two phenomena, the Kerr effect and TPA, is the complex tensor, $\chi^{(3)}$, of the 4th order and 81 components, and it is possible to write n_2 and β in terms of elements of this tensor as [8]:

$$Re\ \chi_{xxxx}^{(3)} = 2n_0^2 \varepsilon_0 c\ n_2, \tag{3.9}$$

$$Im\ \chi_{xxxx}^{(3)} = \frac{n_0^2 \varepsilon_0 c^2}{\omega} \beta, \tag{3.10}$$

where c is the speed of light in a vacuum.

Equation (3.7) assumes that the TPA is a unique process occurring in the interaction of light with the material. However, there is another process that can occur, the *Free-Carrier Absorption* (FCA). In the FCA, electrons absorb photons and are excited across the energy gap, being able to absorb more photons in the sequence. Considering both the TPA and FCA, Equation (3.7) is written as:

$$\alpha(I) = \alpha_0 + \beta I + \sigma_{FCA} N^*, \tag{3.11}$$

where σ_{FCA} and N^* are the FCA cross section and the numerical density of free carriers.

The Z-Scan (ZS) is an experimental technique that can be employed to measure n_2, β and σ_{FCA} of a material. The Real and Imaginary parts of $\chi^{(3)}_{xxxx}$ can, therefore, be calculated [9, 10].

Knowing the energy-band spectrum of the material and suitably choosing the photon energy of the laser, pulse width and frequency in the ZS experiment, it is possible to measure, β and σ_{FCA} independently [11].

While considering the interaction of light with a small volume, v (linear dimensions much smaller than the light wavelength), containing a large number,, of independent single absorbers/scatterers (named hereafter, *unit element*). The electric field of the light generates a dipole in this unit element, and the total scattering from this volume is proportional to $N = N_0/v$.

The Hyper-Rayleigh Scattering (HRS) is a nonlinear elastic incoherent mechanism where two photons (frequency ω) from the pumping beam in interaction with a scatterer are annihilated, giving rise to a single photon of frequency 2ω [12, 13]. This process is also known as a second-harmonic generation. Solutions showing inversion symmetry, like ferrofluids, would not be allowed to show nonlinear optical processes of even order. However, fluctuations in the local orientation of the nanoparticles are responsible for the existence of the second-harmonic generation [14]. The Hyper-Rayleigh scattered intensity, I(2ω), by a sample illuminated by a light beam of intensity $I(\omega)$ is written as:

$$I(2\omega) = GN\beta_{HRS}^2 I(\omega), \tag{3.12}$$

where G and β_{HRS} are an experimental factor and the first-order hyperpolarizability [15].

Details about the HRS experimental technique may be found in [15–17].

3.3 The Z-Scan technique and setup to measure n_2, β and the σ_{FCA}

As pointed out earlier, when a laser beam interacts with a material, different physical phenomena may occur, depending on various factors, like the light intensity, wavelength, duration of the pulse and frequency (in the case of pulsed beams). Isolating the different phenomena to measure reliable parameters that characterize a particular phenomenon is a challenge that experimentalists face in life. In the case of electronic effects, like those that one is interested here, the most common source of problems in the ZS experiment is to ensure there no contamination in the measurements due to thermal effects. There are orders of magnitudes in the characteristic times related to these phenomena. Electronic effects occur in the time scale of femtoseconds and thermal effects (usual ferrofluids and laser light of wavelength of 532 nm) in time scales of the order of milliseconds.

Figure 3.1 shows a sketch of the ZS experiment used to measure nonlinear optical parameters of ferrofluids in the femtoseconds time-scale domain [18, 19]. In the ZS technique, a light beam originates from a pulsed laser, propagating along the z direction of the laboratory frame axes, and is focused at the $z = 0$ position. The sample, in the slab shape, is scanned along the focus position. The beam transmitted by the sample is divided in two by a beam splitter. Part of the transmitted beam is collected in a photo detector (D_2) to measure the nonlinear optical absorbance (*opened aperture geometry*). The other part of the beam reaches an iris in front of a second photo detector (D1) that registers the transmittance to measure the nonlinear refractive index (*closed aperture geometry*). The technique allows the optical transmittance measurement in both detectors as a function of the sample (z) position.

The experiments described in this chapter were obtained with a Ti: Sapphire laser (Chameleon Ultra II, Coherent Inc.), working at the frequency of 80 MHz, pulse width of 196(3) fs FWHM. Depending on the particular nonlinear absorption we were interested in measuring (FCA or TPA), the photon's wavelength was conveniently chosen (for magnetite nanoparticles, 532 nm or 800 nm, respectively). In the focus of the setup, the on-axis peak intensity is $I_0 = 16$ (1) GW/cm^2.

At these experimental conditions, thermal effects are expected in the experiment, besides the electronic effects of interest. Therefore, the original laser frequency is reduced to 1 kHz by using a pulse picker in the beam trajectory before the sample. The pulse picker reduces the frequency of the high-intensity laser pulses. However, polarization leakage in the device allows the transmission of low-energy pulses. As the frequency of these pulses is high, a background beam reaches the sample, and the integral energy accumulated in the sample is enough to induce thermal effects, introducing objects in the measurement of the nonlinear parameters associated with the TPA. To solve this problem, a shutter is introduced in the setup to prevent the background reaching the sample. As depicted in Fig. 3.1, the sample receives a high-intensity light pulse during 196 fs, about three low-energy pulses (about 10 ms) after the shutter closes, and no light reaches the sample during 2 ms. This sequence is repeated during the ZS measurements. It was experimentally observed that for the ferrofluids investigated and under these experimental conditions, no thermal effects are triggered in the ZS experiment [19]. These cautions are essential to obtain reliable values of the nonlinear optical parameters in the fs time scale.

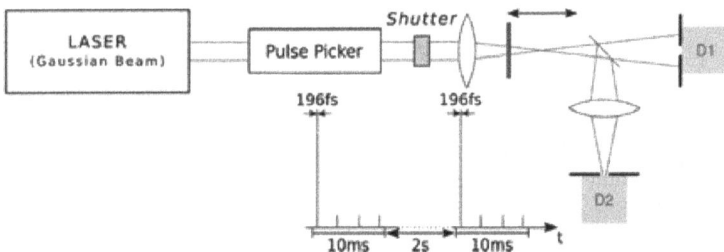

Figure 3.1. Sketch of the ZS experiment employed to measure nonlinear optical parameters of ferrofluid in the femtoseconds time-scale domain. In front of detector D$_1$ an iris limits the detected optical transmittance (this geometry is named *closed aperture geometry*). Detector D$_2$ detects the hole transmittance transmitted by the beam splitter (this geometry is named *opened aperture geometry*). Adapted from [19]. Reproduced with the permission from AIP Publishing.

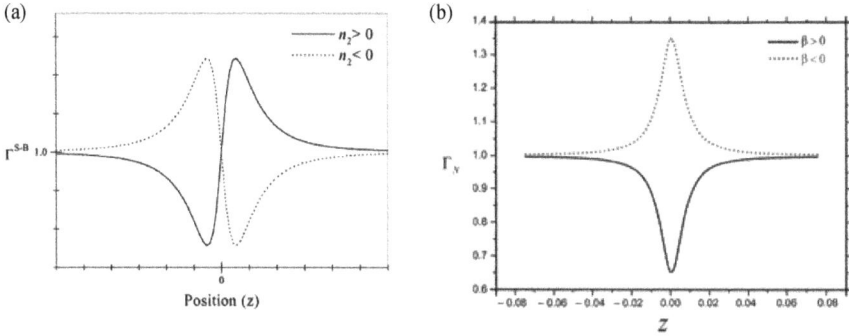

Figure 3.2. Sketches of the normalized optical transmittance, Γ^{S-B}, Γ_N, as a function of the sample position, z, in the ZS experiment. (a) closed aperture geometry; (b) opened aperture geometry.

In the ZS technique, the sample acts like an additional lens placed in the path of the laser beam, whose refractive index depends on the beam intensity. Depending on the sign of n_2, the typical ZS normalized optical transmittance (closed aperture geometry) as a function of the z position of the sample is characterized by a peak to valley ($n_2 < 0$) or valley to peak ($n_2 < 0$), Fig. 3.2a. In the case of the absorption measurement (opened aperture geometry), a peak ($\beta < 0$) or a valley ($\beta > 0$) is formed when the sample is in the vicinity of $z = 0$, Fig. 3.2b.

3.4 Fe_3O_4 nanoparticles

Figure 3.3 shows the typical linear absorbance spectra of Fe_3O_4 nanoparticles in a ferrofluid, with and without an external magnetic field [20]. The absorbance spectrum allows the evaluation of the direct and indirect optical gaps, E_g, according to Equation (3.13) [21, 22]:

$$[\alpha_0 \hbar\omega]^n = C(\hbar\omega - E_g), \tag{3.13}$$

where C is a constant, $n = 2$; $1/2$ correspond to the direct and indirect allowed transitions, respectively. These plots are shown in Figs. 3.4a and 4b. The optical

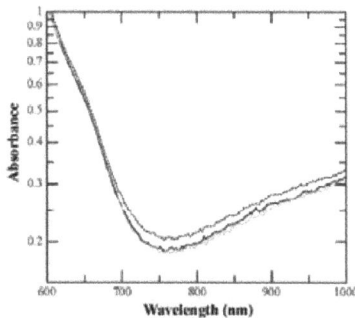

Figure 3.3. Typical linear absorbance spectra of spherical magnetite nanoparticles with diameters of about 7 nm, dispersed in a synthetic ester oil. Solid line: unpolarized light, without magnetic field; Dotted and gray lines: magnetic field, $\vec{H}$, of 2700 Oe, light polarized parallel and perpendicular to $\vec{H}$. Reproduced, with permission, from [20] © The Optical Society.

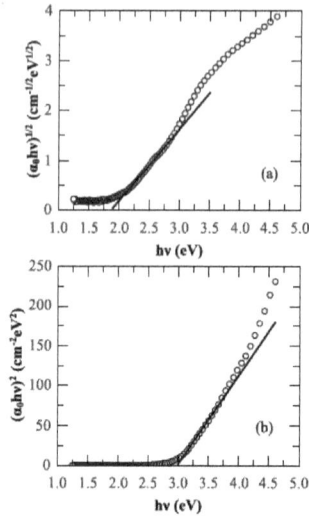

Figure 3.4. Plots of $(\alpha_0 \hbar\omega)^{1/2}$ (a) and $(\alpha_0 \hbar\omega)^2$ (b) as a function of the photon energy, with linear absorption coefficient data from Fig. 3.3. Solid black lines are linear fits to evaluate the gaps in the limits of $(\alpha_0 \hbar\omega)^{\frac{1}{2;2}} \to 0$. Reproduced, with permission, from [24].

gaps in both transitions are evaluated by fitting a linear function in the range of small values of $(\alpha_0 \hbar\omega)$, extrapolating the fits to $(\alpha_0 \hbar\omega)^{\frac{1}{2;2}} \to 0$. The optical gaps obtained through this procedure are 3.0(1) eV and 1.9(1) eV for the direct and indirect transitions. Fontijn and co-workers [23] obtained the electronic structure of the Fe ions in Fe_3O_4 (Fig. 3.5). The transitions evaluated in Fig. 3.4 are identified in the sketch of the electronic structure with arrows. These values of the gaps were shown to be independent on the diameter (in the range from 7 to 16 nm) and on the different coatings of the coating of the nanoparticles [24]. Knowing the electronic structure and energy gaps, one can choose the convenient energy of photons (i.e., laser wavelength) to investigate the particular nonlinear optical parameter one is interested in.

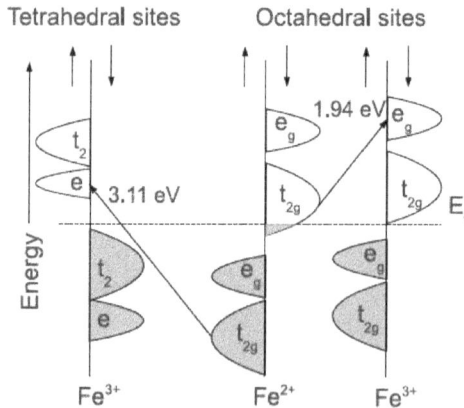

Figure 3.5. Sketch of the electronic structure of the Fe ions in Fe_3O_4. Reproduced, with permission from [23].

3.4.1 The free-carrier absorption cross section, σ_{FCA}

In this particular study, a synthetic isoparaffinic oil-based surfaced ferrofluid was used [11]. The mean diameter of the particles is about 10 nm. The energy of the band gap, obtained from the measurement of the linear absorbance spectrum, is $E_g = 1.9$ eV, as mentioned earlier. The sample is encapsulated between optical glasses in the shape of a slab, and the ZS experiment in the opened aperture geometry is used. Special care must be taken to choose the pulse frequency of the laser, f, pulse duration, Δt, pulse energy, E_p and wavelength of the light, λ, to get rid of thermal effects (i.e., thermal lens and the Soret effect). The optimum experimental conditions were achieved with $\Delta t = 100$ ps, $f = 20$ Hz, $E_p = 3.3$ μJ and $\lambda = 532$ nm, which corresponds to an energy $E_f = 2.33$ eV. The interval between pulses is of 50 ms. In these experimental conditions, photons promote electrons from the valence band to the conduction band, and these electrons can absorb more photons. Figure 3.6 shows the ZS experimental results of the normalized transmittance as a function of the sample position. A simple model for the normalized transmittance, Γ_N, as a function of the sample position, assuming $\beta = 0$ in Equation (3.11), enables writing Equations (3.14, 3.15), used to fit the experimental results [11]:

$$\Gamma_N(z) = \frac{ln[1 + p(z)]}{p(z)}, \tag{3.14}$$

$$p(z) = \frac{\alpha_0 \, \sigma_{FCA} \, F_0 \, L_{eff}}{2\hbar\omega(1+z^2/z_0^2)}, \tag{3.15}$$

where, $L_{eff} = [1 - \exp(-\alpha_0 L)]/\alpha_0$, F_0, L and z_0 are the on-axis fluence at focus, the sample thickness and the Rayleigh length, respectively. It is important to note that this is a one-parameter fit for the experimental data.

The free-carrier absorption coefficient obtained is $\sigma_{FCA} = 3.1(5) \times 10^{-18}$ cm^2.

An aspect that should be emphasized at this point is that the determination of σ_{FCA} is tricky since, besides removing objects brought by undesirable thermal effects in the ZS experiment, wrong interpretations of the experimental results can be easily made. For example, if the experimental parameters are not precisely determined to define which type of process one is investigating, it is easy to try to interpret the nonlinear absorption results (e.g., Fig. 3.6), assuming them as a sign of the parameter β. It has led to claims from some authors giving values for β, that are incorrect.

3.4.2 The nonlinear absorption coefficient, β, and the refractive index, n_2

Since the aim now is to investigate the TPA process occurring in the ferrofluid, the choice of the photon energy of the laser beam is a critical issue. As shown in Fig. 3.5, the energy gap of the direct transition is about 3.0 eV. If one tunes the laser to produce photons of 800 nm (i.e., the energy of 1.55 eV), the linear absorption is small (see Fig. 3.3), and the TPA phenomenon is favored, in the case of a high-intensity laser beam.

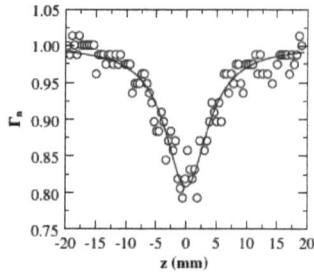

Figure 3.6. ZS in opened-aperture geometry. Normalized transmittance as a function of the sample position. Ferrofluid with Fe_3O_4 nanoparticles, 10 nm in diameter. Solid line is a fit to determine the free-carrier absorption coefficient. Reproduced, with permission, from [11] © The Optical Society.

Figure 3.7 shows the ZS experimental data obtained with a ferrofluid in both closed (a) and opened-apertures (b) geometries [25]. In the ZS experimental technique, the normalized transmittances for the closed (Equation 3.16) and opened-aperture (Equation 3.17) geometries are written as [20]:

$$\Gamma_N(z) = 1 + \frac{8\pi}{\lambda} \frac{n_0 I_0 L_{eff}(z/z_0)}{[(z/z_0)^2 + 9][(z/z_0)^2 + 1]},$$ (3.16)

$$\Gamma_N(z) = \sum_{m=0}^{\infty} \frac{[-\beta I_0 L_{eff}/(1+(z/z_0)^2)]^m}{(m+1)^{3/2}}.$$ (3.17)

These equations are used to determine n_2 and β. With these parameters, it is possible to calculate the σ_{TPA} (Equation 3.8) and (Equation 3.9 and 3.10):

$$Re\chi^{(3)} = -(4.1 \pm 1.8) \times 10^{-20}\ m^2/V^2;$$

$$Im\chi^{(3)} = (1.2 \pm 0.6) \times 10^{-20}\ m^2/V^2;$$

$$|\chi^{(3)}| = (4.3 \pm 1.9) \times 10^{-20}\ m^2/V^2 \sim 3.1 \times 10^{-12} \sim esu;$$

$$n_2 = -(3.5 \pm 1.5) \times 10^{-14}\ cm^2/W;$$

$$\beta = (1.6 \pm 0.8)\ cm/GW;$$

$$\sigma_{TPA} = (50 \pm 2)\ GM^{\#}.$$

[#] 1 GM = 1 Goeppert-Mayer corresponds to 10^{-50} cm⁴ s photon⁻¹.

Since there is no external magnetic field applied to the samples in these experiments, these results are mean values of the parameters, assuming that the nanoparticles are randomly oriented in the volume illuminated by the laser beam. Moreover, these values were shown to be independent of the particle's coatings (dextran sulfate; dextran; uncoated anionic charged; phosphatidylcholine; citric acid; aminosilane; chitosan and oleic acid) [24].

On the other hand, n_2 and β showed a dependence on the particle's diameter, D_{XRD}, measured by using the X-Ray diffraction technique [24]. The dependence of n_2 with D_{XRD} was shown to be not monotonic (Fig. 3.8a), presenting a minimum of

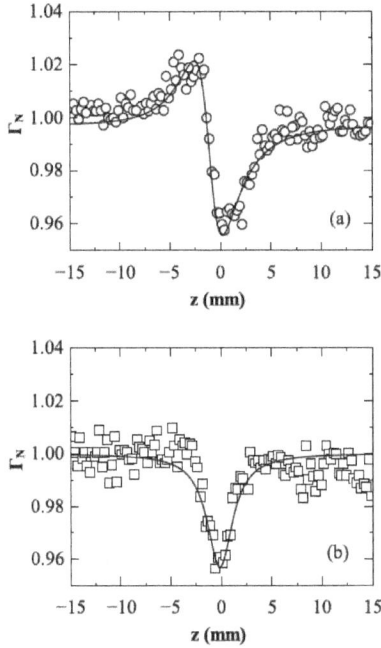

Figure 3.7. Zs experiment in closed-aperture (a) and opened-aperture (b) geometries. Normalized transmittance as a function of the sample position. Ferrofluid with Fe_3O_4 nanoparticles, 10 nm in diameter. Laser wavelength 800 nm, on-axis peak intensity at focus 11 GW/cm^2. Solid lines are fitted (Equations 3.16 and 3.17) to determine n_2 and β. Reproduced with permission of AIP Publishing, from [25].

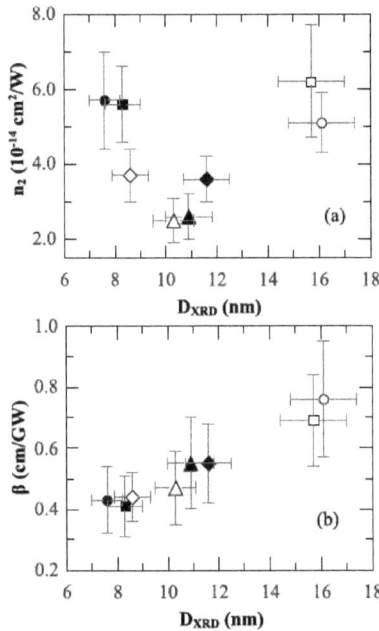

Figure 3.8. (a) Nonlinear refractive index, n_2, and (b) nonlinear absorption coefficient, β, as a function of the magnetite particle's diameter, D_{XRD}. Reproduced, with permission, from [24].

around D_{XRD} = 10 nm. This behavior is not yet clearly understood, and additional experiments are necessary to comprehensively describe the nonlinear refraction of ferrofluids with particles of different dimensions. However, the nonlinear absorption coefficient showed a linear behavior, increasing with the growing particle diameter (Fig. 3.8b), although the volume fraction of nanoparticles in all the samples was the same (2%). The energy gaps evaluated from the linear absorbance spectra measurements and the plots obtained with Equation (3.13) were shown to be about the same for particles of different diameters (differences smaller than 3.5%). Hence, the linear behavior observed in Fig. 3.8b could be due to a change in the electronic-orbitals shapes, modifying the σ_{TPA}, increasing the probability of the TPA process as the particles are bigger. Another possibility that cannot be disregarded is the effect of the shell, always present in the magnetite nanoparticles employed in this study. These particles are of the core-shell type, with a shell of about the same thickness as all the particles. The surface-to-volume ratio of spherical particles is inversely proportional to the particle diameter. The absorption properties of Fe and the electronic orbitals structures, may depend on the environment where the atom is located, in bulk or at the surface [26].

Another approach is to investigate the behavior of n_2 and β as a function of the *numerical density of* Fe_3O_4 *unity absorbers, N*:

$$N = \frac{\rho N_A}{M} \emptyset,$$

(3.18)

where ρ = 5.18 g/cm³, N_A, M = 231.54 g/Mol and $\emptyset$ are the mass density, the Avogadro number, the unit weight and the volume fraction of solid material in the ferrofluid [19]. Figure 3.9 shows n_2 (Fig. 3.9a) and β (Fig. 3.9b) as a function of N. The behavior of these parameters is similar to that obtained as a function of D_{XRD}, i.e., a nonmonotonic and linear, respectively. As described earlier, this absorption refers to the $t_{2g} \to e$ intervalence charge-transfer transition from Fe^{2+} ions (octahedral sites) to Fe^{3+} (tetrahedral sites), as shown in Fig. 3.5.

3.4.3 The Hyper-Rayleigh scattering setup to measure the hyperpolarizability, β_{HRS}

Figure 3.10 shows a sketch of the setup used to measure β_{HRS}. A mode-locked, Q-switched Nd: YAG laser (λ = 1064 nm and frequency ω) is the primary light source, and the detector registers the light at 532 nm (frequency 2ω) scattered by the sample [15]. In this experiment, the sample is oriented by an external magnetic field, H = 800 Gauss, whose direction can be parallel or perpendicular to the incident light-polarization direction. In these experimental conditions, it is possible to measure β_{HRS} in both (orthogonal) relative polarization directions. The mean diameter of the nanoparticles is D_{XRD} ~ 16 nm.

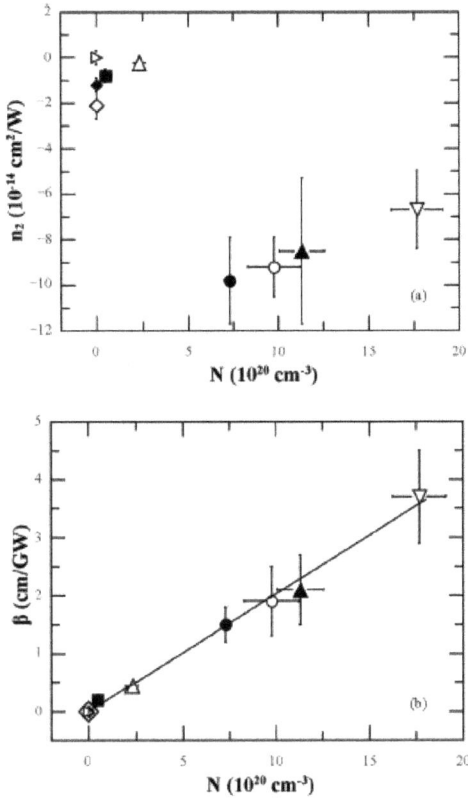

Figure 3.9. Nonlinear refractive index, n_2, and nonlinear absorption coefficient, β, as a function of the numerical density of Fe_3O_4 units, N. Reproduced with permission of AIP Publishing, from [19].

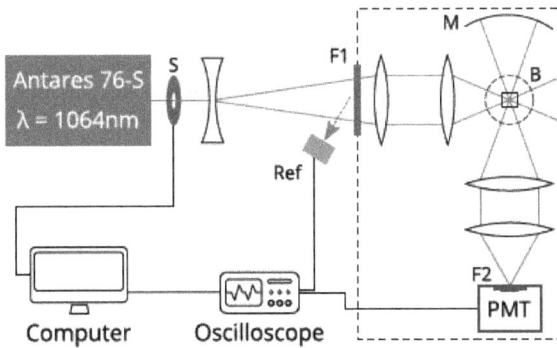

Figure 3.10. Sketch of the setup employed to measure β_{HRS}. F_1 and F_2 are bandpass filters, S, M, B, PMT and Ref are the shutter, mirror, sample and detector, respectively. Reproduced, with permission, from [15] © The Optical Society.

Knowing the incident-light intensity $I(\omega)$ and N and measuring the second-harmonic scattered light intensity $I(2\omega)$, Equation (3.12) allows the determination of β_{HRS} in different polarizations [15] (see Fig. 3.11):

$$\beta_{HRS}^{H=0} = 8.5(1) \times 10^{-28} \ cm^5/esu,$$

$$\beta_{HRS}^{\parallel} = 9.8(2) \times 10^{-28} \ cm^5/esu,$$

$$\beta_{HRS}^{\perp} = 8.1(1) \times 10^{-28} \ cm^5/esu.$$

The anisotropy of β_{HRS}, defined as:

$$A(\%) = \frac{\beta_{HRS}^{\parallel} - \beta_{HRS}^{\perp}}{\beta_{HRS}^{\parallel}} \times 100, \tag{3.19}$$

is about 17%.

The easy-magnetization axis of magnetite is along the <111> crystallographic direction, which corresponds to the main diagonal of the cube formed in the unitary cell of the spinel structure [27]. The magnetic moment of the nanoparticle is oriented in this direction. The measured hyperpolarizability $\beta_{HRS}^{\parallel}$ represents the value of this parameter along the <111> direction, and $\beta_{HRS}^{\perp}$ corresponds to a mean value in the plane perpendicular to the <111> direction. Assuming that the x-axis of the laboratory frame axes coincides with the <111> direction, $\beta_{HRS}^{\parallel} = \beta_{HRS}^{xxx}$ and $\beta_{HRS}^{\perp} = \frac{1}{2}(\beta_{HRS}^{yyy} + \beta_{HRS}^{xxx})$.

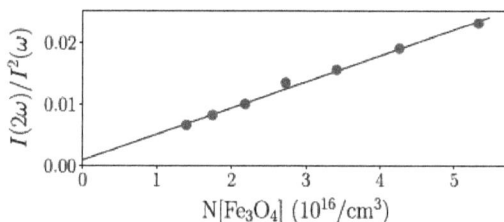

Figure 3.11. Typical result from the Hyper-Rayleigh Scattering technique. $I(2\omega)$ and $I(\omega)$ are the light intensity of the second harmonic and fundamental frequency, respectively. N is the numerical density of unitary scatterers. Reproduced, with permission, from [15] © The Optical Society.

3.5 Manganese-zinc ferrite nanoparticles, MnZFe

The ferrofluids investigated here are synthesized with a basis of Mn, Zn, Fe, O, giving rise to spherical nanoparticles: $Mn_{0.5}Zn_{0.5}Fe_2O_4$. The replacement of 5% of the Fe atoms by Ho during the synthesis changes the particles' shape from spheres to cubes with rounded edges [28]. Ferrofluid with spherical particles will be labeled MZ and the one with cubes, MZH (Fig. 3.12). Typical dimensions of the particles are a diameter of 13 nm and a side of 10 nm. Particles are coated with oleic acid and dispersed in kerosene. The X-Ray diffraction analysis revealed that the structure of the nanoparticles is a single-phase spinel structure, corresponding to the Fd3m space group.

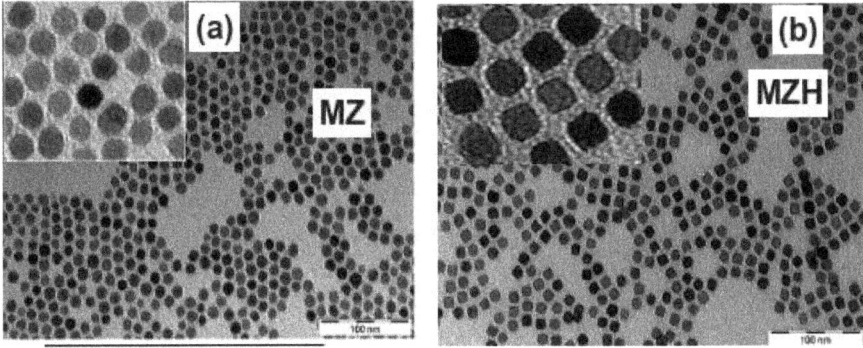

Figure 3.12. Transmission electron microscopic pictures of the MZ and MZH. Scale bars represent 100 nm. Reproduced, with permission, from [28].

3.5.1 The imaginary component of $\chi^{(3)}$, $Im\chi^{(3)}$ and the nonlinear absorption cross section, σ_{TPA}

The ZS (opened-aperture geometry) setup, described in Fig. 3.1, with $\lambda = 800$ nm, was employed in the measurement of the nonlinear optical absorption, β, and the imaginary component of the third-order electrical susceptibility, $Im\chi^{(3)}$, without and with an externally applied magnetic field ($H = 3100$ Oe), with the incident light polarized parallel and perpendicular to the magnetic field. As described earlier, the parameters measured parallel to the magnetic field (i.e., parallel to the direction of the particle's magnetic moment) is the one along the <111> direction of the crystallographic structure of the crystal's unit cell. The value measured in the plane perpendicular to the <111> direction is the mean value of this parameter in that plane.

The numerical results are given in Table 3.1 [28]. In this table, the magnetite experimental values are given as a matter of comparison. Magnetite presents the higher values of the imaginary part of the third-order electrical susceptibility and also the higher anisotropy, A, defined as:

$$A(\%) = \frac{Im\chi_{\parallel}^{(3)} - \langle Im\chi_{\perp}^{(3)} \rangle}{Im\chi_{\parallel}^{(3)}} \times 100.$$

Since the imaginary part of $\chi^{(3)}$ in the MnZFe ferrofluids is directly related to the nonlinear optical absorption. The results shown in Table 3.1 reveal that spherical

Table 3.1. Values of the imaginary component of the third-order electrical susceptibility of the MZ and MZH ferrofluids in different geometries of the orientation of the electric field and the external magnetic field. A represents the anisotropy of $Im\chi^{(3)}$ measured in the direction of the applied magnetic field and perpendicular to it.

	MZ (10^{-21} m²V⁻²)	MZH (10^{-21} m²V⁻²)	Fe_3O_4 (10^{-21} m²V⁻²)
$Im\chi^{(3)}$	1.6(3)	1.2(3)	2.0(5)
$Im\chi_{\parallel}^{(3)}$	2.8(5)	1.6(4)	3.1(5)
$\langle Im\chi_{\perp}^{(3)} \rangle$	1.5(3)	1.0(2)	1.4(5)
$A(\%)$	46	38	55

particles absorb more than cubic particles and also exhibit a higher anisotropy of $Im\chi^{(3)}$. This fact could be useful in designing optical elements required for special needs.

The TPA cross section, σ_{TPA} was calculated as a function of the applied magnetic field, $\vec{H}$ in both relative orientations of the light-polarization direction and $\vec{H}$, parallel and perpendicular. Figure 3.13 shows the dependence of with [29]. For spherical and cubic particles, the parallel configuration showed the σ_{TPA} increasing as a function of H. In the perpendicular configuration, σ_{TPA} is almost independent on H (within the experimental errors). These results were interpreted as the formation of linear clusters of nanoparticles in the direction of the magnetic field, a hypothesis confirmed by small-angle X-ray scattering experiments. The anisotropies of σ_{TPA}, at the highest applied magnetic field, defined as $A(\%) = 100 \times (\sigma_{TPA}^{\parallel} - \sigma_{TPA}^{\perp})/\sigma_{TPA}^{\parallel}$, are: $A^{MZ} = 23(6)\%$ and $A^{MZH} = 27(6)\%$.

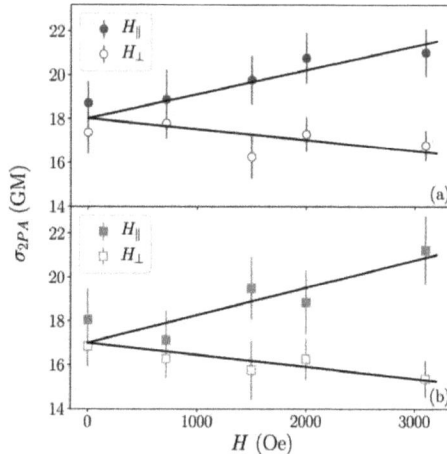

Figure 3.13. Two-photon absorption cross-section as a function of the external applied magnetic field. (a) spherical nanoparticles; (b) cubic nanoparticles. Samples MnZFe. Reproduced, with permission, from [29].

3.5.2 *The Hyper-Rayleigh Scattering to measure the hyperpolarizability,*

The structure used to measure the first-order hyperpolarizability was the same as described earlier. Different concentrations of nanoparticles in the ferrofluids were prepared to account for the variation on the parameter N (see Equation 3.18), which informs about the number of scatterers in the samples. The experiment consists in illuminating the sample with a laser along the light of frequency ω and detecting the light scattered at frequency 2ω. A magnetic field of $H = 3100$ Oe was applied to orient the magnetic moment of the particles and the measurements were performed in both the relative orientation of the light electric field, parallel and perpendicular to $\vec{H}$. Figure 3.14 shows the typical HRS measurements of $I(2\omega)/I^2(\omega)$ as a function of N, with and without H.

Table 3.2 shows the values of the first-order hyperpolarizability measured in the geometries parallel and perpendicular to the applied magnetic field and the values

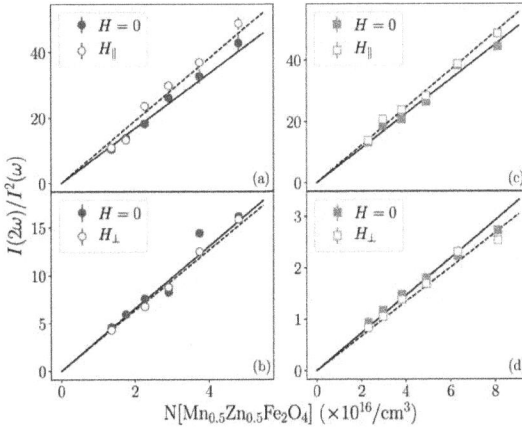

Figure 3.14. Typical result from the Hyper-Rayleigh Scattering technique. $I(2\omega)$ and $I(\omega)$ are the light intensity of the second harmonic and fundamental frequency, respectively. N is the numerical density of unitary scatterers. Spheres and squares represent the MZ and MZH samples, respectively. Reproduced, with permission, from [29].

Table 3.2. First-order hyperpolarizability of MZ and MZH samples. Symbols $\parallel$ and $\perp$ refer to the orientation of the electric field of the laser light parallel and perpendicular to the external applied magnetic field. A shows the anisotropy calculated according to Equation (3.19). The last column brings the magnetite values of β_{HRS}.

	MZ	MZH	Fe_3O_4
$\beta_{HRS}^{\parallel}$ (10^{-28} cm^5/esu)	10.1(2)	8.1(2)	9.8(2)
$\beta_{HRS}^{\perp}$ (10^{-28} cm^5/esu)	9.3(3)	7.4(2)	8.1(1)
$A(\%)$	8(3)	8(2)	17

measured for the magnetite ferrofluid. It can be seen that, despite the change in the shape of the particles, the anisotropy of β_{HRS} remains the same. The introduction of Ho in the crystalline structure in this concentration equally reduced the values of β_{HRS}, keeping the same anisotropy. Comparing these values with those from magnetite, the results indicate that magnetite nanoparticles still present the higher values of β_{HRS}, and also higher anisotropy. This information may be useful in the design of optical devices using optical fibers with ferrofluid incorporated.

3.6 Rotation of the magnetic moments of the particles

This problem has been treated theoretically and experimentally in various scientific publications [30–37]. The theoretical tools are given to the reader who wants to perform computer simulations of the movements of the particles and their magnetic moments, for example, to compare with experimental results of ferromagnetic resonance. The focus is on the method and not on the results of simulations. The present text differs from the earlier theoretical ones in two crucial respects: (a) ferrofluids are always polydisperse in particles' diameter, which is usually not taken into account in theoretical works, and (b) realistic values for the parameters, like magnetization, anisotropy constant, liquid's viscosity, applied field, temperature, etc., used in the simulations, instead of arbitrary values in nondimensional units. It can be seen how to obtain the equations of motion for the rotations of the magnetic moments and the symmetry axis of the particles. These equations are stochastic differential equations with multiplicative noise. They may be used as a basis for simulations. Therefore, the roles of stochastic calculus must be used in their interpretation [6, 38].

The dynamics of the magnetic moments of super-paramagnetic particles in ferrofluids may be studied by several experimental and theoretical procedures. Among the experimental tools, ferromagnetic resonance is particularly appropriate for the observation of several properties of the particles. Moreover, the absorption lines of ferromagnetic resonance may be calculated theoretically once a model for the dynamics of the magnetic moments is surmised, and therefore, a comparison between theoretical and experimental results is a good test for the assumed model.

When a weak periodic field $\vec{H}_1(t)$ is applied perpendicular (such as, in the x direction) to a strong constant field $\vec{H}_0$ (in the z-direction) on the sample, in the appropriate experimental setup, the absorption power, as a function of the applied frequency, is related to the imaginary part of the complex susceptibility by:

$$P^{\dagger}(\omega) = C\omega\chi''_{xx}(\omega)$$

where the constant C depends on experimental details (field intensity, geometry). The complex susceptibility

$$\chi_{xx}(\omega) = \chi'_{xx}(\omega) + i\chi''_{xx}(\omega)$$

is the Fourier-Laplace transform of the time-dependent "response function" $\phi_{xx}(t)$ which may be obtained from a numerical simulation procedure by "Kubo formula"

$$\phi_{xx}(t) = -\frac{1}{k_BT}\langle\vec{\mu}_x(0)\,\dot{\vec{\mu}}_x(t)\rangle$$

where $\dot{\vec{\mu}}_x = \dfrac{d\vec{\mu}_x}{dT}$ and k_B is the Boltzmann constant.

3.6.1 Some assumptions about the magnetic particles

For realistic ferrofluids, the following assumptions about the magnetic particles are at least good approximations:

(a) The magnetic particle is approximately spherical and has a symmetry axis of easy magnetization, which will be characterized by the unit vector $\hat{c}$;

(b) The particles' diameters d are distributed approximately as a log-normal distribution;

(c) The absolute values of the particles' magnetic moments are constant and proportional to the particles' volumes, $\mu = M_s V$, where M_s is the saturation magnetization of the material, but the "vector" magnetic moment, $\vec{\mu} = \mu\hat{e}$, where $\hat{e}$ is a unit vector that can rotate inside the particle under the influence of an interaction potential, modeled by $U = -A(\hat{e}\cdot\hat{c})^2$ and A is a constant, called "anisotropy constant", whose value depends linearly on volume V, i.e., $A = KV$, where K is a material's constant;

(d) The internal (Néel) dissipation constant depends on the particle's volume V in the form of an Arrhenius-type viscosity function, $\gamma = \gamma_0 \, exp \, (E \, V/k_B T)$, where E is the materials' *activation energy;*

(e) The magnetic moment, $\vec{\mu}$, rotates much faster than the particle, so that the particle's angular velocity may be neglected in estimation of the torques (noise and dissipative torques) on $\vec{\mu}$;

(f) The particle's rotational inertia is negligible in the equations of motion compared to the Brownian and dissipative terms.

3.6.2 *The equations of motion*

The equations of motion for $\vec{\mu}$ are obtained from the classical equation for the rotation of angular momentum, $\vec{j}$, caused by a torque, $\vec{N}$,

$$d\vec{j}/dt = \vec{N}.$$

A magnetic moment, $\vec{\mu}$, is associated with an angular momentum by $\vec{\mu} = g\,\vec{j}$, where g is the gyromagnetic tensor. In many situations, it is a good approximation to assume $\vec{\mu}$ to be parallel to $\vec{j}$, i.e., to surmise that g is a scalar. Therefore,

$$d\vec{\mu}/dt = g\vec{N}.$$

The torque $\vec{N}$ has several origins:

(a) Interaction with the applied field, in the form $\vec{\mu} \cdot \vec{H}$;

(b) Torque caused by the anisotropy potential $U = -A(\hat{e} \cdot \hat{c})^2$, having the form 2A $(\hat{e} \cdot \hat{c}) \, (\hat{e} \times \hat{c})$;

(c) Néel type of internal noise, caused by atomic vibrations (phonons), in the form $\alpha\vec{\mu} \times \vec{\xi}(t)$, where $\vec{\xi}(t)$ may be simulated as normalized white noise;

(d) Dissipative torque, caused by interaction with the phonons in reaction to the rotation of $\vec{\mu}$, in the form $\gamma\vec{\mu} \times d\hat{e}/dt$; here, hypothesis (e), above, has been taken into account.

Therefore, the equation for $\boldsymbol{\mu}$ may be written as

$$d\vec{\mu}/dt = g\vec{\mu} + \{\vec{H} + (2A/\mu)\,(\hat{e} \cdot \hat{c})\hat{c} - \gamma \, d\hat{e}/dt + a \, \vec{\xi}(t)\}. \qquad (3.20)$$

Substituting $d\hat{e}/dt$ at the right-hand-side of Equation (3.20) by the whole of the right-hand-side of the same equation, divided by μ, we get:

$$d\hat{e}/dt = g\hat{e} \times \{\vec{H} + (2A/\mu)\,(\hat{e} \cdot \hat{c}) \, \hat{c} + \alpha \, \vec{\xi}(t) - \gamma \, g \, \hat{e} \times [\vec{H} + (2A/\mu)\,(\hat{e} \cdot \hat{c}) \, \hat{c} - \gamma \, (d\hat{e}/dt) + a \, \vec{\xi}(t)]\}$$

Since $\hat{e}$ is unitary and $d\hat{e}/dt$ is perpendicular to $\hat{e}$, the vector identity may be used

$$\hat{e} \times (\hat{e} \times d\hat{e}/dt) = - d\hat{e}/dt$$

which, after some algebra, leads to

$$d\hat{e}/dt = (g/(1+g^2\,\gamma^2))\,\hat{e} \times \{\vec{H}_{eff} - \gamma\,g\,\hat{e} \times \vec{H}_{eff}\}, \qquad (3.21)$$

where the effective $\vec{H}_{eff}$ field is

$$\vec{H}_{eff} = \vec{H} + (2A/\mu)\,(\hat{e}\,.\,\hat{c})\hat{c} + \alpha\,\vec{\xi}(t).$$

Equation (3.21) for $d\hat{e}/dt$ is the well-known "Landau-Lifshitz" equation.

The particle itself also rotates inside the liquid. Its rotation about the symmetry axis, $\hat{c}$, is irrelevant for the motion of $\vec{\mu}$. For this reason, only the rotation of itself will be considered. Due to the hypothesis (f), above, the equation of motion for is simply a balance between torques, i.e.,

$$\sum \vec{N}_j = 0$$

where,

(e) $\vec{N}_1 = \alpha_1\,\hat{c} \times \vec{\xi}_1(t)$ is the Brownian torque caused by the liquid's molecular collisions with the particle;

(f) $\vec{N}_2 = -\gamma_1\hat{c} \times d\hat{c}/dt$ is the dissipative torque opposing to the particle's rotation;

(g) $\vec{N}_3 = - \vec{\mu} \times \{(2A/\mu)\,(\hat{e}\,.\,\hat{c})\,\hat{e} - \gamma\,d\hat{e}/dt + a\,\vec{\xi}_1(t)\}$ is the "reaction" on the particle to the torque caused by the particle on the magnetic moment.

Therefore, the balance equation of torques becomes:

$$\hat{c} \times \{\alpha_1\vec{\xi}_1(t) - \gamma_1\,d\hat{c}/dt\} - \vec{\mu} \times \{(2A/\mu)\,(\hat{e}\,.\,\hat{c})\,\hat{c} - \gamma\,d\hat{e}/dt + a\,\vec{\xi}_1(t)\} = 0.$$

Multiplying this equation vectorially at the left by and using the vector identity

$$\hat{c} \times (c\,d\hat{c}/dt) = - d\hat{c}/dt$$

we get.

$$d\hat{c}/dt = (\hat{c}/\gamma_1)\ \times\{\hat{e} \times [(2A/\mu)\,(\hat{e}\,.\,\hat{c})\hat{c} - \gamma\,d\hat{e}/dt + a\,\vec{\xi}_1(t)] + a_1\vec{\xi}_1(t)\}, \qquad (3.22)$$

where, in the last case, we substituted $- \hat{c} \times (\hat{c} \times \vec{\xi}_1(t))$ by $\hat{c} \times \vec{\xi}_1(t)$ because both expressions are normalized white noise, perpendicular to $\hat{c}$ having, therefore, the same statistical distributions, and the second expression is simpler. In these equations

$\vec{\xi}$ and $\vec{\xi}_l$ are the Néel and Brownian torques, respectively. They may be modeled as white noise, having the following statistical properties:

$$< \xi_i(t) > = 0,$$

$$< \xi_i(t_1)\xi_j(t_2) > = \delta_{ij}\delta(t_1 - t_2).$$

where ξ_i and ξ_j are the Cartesian components of $\vec{\xi}$ and $\vec{\xi}_l$.

Equations (3.21) and (3.22) for $d\hat{e}/dt$ and $d\hat{c}/dt$ for form a coupled system of stochastic differential equations. Realizations of the stochastic processes $\hat{\mu}$ and $\hat{c}$ may be obtained by simultaneous simulations of both equations. Since they have terms containing "multiplicative white noise" the reader must be forewarned that the simulations must be realized following the rules of Stratonovich calculus [5, 6].

3.7 Conclusions

Oriented magnetic nanoparticles in suspensions were analyzed under the action of strong electric fields. Typical intensity of the light beam interacting with the ferrofluid is of the order of GW/cm^2. In this condition, nonlinear optical effects are evidenced, and accessible in experiments. Techniques as the Hyper-Rayleigh scattering and the Z-Scan allowed the measurement of the two-photon absorption cross section, the first-order hyperpolarizability, the Real and Imaginary parts of the third-order electrical susceptibility. Moreover, anisotropies in the values of these parameters were measuremed in the directions parallel and perpendicular to that of the external magnetic field employed to orient the magnetic nanoparticles. The knowledge of these parameters is interesting not only from the fundamental point of view, but also, for technological applications of magnetic colloids.

Acknowledgments

Financial support from INCT/CNPq (Conselho Nacional de Desenvolvimento Científico e Tecnológico; Grant number: 465259/2014-6), INCT/FAPESP (Fundação de Amparo à Pesquisa do Estado de São Paulo; Grant Number: 14/50983-3), INCT/ CAPES (Coordenação de Aperfeiçoamento de Pessoal de Nível Superior; Grant number: 88887.136373/2017-00), FAPESP (Thematic Project; Grant 2016/24531-3), and INCT-FCx (Instituto Nacional de Ciência e Tecnologia de Fluidos Complexos). We deeply thank Mr. Maurício Rheinlander de Pinho Klein for the preparation of the figures.

References

[1] M. Rajnak, J. Kurimsky, B. Dolnik, P. Kopcansky, N. Tomasovicova, E. A. Taculescu-Moaca and M. Timko. Dielectric-spectroscopy approach to ferrofluid nanoparticle clustering induced by an external electric field. Phys. Rev. E 90: 032310 (2014).

[2] T. Cherian, F. Sohrabi, C. Rigoni, O. Ikkala and J. V. I. Timonen. Electroferrofluids with nonequilibrium voltage-controlled magnetism, diffuse interfaces, and patterns. Sci. Adv. 7: eabi8990 (2021).

[3] D. Soga, S. Alves, A. Campos, F. A. Tourinho, J. Depeyrot and A. M. Figueiredo Neto. Nonlinear optical properties of ionic magnetic colloids in the femto and milliseconds time scales: change from convergent-to-divergent lens-type behaviors. Journal of the Optical Society of America B, 24: 49 (2007).

[4] S. Alves, A. Bourdon and A. M. Figueiredo Neto. Generalization of the thermal lens model formalism to account for thermodiffusion in a single beam Z-Scan experiment: determination of the Soret coefficient J. Opt. Soc. Am. B 20: 713 (2003).

[5] R. L. Stratonovich. Introduction to the Theory of Random Noise, Gordon and Breach, New York (1963).

[6] R. L. Stratonovich. A new representation for stochastic integrals and equations. SIAM Journal on Control, 4: 362 (1966).

[7] P. N. Butcher and D. Cotter. The elements of nonlinear optics Cambridge studies in modern optics 9. Cambridge; New York: Cambridge University Press, pg. 29 (1990).

[8] R. W. Boyd. Nonlinear Optics, 3. ed. Burlington, MA: Academic Press (1980).

[9] M. Sheik-Bahae, A. A. Said and E. W. Van Stryland. High-Sensitive, single-beam measurements. Opt. Lett. 14: 955 (1989).

[10] M. Sheik-Bahae, A. A. Said, T. H. Wei and D. J. Hagan. Sensitive, measurement of optical nonlinearities using a single beam. IEEE J. Quantum Electron. 26: 760 (1990).

[11] D. Espinosa, D. Soga, S. Alves, L. De Boni, S. C. Zílio and A. M. Figueiredo Neto. Investigation of the optical absorption of a magnetic colloid from the thermal to the electronic time-scale regime: measurement of the free-carrier absorption cross-section. J. Opt. Soc. Am. B 29: 280 (2012).

[12] K. Clays and A. Persoons. Hyper-Rayleigh scattering in solution. Phys. Rev. Lett. 66: 2980 (1991).

[13] R. W. Terhune, P. D. Marker and C. M. Savage. Measurements of nonlinear light scattering. Phys. Rev. Lett. 14: 681 (1965).

[14] K. Clays, E. Hendrickx, M. Triest and A. Persoons. Second-order nonlinear optics in isotropic liquids: Hyper-Rayleigh scattering in solutions. J. Mol. Liq. 67: 133 (1995).

[15] E. S. Gonçalves, R. D. Fonseca, L. De Boni and A. M. Figueiredo Neto. Tuning hyper-Rayleigh scattering amplitude on magnetic colloids by means of an external magnetic field. JOSA B 35: 2681 (2008).

[16] P. L. Franzen, L. Misoguti and S. C. Zilio. Hyper-Rayleigh scattering with picosecond pulse trains. Appl. Opt. 47: 1443 (2008).

[17] R. D. Fonseca, M. G. Vivas, D. L. Silva, G. Eucat, Y. Bretonnière, C. Andraud, L. De Boni and C. R. Mendonça. First-order hyperpolarizability of triphenylamine derivatives containing cyanopyridine: molecular branching effect, J. Phys. Chem. C 122: 1770 (2018).

[18] A. L. Sehnem, D. Espinosa, E. S. Gonçalves and A. M. Figueiredo Neto. Thermal lens phenomenon studied by the Z-Scan technique: measurement of the thermal conductivity of highly absorbing colloidal solutions. Braz. J. Phys. 46: 547 (2016).

[19] D. Espinosa, E. S. Gonçalves and A. M. Figueiredo Neto. Two-photon absorption cross section of magnetite nanoparticles in magnetic colloids and thin films. J. Appl. Phys. 121: 043103 (2017).

[20] D. H. G. Espinosa, C. L. P. Oliveira and A. M. Figueiredo Neto. Influence of an external magnetic field in the two-photon absorption coefficient of magnetite nanoparticles in colloids and thin films. J. Opt. Soc. Am. B 35: 346 (2018).

[21] J. Tauc, R. Grigorovici and A. Vancu. Optical properties and electronic structure of amorphous germanium. Phys. Status Solidi 15: 627 (1966).

[22] E. A. Davis and N. F. Mott. Conduction in non-crystalline systems V. Conductivity, optical absorption and photoconductivity in amorphous semiconductors. Philos. Mag. 22: 0903 (1970).

[23] W. F. J. Fontijn, P. J. van der Zaag, M. A. C. Devillers, V. A. M. Brabers and R. Metselaar. Optical and magneto-optical polar Kerr spectra of Fe_3O_4 and Mg^{2+} - or Al^{3+}-substituted Fe_3O_4, Phys. Rev. B 56: 5432 (1997).

[24] D. Espinosa, L. B. Carlsson, S. Alves and A. M. Figueiredo Neto. Influence of nanoparticle size on the nonlinear optical properties of magnetite ferrofluids. Phys. Rev. E 88: 032302 (2013).

[25] M. Vivacqua, D. Espinosa and A. M. Figueiredo Neto. Application of the Z-scan technique to determine the optical Kerr coefficient and two-photon absorption coefficient of magnetite nanoparticles colloidal suspension. J. Appl. Phys. 111: 113509 (2012).

[26] A. G. Roca, M. P. Morales, K. O'Grady and C. J. Serna. Structural and magnetic properties of uniform magnetite nanoparticles prepared by high temperature decomposition of organic precursors. Nanotechnology 17: 2783 (2006).

[27] L. R. Bickford. Ferromagnetic resonance absorption in magnetite single crystals. Phys. Rev. 78: 449 (1950).

[28] D. Parekh, H. G. Espinosa, D. Reis, C. L. P. de Oliveira, W. Wlysses and A. M. Figueiredo Neto. Morphological metamorphosis of magnetic nanoparticles due to the presence of rare earth atoms in the spinel structure: From spheres to cubes. Materials Chemistry and Physics 222: 217 (2019).

[29] E. S. Gonçalves, L. H. Z. Cocca, W. W. R. Araujo, K. Parekh, C. L. P. Oliveira, J. P. Siqueira, C. R. Mendonça, L. De Boni and A. M. Figueiredo Neto. Influence of magnetic field on the two-photon absorption and hyper-rayleigh scattering of manganese–zinc ferrite nanoparticles. J. Phys. Chem. C 124: 6784 (2020).

[30] C. Scherer. Computer simulation of the stochastic dynamics of super-paramagnetic particles in ferrofluids. Braz J. Phys. 36: 676 (2006).

[31] C. Scherer and G. Matuttis. Rotational dynamics of magnetic particles in suspensions. Phys. Rev. E 63: 0115504 (2001).

[32] C. Scherer and T. F. Ricci. Rotational dynamics of the magnetic moments in ferrofluids. Bras. J. Phys. 31: 380 (2001).

[33] M. I. Shliomis and V. I. Stepanov. Anomalous diffusion of a dipole interacting with its surroundings. Adv. Chem. Phys. Series 87: 1 (1994).

[34] V. P. Shilov, J. C. Bacri, F. Gazeau, F. Gendron, R. Perzynski and Y. L. Raikher. Dynamic optical probing of the magnetic anisotropy of nickel-ferrite nanoparticles. J. Appl. Phys. 85: 6642 (1999).

[35] U. Wieddwald, M. Spasova, M. Farle, M. Hilgendorff and M. Giersig. Ferromagnetic resonance of monodisperse Co particles. J. Vac. Sci. Technol. A 19: 1773 (2001).

[36] Q. A. Pankhurst, J. Connolly, S. K. Jones and J. Dobson. Applications of magnetic nanoparticles in biomedicine, J. Phys. D: Appl. Phys. 36 R167 (2003).

[37] P. C. Morais, G. R. R. Gonçalves, A. F. Bakuzis, K. Skeff Neto and F. Pelegrini. Experimental evidence of dimer disruption in ionic ferrofluid: A ferromagnetic resonance investigation. J. Magn. Magn. Mat. 225: 84 (2001).

[38] P. E. Kloeden and E. Platen. Numerical Solution of Stochastic Differential Equations. Springer, New York (1999).

4

Thermal Conductivity of Magnetic Fluid
Effect of Magnetic Field and Concentration on Microstructure Evolution

Kinnari Parekh,* Jaykumar Patel and *R.V. Upadhyay*

◇◇

4.1 Introduction

Since the Nobel Prize winner Richard P. Feynman presented the concept of micromachines in his seminal talk, "There is Plenty of Room at the Bottom—An Invitation to Enter a New Field of Physics," in December 1959 at the annual meeting of the American Physical Society at the California Institute of Technology [1], miniaturization has been a significant trend in modern science and technology. Almost 40 yr later, another Nobel Prize winner, H. Rohrer, presented the chances and challenges of nano-age and declared that nanoscience and nanotechnology had entered the limelight in the 1990s from virtual obscurity in the 1980s [2]. *Nano* is a prefix meaning one-billionth, so a *nanometer* is one-billionth of a meter. *Nanotechnology* is the creation of functional materials, devices and systems by controlling matter at the nanoscale level, and nanoscience exploits the novel properties and phenomena that emerge at the nanoscale. The emergence of nanofluid as a new field of nanoscale heat transfer in liquids is related directly to miniaturization trends and nanotechnology. The thermal transport properties of any liquid are essential to study when one thinks of using it in a device as a coolant media to dissipate the heat generated in the device. Increasing the thermal transport properties will not only help improve the device's efficiency but also help minimize the device's size.

Dr. K. C. Patel R & D Center, Charotar University of Science & Technology, Changa-388421, Gujarat, India.

* Corresponding author: kinnariparekh.rnd@charusat.ac.in

Liquid metals are thermally conductive, and their heat transfer characteristics have attracted great interest. Liquid metals are heat transfer fluids in specialized branches of engineering involving very high heat fluxes [3]. As an example, in nuclear engineering, there is a requirement to obtain high rates of heat extraction from reactors. Liquid metal is also used in a gas turbine, where the need for effective blade-cooling systems remains as pressing as ever to achieve the greatest thermodynamic advantage. Liquid metals usually have very high thermal conductivity, which marks them from other conventional heat transfer fluids, i.e., water, oils and glycols. Table 4.1 shows the thermo-physical properties of liquid metals.

Replacing the conventional heat transfer liquids with liquid metal is impossible due to the unique characteristics of each liquid metal. For example, liquid mercury is toxic, corrosive to certain metals, incompatible with metal oxides and expensive. At the same time, sodium spontaneously ignites in contact with air and is potently reactive with water, so it must be stored and used in an inert environment. Similarly, galinstan tends to wet and adhere to many materials, such as gallium, which limits its use compared to mercury. Surfaces must be coated with gallium oxide, to prevent the metal from sticking. The high cost and the aggressive properties of galinstan are major obstacles to its use. Hence, only in certain special applications do liquid metal coolants appear as a suitable heat-transfer media. In most engineering components, conventional heat transfer fluids are more suitable and improving their thermal conductivity is thus important and necessary.

With the ever-increasing demand for energy use in various sectors, the smart and automatic performance of heat transfer devices is a need for future generation coolant systems. Magnetic fluid, also known as a ferrofluid, is a suspension of magnetic nanoparticles in a liquid medium, giving an extraordinarily distinct response when exposed to a magnetic field. Despite being a liquid, a magnetic fluid can defy gravity with the help of a magnetic field or the flow pattern can be tuned by changing magnetic field strengths [5]. Thus, using a magnetic fluid in heat transfer devices will add a new dimension to the system to make it smart and automatic.

The thermal conductivity and viscosity of the magnetic fluid can be enhanced drastically through external magnetic fields. This is because magnetic particles suspended in the magnetic fluid form some chainlike or aggregated structures under an externally applied magnetic field. This structure formation determines the energy transport processes of the suspended nanoparticles inside the magnetic fluid. Alteration of the magnetic field direction changes the orientation of these structures towards the direction of the external field. These oriented structures considerably affect the energy transport process of the magnetic fluid, which makes the magnetic fluid capable of being using as a smart fluid for possible thermal applications.

Table 4.1. Thermo-physical properties of liquid metals [4].

Liquid Metal	Melting Point (°C)	Density (kg/m³)	Thermal Conductivity (W/m-K)	Dynamic Viscosity (mPa-s)
Mercury	−38.8	13.534	8.3	1.526 (@ 300 K)
Sodium	97.7	0.968	142	0.7 (@ 400 K)
Galinstan	−19.6	6.441	16.5	2.4 (@ 300 K)

The tuning of the thermal transport phenomenon under the influence of the magnetic field is mainly affected by the internal structure formation of these particles in the absence or the presence of the magnetic field. Most published research papers outline the thermal conductivity study as a function of different parameters [6–11], such as particle size, size distribution, the concentration of particles, the direction of the magnetic field as compared to the direction of heat transfer, etc. Despite this, practical smart and automatic cooling devices using a magnetic fluid coolant are yet to be developed and more significant work is required for its realization. Understanding the interaction at a microscopic level under different volume fractions of magnetic particles, magnetic field strength, temperature, etc., sheds light on the evolution of structure formation in the system, which ultimately affects its thermal transport properties. Therefore, studying micro-structure formation under different situations is enlightening and worthwhile for its practical utility.

Li et al. [12] 2005 showed the importance of the applied magnetic field's direction during the magnetic fluid's thermal conductivity. When the fluid sample is subjected to a perpendicular magnetic field, little change in the sample's thermal conductivity was found, whereas a parallel field shows significant enhancement in thermal conductivity. The results were explained based on the formation of chainlike aggregation structures along the orientation of the external magnetic field. On the other hand, no chain alignments of the magnetic particles along the temperature gradient (i.e., perpendicular to the magnetic field) appear. Such morphology suppresses energy transport inside the fluid along the temperature gradient. Therefore, the applied magnetic field causes the anisotropy of the thermal conductivity of the magnetic fluid. Therefore, the perpendicular magnetic field weakly affects a thermal process in the magnetic fluid. In the case of a parallel magnetic field, the chain formation takes place in the direction of the temperature gradient, thus providing more effective bridges for energy transport and enhancing the thermal process in magnetic fluids.

The enhancement in thermal conductivity also depends on the concentration of magnetic particles. The higher the particle concentration, the more the enhancement. The magnetic field effect on the thermal conductivity of the magnetic fluid at a higher particle concentration is stronger than that at a lower particle concentration. The ratio of the thermal conductivity of magnetic fluid under the field strength of 240 G to that in the absence of the magnetic field varies from 1.12 to 1.25 if the volume fraction of the nanoparticles increases from 1.0 volume to 5.0 volume % [13].

Another important point for a magnetic material is "Remanence" or "Hysteresis". Thus, to see the effect of the hysteresis on magnetic fluid Philip et al. [14] measured the thermal conductivity of magnetic fluid having 2.6 volume % of Fe_3O_4 nanoparticles as a function of the applied magnetic field while increasing as well decreasing the field. They showed that the enhancement starts above 20 G. Further increase in a magnetic field leads to drastic enhancement in thermal conductivity. The maximum 128% enhancement is observed at a magnetic field of 94.5 G, above which the thermal conductivity value decreases slightly. While lowering the magnetic field, the thermal conductivity value shows a small hysteresis, but returns to the original value when the magnetic field is turned off. After repeating the magnetic cycles several times, the thermal conductivity returns to the original value. They repeated this experiment in various magnetic fluid samples with aqueous and non-aqueous carrier fluids and obtained similar results.

Shima et al. [15] synthesized and measured the thermal conductivity of nanofluids as a function of the applied magnetic field parallel to the temperature gradient at 0.078 volume % of magnetite nanoparticles. In the low magnetic field region, the thermal conductivity increases rapidly, reaching up to 300% of enhancement at 82 G and then decreases and becomes stable for a higher field value. According to the authors, the nanoparticles are well suspended and conduction is through series modes in the lower field limit, whereas in the upper limit, the conduction path is through the dispersion of particles. The observed decrease in thermal conductivity above 82 G had been explained using the concept of "zippering" of chains. Mapping the magnetic field within the probe area shows a fairly uniform field which ruled out the possibility of induced body forces. However, the linear and thick aggregates with high aspect ratio due to zippering can collapse to the bottom of the cell and hence cannot be seen by the hotwire.

A significant contribution was made by Shima et al. [16] by measuring the effective thermal conductivity of two nanofluids having particle diameters of 3.6 and 6.1 nm in the presence of an external magnetic field parallel to the temperature gradient, at 5.5 volume % of Fe_3O_4. The thermal conductivity grows with an increase in applied field strength as the heat conduction is parallel to the direction of the chain of particles. They observed a lesser value of thermal conductivity for small size particles compared to larger size particles which they feel is due to the lower aspect ratio of the chains for smaller size particles which resulted in lower enhancement of thermal conductivity.

To obtain better insight into the effect of magnetic field orientation (i.e., the orientation of nano chains with respect to the heat flow direction) on thermal conductivity enhancement, Shima et al. [17] carried out thermal conductivity measurements under different magnetic field orientations with respect to the thermal gradient. They studied the variation of thermal conductivity enhancement with magnetic field strength for hexadecane-based Fe_3O_4 nanofluids with 6.8 volume % under different field orientations. The maximum enhancement in thermal conductivity is observed when the field direction is parallel to the thermal gradient, whereas practically no enhancement is observed when the field was perpendicular to the thermal gradient. A gradual reduction in the thermal conductivity enhancement is observed as the field direction is shifted from the parallel to the perpendicular direction for the thermal gradient. Hence the enhancement of thermal conductivity of magnetic fluid when subjected to an external magnetic field depends on the volume fraction of magnetic nanoparticles, magnetic field, particle size and the direction of the applied magnetic field to the temperature gradient direction.

In addition to the parameters discussed above, the influence of time on the field-induced structure also plays an important role in determining thermal conductivity enhancement. Gavili et al. [18] studied the effect of time on the thermal conductivity of the water-based magnetite magnetic fluid at a constant magnetic field of different strengths. They reported about 200% enhancement in thermal conductivity after 120 min of applying a 0.1 T magnetic field. Kudelcik et al. [19] recently studied the structural change in transformer oil-based magnetic fluid by acoustic spectroscopy.

They observed that the process of aggregation in the magnetic fluid took a long time. Hence, time-dependent field-induced aggregation is also a decisive parameter to obtain the maximum enhancement in the thermal conductivity of magnetic fluids.

The field-dependent thermal conductivity can be explained using the concept of dipolar interaction. The interaction potential, U, between two identical isolated magnetizable particles scales with the square of the field strength, is attractive or repulsive depending on the head-to-tail or side-by-side alignment of dipoles, respectively. Under weak fields, the inter-particle interaction strength will be comparable to thermal energy, therefore their attractive dipolar interaction is balanced by thermal forces. However, in most practical situations, the dipolar interaction strength greatly exceeds thermal energy and aggregation proceeds as a non-equilibrium transport limited process.

Halsey and Toor (HT) [20, 21] reported a long-range interaction of dipolar chains due to strong Landue-Peierls fluctuations. According to the HT model, longitudinal and transverse particle fluctuations of wave vector **k** in a dipolar chain create local variations in the lateral field. The total interaction energy of two chains is the sum of the interaction energy between their fluctuations and the energy of deformation required for the fluctuation mode. They show that the mean square lateral field for fluctuating chains decays as a power law and is independent of field strength. Later, the HT model was modified [22, 23] and showed that the interaction energy per unit length 'a' accounts for the dependence on field strength.

Fermigier and Gast [24] calculated the time for the lateral alignment of chains (t_{lat}) and tip-to-tip alignment of chains (t_{tip}) for the suspension of latex particles. In their study, the two-dimensional aggregation kinetics at sufficiently high particle surface fractions ($\varphi_e \sim 0.03$) rapidly forms long, concentrated chains in the field direction. The interaction spacing was small to observe lateral chain coalescence through a 'zippering' motion.

Martin et al. [25] used computer simulations of dipolar suspension aggregation to show that coarsening can occur in the absence of thermal fluctuations for concentrated suspensions ($\varphi \sim 0.05 - 0.5$) because they have a greater tendency to form highly irregular chains. Martin et al. showed that a single particle or group of particles tend to cling to the sides of chains which causes them to interact strongly in the absence of thermal fluctuations, driving them toward further coarsening.

Apart from the physical parameters like material thermal conductivity, particle size, shape, concentration, temperature and interparticle interactions, the dispersion of nanoparticles into a liquid medium has a few more effects that can modulate the thermal conductivity of nanofluids. They are (i) ordered nanolayer of liquid molecules, (ii) percolation-like behavior, (iii) interfacial thermal resistance, (iv) Brownian motion of nanoparticles, (v) nanoconvection, (vi) surface charge model, (vii) molecular level mechanism (viii) heat transfer models, (ix) pool boiling heat transfer mechanism and (x) mechanism of enhanced heat transfer of nanofluid. The dependency of these parameters on thermal conductivity is not established very well, so a discrepancy is observed between the theory and experimental value. Understanding the origin of these effects is a challenging task that needs to be addressed soon.

The experimental observations in this direction mainly focus on optical trapping, video-microscopy, light scattering and optical transmission [22–27]. Several models have been developed to explain the long-range interaction of dipolar chains to form aggregates due to strong thermal fluctuations [25, 20–21, 27] or perturbations in the local lateral field in the absence of thermal fluctuations [26]. The present work reports the use of these models to explain the results of the development of thermal conductivity of magnetic fluid under prolonged magnetic field exposure. Most of the literature reports the study of volume fraction and magnetic field dependent thermal conductivity on magnetite magnetic fluid, which is mainly light hydrocarbon oil (kerosene or hexane). In the present work, we report the investigation of thermal conductivity as a function of constant magnetic field exposure time on Mn-Zn ferrite dispersed in transformer oil. Here, the field-dependent study of thermal conductivity and viscosity as a function of time for the transformer oil-based Mn-Zn ferrite fluid are described. To the best of the authors' knowledge, this kind of study has not been reported in literature. However, this parameter is one of the important criteria to investigate when the device holds a constant magnetic field over a long period. The experimental results are interpreted using the concept of structure formation and its alignment under different conditions. The results are supported using optical microscopy images. The study will help to design magnetic field-assisted heat transfer devices using oil-based magnetic fluid.

4.2 Experimental

Mn-Zn ferrite nanoparticles were synthesized using the chemical co-precipitation method, followed by surface coating with oleic acid and dispersion in transformer oil (TASHOIL-50) [9]. The solid volume fraction of particles in a fluid is 10.3%. Bruker X-Ray Diffractometer (XRD) model D2 Phaser was used to measure the crystalline nature of the particles. The sample measurement was carried out in the angular range of 10–80° in steps of 0.05° using Cu Kα radiation. Zetasizer Nano-S90 model ZEN1690 with red laser light was used to measure the hydrodynamic size of the coated particles. Transmission Electron Microscopy (TEM) model JEOL Tecnai was used to investigate the morphology of the particles. Room temperature magnetic measurement of the fluid was investigated using Lakeshore Vibrating Sample Magnetometer (VSM) model 7404 with a maximum field up to 1 T.

The thermal conductivity of magnetic fluid was measured at 303 K using thermal properties analyzer (Decagon KD2 Pro) that works on the principle of the transient hot-wire technique. KD2 Pro Thermal Properties Analyzer (Decagon Devises, USA) consists of a handheld controller and sensors inserted into the medium. The instrument meets the standards of ASTM D5334 and IEEE 442-1981. The single-needle sensor with 1.3 mm diameter and 60 mm length was used (KS-1) to study thermal conductivity. The sensor integrates a heating element and a thermo-resistor, while a microprocessor controls and conducts measurements. The nanofluid sample (20 ml) was filled in a cylindrical glass container of 25 mm in diameter and 70 mm in height. The probe was calibrated using standard fluid samples. A single reading generally takes 90 s. The first 30 s are used to ensure temperature stability, after which the probe is heated for 30 s using a known amount of current and the last 30 s measures the change in temperature after heating. Thermal conductivity is measured according to Equation (4.1).

$$\lambda = \frac{Q}{4\pi(T_2 - T_1)} \, ln\left(\frac{t_2}{t_1}\right) \tag{4.1}$$

where, λ is the thermal conductivity, Q is the applied electric power per unit length of the wire, and $T_2 - T_1$ is the temperature rise of the wire between time t_1 and t_2. Field-induced thermal conductivity was measured by placing a vial containing samples in the magnetic field generated using permanent rare earth-based magnetic slabs (50 mm x 100 mm x 100 mm). Figure 4.1 shows the photograph of the arrangement of the sample in the center of magnetic fields. The 10 mm long needle was completely inserted in the sample, ensuring it remained at the center of the vial. The direction of the magnetic field was parallel to the direction of heat transfer. The data were recorded at every 20 min interval with the following protocol: Initially, no field was applied, but data were recorded for 2 hr, then a particular field was applied for 24 hr (0.1, 0.12 and 0.23 T) and finally the field was removed and data were recorded for another 2 hr.

The rheological properties of the magnetic fluid were studied using Physica MCR 301 rheometer from Anton Paar equipped with an MRD set up in the strain-controlled mode. The measuring system used was a parallel plate geometry with a plate diameter of 20 mm. For steady-state measurements, the sample was placed in the measuring system of the rheometer with the field set to zero. The sample was pre-sheared at a rate of 50 s^{-1} for 30 s and then allowed to rest for 30 s for an equilibrium

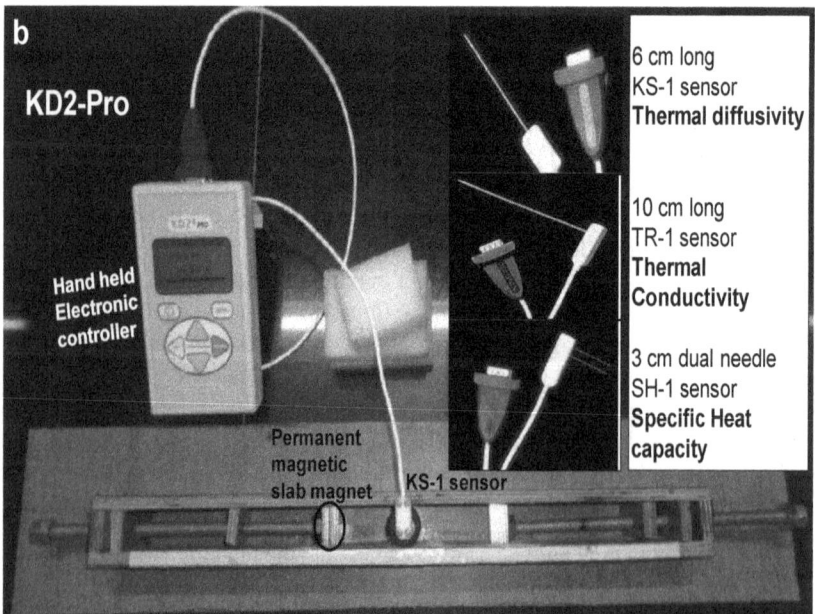

Figure 4.1. Photograph of Transient Hot Wire technique to measure the thermal conductivity using KD2-Pro thermal property analyzer.

structure to be established. The viscosity was then measured as a shear rate from 0.1 to 600 s^{-1}. For time-dependent relaxation, the measurement was performed for a constant shear rate of 100 s^{-1} under a fixed magnetic field (0.1, 0.12 and 0.23 T) to study the behavior of structure formation in the sample.

An optical microscope (MAGNUS MLM) with a 40X numerical aperture (NA) = 0.65 air objective with 10X eyepieces attached to a digital camera (Sony DSC-WX80) was used to record the dynamics of the magnetic-field-induced structure formations. The fixed magnetic field of 0.1, 0.12 and 0.23 T was applied to the sample at 305 K using rare earth slab magnets and the time evolution of structure formation was recorded for each field.

4.3 Results and discussion

4.3.1 Structural and magnetic characterization

The crystalline structure of these particles measured using an X-ray diffractometer shows characteristic peaks of spinel ferrite structure (Fig. 4.2a). The crystallite size of these particles calculated using Scherer's formula for the most intense (311) reflection plane is 8.0 ± 0.5 nm. However, the hydrodynamic particle size (D_h) measured using DLS by fitting the data with a log-normal diameter distribution curve is 11.6 nm (Fig. 4.2b). Hydrodynamic diameter considers the 2 nm thick layer of oleic acid coating the particles. The magnetic response of the fluid is shown in Fig. 4.2c (symbols) [9]. The solid line in the figure represents Langevin's theory incorporated with the log-normal size distribution function as described in Equation (4.2).

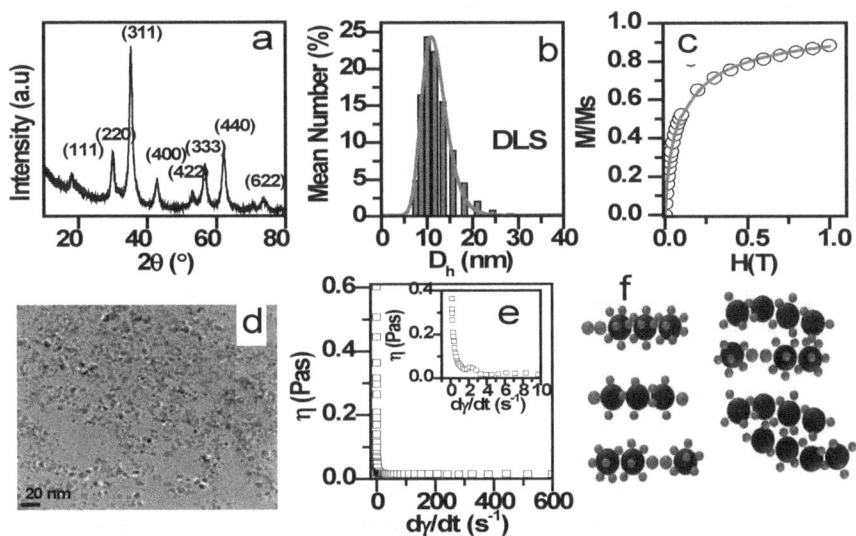

Figure 4.2. (a) XRD pattern of particles (b) Mean number distribution as a function of hydrodynamic size, diameter distribution obtained using log-normal moment distribution function. (c) Reduced magnetization M/Ms as a function of the magnetic field strength of magnetic fluid measured at room temperature. The data (symbols) were fitted with Langevin's theory (line) (d) TEM image of particles (e) Viscosity as a function of shear rate. The inset figure shows the zoom image of zero shear viscosity (f) Schematic of structure formation of particles in magnetic fluid due to the particle size distribution.

$$M(H,T) = n \int_{\mu_{min}}^{\mu_{max}} \mu L(\xi) P(\mu) d\mu$$

$$= n \int_{\mu_{min}}^{\mu_{max}} \mu \left\{ Coth(\frac{\mu H}{k_B T}) - \frac{k_B T}{\mu H} \right\} \left(\frac{1}{\sqrt{2\pi}\sigma_\mu \mu} \right) \exp\left[\frac{\ln\left(\frac{\mu}{\mu_m}\right)^2}{2\sigma_\mu^2} \right] d\mu \qquad (4.2)$$

Here, n is the number density of the particles, and $L(\xi)$ is Langevin's function with ξ as Langevin's parameter ($=\mu H/k_B T$). $P(\mu)d\mu$ is the log-normal moment distribution function. μ represents the particle magnetic moment. Similarly, μ_m represents the mean magnetic moment and σ_μ represents the distribution in $\ln(\mu)$. k_B is the Boltzmann constant and T is the absolute temperature. The best-fit parameters indicate the mean magnetic moment of the particle is $0.9 \pm 0.02 \times 10^{-19}$ A/m^2 with moment distribution, σ_μ as 1.2 ± 0.05 and 23 kA/m fluid magnetization.

Figure 4.2d shows TEM images of the nearly spherical particles with the median diameter, D = 9.5 nm and σ = 0.43 [9]. The viscosity as a function of shear rate is plotted in Fig. 4.2e. It is seen that the fluid is Newtonian above 50 s^{-1} shear rate, whereas at a low shear rate, it shows some pre-structure formed in the system (inset Fig. 4.2e). It is believed that this type of structure forms in the system because of the presence of large particles and high polydispersity in a sample [28] (as observed in Fig. 4.2b). Using this diameter, the characteristic coupling coefficient ($\lambda_{dcc} = \mu_0 \mu^2/4\pi D_h^3 k_B T$) is calculated where, μ_0 is the permeability of free space. The coupling coefficient is a dimensionless parameter that gives a measure of dipolar contact energy compared to thermal energy, $k_B T$. Figure 4.2b shows that 21% of the particles give dcc ≥ 2, while 79% of particles have $\lambda_{dcc} < 2$. Dipole-dipole interaction among particles having $1 < \lambda_{dcc} < 2$, is very weak and cannot ensure their assembly into long-range aggregates. However, small aggregates such as dimmers or trimmers may form. At zero magnetic field, there exist chains for $\lambda_{dcc} \gg 1$ in a magnetic fluid [5]. The characteristic interaction energy of large particles is much higher than thermal energy $k_B T$ as $\lambda_{dcc} \gg 1$; hence, this can result in the formation of various aggregates. Thus, the large particles capture the smaller particles and make clusters of small and large particles. The schematic picturization of possible structures can be seen in Fig. 4.2f.

Such clusters are responsible for significant changes in the physical properties of magnetic fluid in the absence of a magnetic field. Both experimental studies and computer simulation, as reported earlier [29–34], agree that strong dipole-dipole interaction results in short flexible chains; the number of chains and its length grow with an increase in particle concentration and the strength of the external magnetic field. Polydispersity in size results in types of aggregates other than a linear chain-like configuration [29, 35–36]. The interaction between the small particles is presumed to be very weak and can be ignored. At the same time, the dipole-dipole interaction is high enough for the large particles that can be assembled into linear chain-like clusters.

Moreover, the interaction between large and small particles is assumed to be moderate with the characteristic energy of the order of k_BT. Therefore, small particles may form "clouds" around a single large particle, extending as a chain [31]. In order to escape from being agglomerate, the effective repulsive force is created by coating the surface of nanoparticles with surfactant molecules, which prevents the particles from coming closer due to the Brownian motion. In addition to the magnetic dipolar interaction, the structural difference between transformer oil (light mixtures of alkanes in the C_{15} to C_{40} range and cyclic paraffin) and oleic acid ($C_{17}H_{33}COOH$) coated particles reduces the compatibility of the dispersed phase with the dispersion medium [37]. Incompatibility of carrier molecules with the surfactant tail leads to coil up or collapse to minimize the Gibbs free energy [38]. It leads to particle aggregation or short oligomers in the system. Such carrier-induced particle aggregation behavior is observed in studying acoustic wave propagation in transformer oil-based magnetic fluid [37]. The presence of aggregates also reduces the thermal conductivity of magnetic fluid [9]. The application of the magnetic field shows enhancement in thermal conductivity, but it was observed that to reach the point value of thermal conductivity in transformer oil-based fluid, large magnetic field strength is needed to be held for a longer duration [9].

4.3.2 Thermal Conductivity measurement

Figure 4.3 shows thermal conductivity enhancement as a function of time at a fixed applied magnetic field (0.1 T, 0.12 T and 0.23 T). The thermal conductivity of transformer oil at 303°K is 0.128 W/m-K. Thermal conductivity enhancement increases, reaching a peak and then decreasing for instantaneous application of 0.1, 0.12 and 0.23 T magnetic fields. The following features can be observed from the figure.

➤ Maximum enhancement in thermal conductivity is 78, 102 and 129%, respectively for 0.1 T, 0.12 T and 0.23 T field (Fig. 4.4).

➤ The time needed to achieve a peak is very long (a few hours) and decreases as the field intensity increases (Fig. 4.4).

➤ The decrease in thermal conductivity is faster when the field is high compared to the lower magnetic field and is not superimposed on the original value.

➤ After switching OFF the field, thermal conductivity enhancement does not show the original value indicating remanence in thermal conductivity.

The constant exposure of a specific magnetic field causes magnetic particles to align in the direction of the field, as shown in Fig. 4.3 (optical microscopy image). As time passes, the extension of the chain-like structure takes place and reaches its maximum alignment after several hours by increasing the aspect ratio. The expansion in the aspect ratio results in an increase in thermal conductivity with time. After a critical aspect ratio value, the structure becomes energetically unstable, changing the structure's shape from a long chain-like to a columnar structure. At the maximum aspect ratio, thermal conductivity enhancement shows a peak and then decreases because of the decrease in the aspect ratio of the aggregates due to the coalescence of a structure. Literature on thermal conductivity of kerosene and water-based magnetic

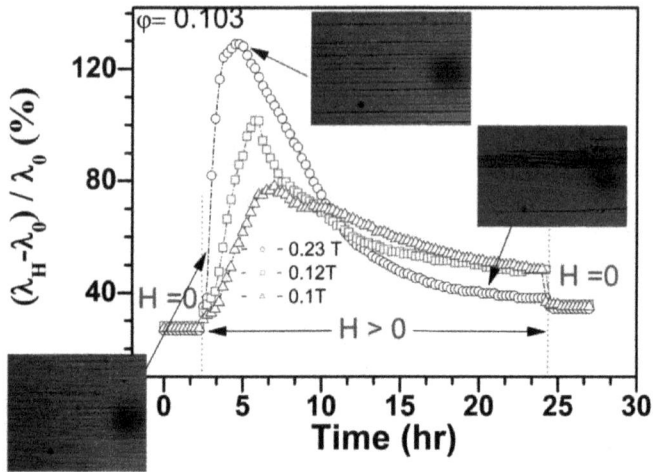

Figure 4.3. Thermal conductivity enhancement as a function of time at a fixed applied magnetic field (0.1 T, 0.12 T and 0.23 T). The inset figure shows an optical image of fluid taken at a different time for a 0.23 T field.

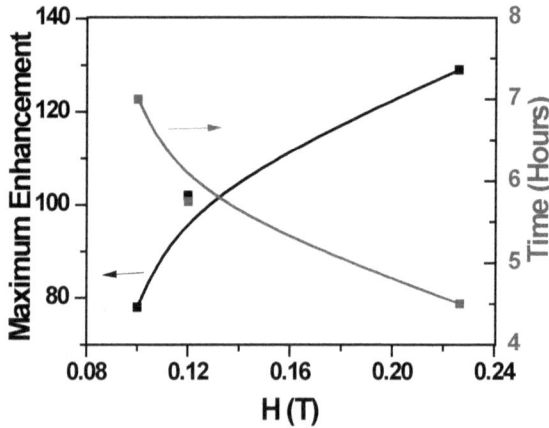

Figure 4.4. Plot of thermal conductivity enhancement and time required to achieve a maximum enhancement as a function of the specified magnetic field.

fluid [8, 10] showed a peak within a few minutes at relatively low field strength. In the present case, it takes a very long time (a few hours). We correlate the shift in peak time due to (i) high viscosity of transformer oil as compared to kerosene, (ii) low magnetic moment of Mn-Zn ferrite particles as compared to magnetite, and (iii) presence of short oligomer in the system even in the absence of magnetic field.

For a specific volume fraction, the increase in magnetic field strength increases the aspect ratio of the aggregates, escalating thermal conductivity enhancement. However, the increase in the field also expands the drag force, i.e., Kelvin body force and hence kinetics of the structure formation enhances , which results in a reduced time to have a maximum aspect ratio and hence peak shifts towards left with the increasing field. For the direct application of a magnetic field, the process of linear extension and the coalescence of small particle chains start together. Therefore, sidewise joining supersedes the chain length extension and creates spherical drop-like structures. This will result in a decrease in the maximum value of thermal conductivity.

Sithara et al. [39] reported that when the particle size (9.6 and 8.3 nm) is relatively smaller than the 10 nm size of the particles and size distribution is also narrow, then the 127% (at 160 G) and 42% (at 70 G), respective enhancement in thermal conductivity within the presence of magnetic field is observed. Whereas, for 10.5 nm size particles, no significant enhancement with the magnetic field was observed. They explained their results using the development of chain formation of particles around the large particles, which act as nucleation centers at low fields and form thick columnar structures at higher fields, which is responsible for reducing the heat transport efficiency. A recent study mentions that adding pristine MCNTs to the magnetic fluid increases thermal conductivity by 12.47% [40].

Similarly, Hakim Abbasov [41] proposed a model in which he considered the total volume of magnetic fluid under applied magnetic fields splits into two parts: a volume with aggregated magnetic nanoparticles (chainlike clusters) and a volume with non-aggregated uniformly dispersed nanoparticles. With the external magnetic field increasing, the aggregation degree—the percentage of the volume with the aggregated magnetic nanoparticles increases. Thus, the contribution of chainlike clusters to the thermal conductivity of ferrofluid depends on the magnetic field strength and the direction of these chains relative to the temperature gradient; when the chain direction is parallel to the temperature gradient, higher thermal conductivity enhancement and almost no change in thermal conductivity was observed when the chain direction is perpendicular to the temperature gradient.

4.3.3 *Viscosity measurement*

The relaxation in relative viscosity, the ratio with field viscosity to zero field viscosity ($\eta_H/\eta_{H=0}$), is carried out under ON-OFF conditions at a magnetic field strength of 0.23 T/0.12 T/0.1 T (ON) and zero fields (OFF), respectively. The field was kept ON for 20 min and data were recorded for every 10 s.

Figure 4.5 shows the relative viscosity ($\eta_H/\eta_{H=0}$) as a function of time measured at 100 s^{-1} shear rate at 303 °K. The blue symbol in Fig. 4.5 is for H = 0.23 T, the red symbol for 0.12 T and the black symbol for the 0.1 T field. The viscosity increases instantaneously with the application of a magnetic field. However, with time for a given magnetic field, it shows a slight decrease indicating that the magnetic structures are disturbed by shear underflow.

It is seen that the increase in relative viscosity is a function of external magnetic field strength. Under the influence of an external magnetic field, the particle's magnetic moment aligns in the direction of the field , and the particle will rotate around the field's direction. The direction of a magnetic field is perpendicular to vorticity, so the viscous friction tilts the magnetic moment against the field direction. The resulting finite angle between the magnetic moment and the field direction will give rise to a magnetic torque counteracting the viscous torque that tries to realign the moment along the field direction. Hindrances to the free rotation of the particles in the flow occur due to the counteraction of the torques.

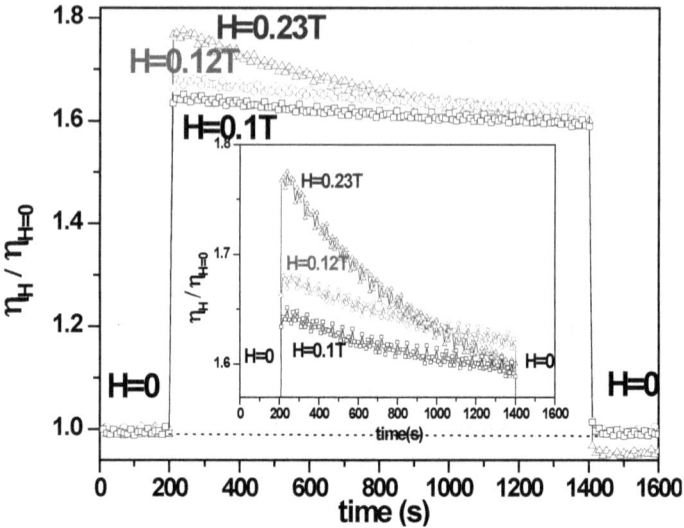

Figure 4.5. Field-dependent relaxation in relative viscosity ($\eta_H/\eta_{H=0}$) as a function of time measured at 303 °K. Square symbol for H = 0.23 T, circle symbol for 0.12 T and triangle symbol for 0.1 T field.

Furthermore, it also increases the viscosity of the fluid. The structure's realignment continues during constant magnetic field exposure, giving rise to a continuous decrease in relative viscosity, irrespective of field strength. The relative viscosity (η/η_0) value drops to zero immediately after field removal or even lower than zero field viscosity for a 0.23 T field. It shows that the structure breaks after removing the field and the suspension becomes homogeneous.

Similar structure evolution in a superparamagnetic latex suspension was observed by Fermigier and Gast [24]. They reported that when a magnetic field is applied to a system of low volume fraction, particles experience an attractive dipolar interaction and form linear chains by aggregating tip-to-tip parallel to the field direction. The long-range dipolar interaction modifies the flux, and lateral interaction alters the aggregation mechanism at sufficient high surface fractions above a few percent. Chains start to coalescence sideways, resulting in a thickening of chains to create fibers of particles.

Interaction between particles and chains

When isolated magnetizable particles experience the external magnetic field, H, it develops a dipole moment μ, which is proportional to the external field given as

$$\mu = \mu_0 \frac{4}{3} \pi a_3 \chi H ... \tag{4.3}$$

where, μ_0 is the magnetic permeability of the vacuum, χ is the magnetic susceptibility and 'a' is the particle radius. The interaction potential, U, between two identical dipole moments is given as

$$U(r, \theta) = \frac{\mu^2}{4\pi\mu_0} \frac{1-3\cos^2 \theta}{r^3} ... \tag{4.4}$$

where, r is the center-to-center distance and θ is the angle between the dipole moment vector and the center-to-center vector. The interaction thus scales with the square of the field strength and is attractive or repulsive depending on the dipoles' head-to-tail or side-by-side alignment. The maximum attraction, at $r = 2a$ and $\theta = 0$, is gauged by the dimensionless dipole strength λ defined as

$$\lambda = \frac{\pi \mu_0 \, a^3 \, \chi^2 \, H^2}{9 k_B \, T} \ldots \tag{4.5}$$

where, k_B is the Boltzmann constant and T is the absolute temperature. Under weak field conditions, the inter-particle interaction strength will be comparable to thermal energy and aggregation will commence. A range of weak external fields exists where the aggregates reach an equilibrium size, their attractive dipolar interaction is balanced by thermal forces. However, in most practical situations, the dipolar interaction strength greatly exceeds thermal energy and aggregation proceeds as a non-equilibrium transport limited process. At a low solid volume fraction, particles experience an attractive dipolar interaction when the magnetic field is applied and form linear chain-like aggregates aligned parallel to the external field. For large volume fractions, the induced dipolar interaction is much stronger than thermal energy. Thus, irreversible aggregates form, which grow continuously until they reach the walls of the sample vial.

If there is no interaction between the particles, the particle motion has a purely Brownian origin and the characteristic time, t_B, is defined as

$$t_B = \frac{1}{6} \frac{a^2}{D\varphi} \ldots \tag{4.6}$$

where, D is the particle diffusion coefficient ($= k_B T / 6\pi \eta a$) and φ is the solid volume fraction. The long-range inter-particle interactions modify the dipolar interaction by the stability factor, W, as

$$t_{BW} = t_B W = \frac{9.16}{6} \frac{a^2}{D\varphi \lambda^{\frac{4}{3}}} \ldots \tag{4.7}$$

The time for the lateral alignment of chains (t_{lat}) and tip-to-tip alignment of chains (t_{tip}) can be calculated using the formula given for the suspension of latex particles [24]

$$t_{lat} = \frac{6\pi \eta a^4 \sqrt{9.16\pi \eta a}}{k_B T \varphi_e^2 \sqrt{t k_B T \varphi \lambda^{\frac{4}{3}}}} \ldots \tag{4.8}$$

and

$$t_{tip} = \frac{\left(\dfrac{t k_B T \varphi \lambda^{\frac{4}{3}}}{9.16\pi \eta a} \right)^{\frac{5}{2}} 6\pi \eta}{k_B T \lambda a^2} \ldots \tag{4.9}$$

where, η is a fluid viscosity, φ_e is an effective solid surface fraction defined as $(N_0 \pi a^2/A)$ with N_0 is the total initial number of particles, A is an area of view and λ is dimensionless dipole strength. The time for the lateral alignment of chains (t_{lat}) and tip-to-tip alignment of chains (t_{tip}) for the present system is calculated using Equations 8 and 9 and the same is shown in Table 4.2. It is can be seen from the table that when the magnetic field is low, the transport time t_{tip} along the aggregates is faster as compared to the diffusion time t_{lat} for the lateral alignment of the aggregates. On increasing the magnetic field, the lateral alignment time is faster due to the time required to travel an average chain sufficiently long to join tip-to-tip. It makes the tip-to-tip process very slow and initiates significant lateral interaction. Lateral interaction is also responsible for the cross-linking of chains resulting in a fully inter-connected fibrous structure.

Table 4.2. The time for lateral alignment and tip-to-tip alignment of chains for various field strengths.

H(T)	$t_{lat}(s)$	$t_{tip}(s)$
0.1	3.36E-05	2.74E-05
0.12	2.66E-05	6.21E-05
0.23	1.15E-05	1.17E-03

Figure 4.5 shows that only in a 0.23 T field the relative viscosity decreases under constant magnetic field exposure. Whereas for lower field strength, such a decrease is not observed. It may be correlated to the lateral alignment of chains at 0.23 T field compared to the tip-to-tip alignment of chains at 0.12 and 0.1 T field.

4.3.4 *Optical microscopy*

The kinetics of structure formation is visualized physically using optical microscopic images taken at various field strengths of 0.1 T, 0.12 T and 0.23 T as a function of time. Figure 4.6 shows the images for all fields for 1 min, 5 min and 20 min. It can be seen that on application of the magnetic field, short chain-like structures formed within 1 min, which grows in length and width with increasing exposure time. After 20 min of exposure to the field, a long rod-like structure is formed. However, with increasing field strength, more fluctuating chains are observed (first column), interacting more strongly than rigid chains [24]. Hence, chain length increases faster, leading to a faster enhancement compared to a lower field. As time progresses, the tip-to-tip interaction increases at a given field, making the structure longer in the direction of the field. With further increase in time, the probability of lateral alignment of the chain increases, making fractals in the sample (last column). Fractals have a lower value of thermal conductivity than chain-like structures.

Moreover, the inter-chain distance decreases with increasing field strength. At 0.1 T, the chains are thicker and the average inter-chain distance is 51 µm. The chain covers the length of the view area. However, the thickness decreases with increasing field strength and the interchain spacing decreases. For 0.12 T, the interchain spacing varies from 20 to 37 µm, and for 0.23 T, it varies from 9 to 37 µm. In addition, the number of short-chain grows with increasing field strength.

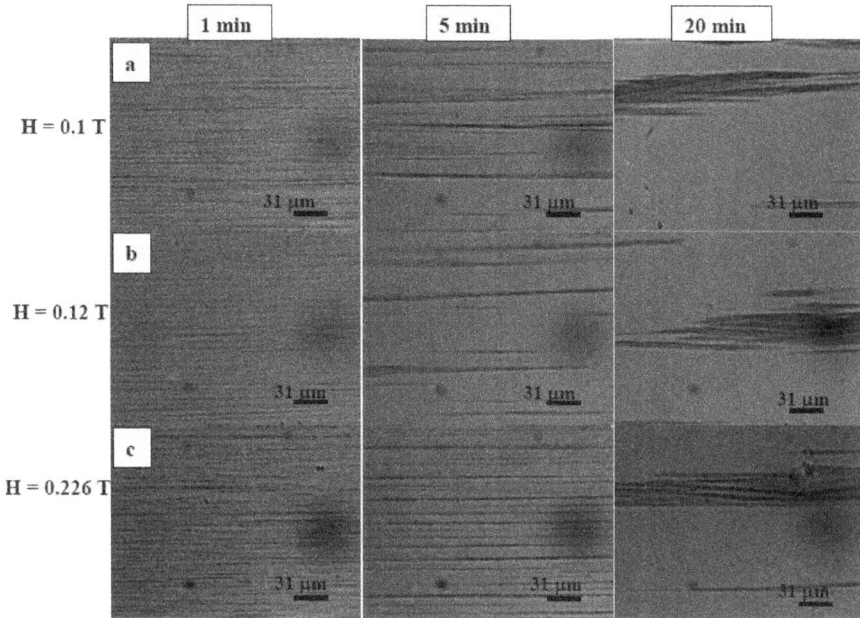

Figure 4.6. Optical microscopic images taken at different times after application of 0.1 T, 0.12 T and 0.23 T magnetic field.

Although chains are well spaced for the 0.1, T applied field. The enhancement is lower than the higher field strength. The extended and rigid chains create a bridge for enhanced heat transfer. However, at the same time, a channel with only a carrier is created, which forms a low conducting path and hence reduces overall thermal conductivity. With increasing field strength, the drag force, i.e., the Kelvin body force, is higher than the viscous force; this causes the formation of many short chains, which joins tail to tail rather than lateral aggregation. The short chains are not stable and fluctuate. Due to this, the heat transfer escalates by increasing the number of fluctuating chains.

4.4 Conclusion

The influence of structure formation on thermal conductivity of transformer oil-based Mn-Zn ferrite fluid studied as a function of time for three different magnetic fields (0.1 T, 0.12 T and 0.23 T). The field was directly applied to the system, and thermal conductivity enhancement's time evolution was investigated. With direct application of the magnetic field, thermal conductivity increases, reaching a peak value and decreases. The time required to achieve maximum enhancement and its peak value is field-dependent. The relative viscosity relaxation also shows the re-organization of the structures in steady-state conditions. The results are explained using the structure formation due to Kelvin's body force and drag force. With increasing field intensity, thermal conductivity and viscosity grow due to an increase in the Kelvin body force and chain formation kinetics. Increasing the magnetic field at a fixed volume fraction will escalate the number of fluctuating chains, enhancing thermal conductivity and viscosity. For a low magnetic field, the tip-to-tip alignment of chains is more favorable than the lateral alignment, which starts at a higher field strength, resulting in cross-linking of chains with time and creating fractals-like structures. Since fractals have low thermal conductivity, the enhancement in thermal conductivity drops after this time.

Acknowledgment

The authors thank the Department of Science & Technology, Government of India, New Delhi, for providing financial support under the "Technology Development Program" vide; DST/TSG/2011/161-G.

References

[1] http://nano.xerox.com/nanotech/feynman.html.
[2] H. Rohrer. The nanoworld: Chances and challenges. Microelectronic Engineering 32(1-4): 5–14 (1996).
[3] A. Miner and U. Ghoshal. Cooling of high-power-density microdevices using liquid metal coolants. Applied Physics Letters 85(3): 506–508 (2004).
[4] Z. Han Nanofluids with enhanced thermal transport properties. Thesis submitted to the University of Maryland at College Park, College Park, Maryland (2008).
[5] R. E. Rosensweig. Ferrohydrodynamics, 1st ed. (Dover Publication, New York, 1997).
[6] Nkurikiyimfura I., Y. Wang and Z. Panl. Heat transfer enhancement by magnetic nanofluids—A review. Renewable and Sustainable Energy Reviews 21: 548–561 (2013).
[7] K. Parekh and H. S. Lee. Experimental investigation of thermal conductivity of magnetic nanofluids. AIP Conf. Proc. 1447: 385–386 (2012).
[8] A. Gavili, F. Zabihi, T. D. Isfahani and J. Sabbaghzadeh. The thermal conductivity of water base ferrofluids under magnetic field. Exp Therm Fluid Sci. 41: 94–98 (2012).
[9] J. Patel, K. Parekh and R. V. Upadhyay. Maneuvering thermal conductivity of magnetic nanofluids by tunable magnetic fields. J. Appl. Phys. 117: 243906 (2015).
[10] P. D. Shima and John Philip. Tuning of thermal conductivity and rheology of nanofluids using an external stimulus. J. Phys. Chem. C 115: 20097–20104 (2011).
[11] H. L. Fu and L. Gao. Theory for anisotropic thermal conductivity of magnetic nanofluids. Physics Letters A 375 3588–3592 (2011).
[12] Q. Li, Y. Xuan and J. Wang Experimental investigations on transport properties of magnetic fluids. Experimental Thermal and Fluid Science 30(2): 109–116 (2005).
[13] Q. Li, Y. Xuan and J. Wang. Experimental investigations on transport properties of magnetic fluids. Experimental Thermal and Fluid Science 30(2): 109–116 (2005).
[14] J. Philip, P. D. Shima and B. Raj. Nanofluid with tunable thermal properties. Applied Physics Letters 92(4): 043108_1–043108_3 (2008).
[15] P. D. Shima, J. Philip and B. Raj. Magnetically controllable nanofluid with tunable thermal conductivity and viscosity. Applied Physics Letters 95(13): 133112_1–133112_3 (2009).
[16] P. D. Shima, J. Philip and B. Raj. Role of microconvection induced by Brownian motion of nanoparticles in the enhanced thermal conductivity of stable nanofluids. Applied Physics Letters 94(22): 223101_1–223101_3 (2009).
[17] P. D. Shima and J. Philip. Tuning of thermal conductivity and rheology of nanofluids using an external stimulus. The Journal of Physical Chemistry C 115(41): 20097–20104 (2011).
[18] A. Gavili, F. Zabihi, T. D. Isfahani and J. Sabbaghzadeh. The thermal conductivity of water base ferrofluids under magnetic field. Exp. Therm. Fluid Sci. 41: 94–98 (2012).
[19] J. Kudelcık, P. Bury, S. Hardon, P. Kopcansky and M. Timko. Influence of nanoparticles diameter on structural properties of Magnetic fluid in magnetic field. J. Electr. Eng. 66: 231–234 (2015).
[20] T. C. Halsey and W. Toor. Fluctuation-induced couplings between defect lines or particle chains. J. Stat. Phys. 61: 1257 (1990).

[21] T. Halsey and W. Toor. Structure of electrorheological fluids. Phys. Rev. Lett. 65: 2820 (1990).

[22] J. E. Martin, J. Odinek and T. C. Halsey. Evolution of structure in a quiescent electrorheological fluid. Phys. Rev. Lett. 69: 1524 (1992).

[23] J. E. Martin, J. Odinek, T. C. Halsey and R. Kamien. Structure and dynamics of electrorheological fluids. Phys. Rev. E 57: 756 (1998).

[24] M. Fermigier and A. Gast. Structure evolution in a paramagnetic latex suspension. J. colloid Inter. Sci. 154: 522 (1992).

[25] J. E. Martin, K. M. Hill and C. P. Tigges. Magnetic-field-induced optical transmittance in colloidal suspensions. Phys. Rev. E 59: 5676 (1999).

[26] E. M. Furst and A. P. Gast. Dynamics and lateral interactions of dipolar chains. Phys. Rev. E. 62 6916 (2000).

[27] J. Liu, E. M. Lawrence, A. Wu, M. L. Ivey, G. A. Flores, K. Javier, J. Bibette and J. Richard. Field-induced structures in ferrofluid emulsions. Phys. Rev. Lett. 74: 2828 (1995).

[28] W. Toor. Structure formation in electrorheological fluids. J. Colloid Interface Sci. 156: 335 (1993).

[29] S. Odenbach. Magnetoviscous effect in ferrofluids. Berlin: Springer-Verlag 185–201 (2002).

[30] A. O. Ivanov and S. S. Kantorovich. Structure of chain aggregates in ferrocolloids. Colloid J. 65 166–176 (2003).

[31] J. B. Mathieu and S. Martel. Aggregation of magnetic microparticles in the context of targeted therapies actuated by a magnetic resonance imaging system. J. Appl. Phys. 106: 044904 (2009).

[32] A. Y. Zubarev and L. Y. Iskakova. Structural transformations in polydisperse ferrofluids. Colloid J. 65: 711–719 (2003).

[33] H. Morimoto and T. Maekawa. Cluster structures and cluster–cluster aggregations in a two-dimensional ferromagnetic colloidal system. J. Phys. A: Math. Gen. 33: 247–258 (2000).

[34] N. S. S. Mousavi, S. D. Khapli and S. Kumar. Direct observations of field-induced assemblies in magnetite ferrofluids. J. Appl. Phys. 117: 103907 (2015).

[35] S. Kantorovich and A. O. Ivanov. Formation of chain aggregates in magnetic fluids: an influence of polydispersity. J. Mag. Mag. Mater. 252: 244–246 (2002).

[36] M. Aoshima and A. Satoh. Two-dimensional Monte Carlo simulations of a colloidal dispersion composed of polydisperse ferromagnetic particles in an applied magnetic field. J. Colloid Interface Sci. 288: 475–488 (2005).

[37] K. Parekh, J. Patel and R. V. Upadhyay. Ultrasonic propagation: A technique to reveal field induced structures in magnetic nanofluids. Ultrasonics 60: 126–132 (2015).

[38] G. Cao. Nanostructures & nanomaterials synthesis, properties and applications. Imperial College Press, London p-43 (2004).

[39] S. Vinod and J. Philip. Experimental evidence for the significant role of initial cluster size and liquid confinement on thermo-physical properties of magnetic nanofluids under applied magnetic field, Journal of Molecular Liquids 257: 1–11 (2018).

[40] Qian Li, Juying Zhao, Licong Jin and Decai Li. Experimental study on thermal conductivity and magnetization behaviors of kerosene-based ferrofluid loaded with multiwalled carbon nanotubes. ACS Omega 5: 13052−13063 (2020).

[41] Hakim Abbasov. Modeling of anisotropic thermal conductivity of ferrofluids. J. Dispersion Sci. Tech. 41(7): 1030–1036 (2019). https://doi.org/10.1080/01932691.2019.1614040.

5

Preparation, Colloidal Structure and Thermodiffusion of Ferrofluids Based on Charged Nanoparticles Dispersed in Polar Solvents and Ionic Liquids

C.L. Filomeno,[1,2] *T. Fiuza,*[1,2] *M. Kouyaté,*[1] *J.C. Riedl,*[1]
M. Sarkar,[1] *G. Demouchy,*[1,3] *J. Depeyrot,*[2] *E. Dubois,*[1]
G. Mériguet,[1] *R. Perzynski*[1,*] and *V. Peyre*[1]

〜〜〜

5.1 Introduction

Magnetic Fluids or Ferrofluids (FF) [1–5] are colloidal dispersions of magnetic nanoparticles (NPs) which are flowing in a gradient of field towards the region where the field is the strongest. In a homogeneous field they present a superparamagnetic behavior at room temperature and an under-field optical anisotropy, related to the dispersed NPs shape and structure. A number of technological applications are derived

[1] Sorbonne Université, CNRS, Lab. PHENIX, case 51, 4 place Jussieu, F-75005, Paris, France.
[2] Univ. de Brasília, Grupo de Fluidos Complexos - Inst. de Quimica e Inst. de Fisica, Brasília D.F., Brazil.
[3] Dpt. de Physique - Univ. Cergy-Pontoise 33 Bd du port - F95011 Cergy-Pontoise Cedex France.
* Corresponding author: regine.perzynski@sorbonne-universite.fr

from these FF properties either in dilute dispersions (optical switches or modulators) [6] or in more concentrated ones (magnetic seals, dampers, sensors) [1,7]. Such dispersions also find applications in the biomedical field (MRI markers, magneto-thermic treatments, cell labelling) [8–12] or in the field of energy (thermocells, low grade energy harvesting, thermionic capacitors) [13–16] or for heat transfer fluids [17]. Contrary to magneto-rheological fluids which are based on much larger magnetic particles and present under-field inhomogeneities [18], the FF's colloidal stability is of paramount importance for the efficiency of the developed applications. Indeed, the NP aggregation [19] or the (sometimes observed) phase separation in two liquid phases [20] is then prohibited. To ensure the colloidal stability of the dispersion an interparticle repulsive force must exist, which surpasses the van der Waals attraction and the dipole-dipole interaction, which is anisotropic under an applied magnetic field and which is on average attractive in zero and low fields. This repulsion can be brought by a surfactant coating of the NPs or by an electrostatic repulsion between charged NPs. In view of thermoelectric applications [21–26], we focus here on the development and the colloidal stabilization of FFs based on electrostatically charged NPs dispersed either in polar solvents or in room temperature Ionic Liquids (named here ILs). As polar solvents, we focus here on water and dimethyl sulfoxide (DMSO), both frequently used solvents [27, 28]. For their parts, ILs are a wide class of materials only constituted of ions [29]. They are liquid at room temperature. They present a high thermal stability associated to a reduced flammability and vapor pressure and can allow to extend standard applications towards more demanding conditions. We probe several ILs hereafter, but mainly focus on two reference systems, namely ethyl ammonium nitrate (EAN) and 1-ethyl 3-methylimidazolium bistriflimide ($EMIM^+$-$TFSI^-$). In the text below, we present how the stability of FFs based on charged NPs in polar solvents and in ILs is theoretically described in these two kinds of solvents. We then focus on the way of preparing such stable FFs and on the different important parameters to ensure their stability. Later, we show how the colloidal structure can be probed and quantified by Small Angle Scattering (SAS) in order to obtain the second virial coefficient A_2 of a NP's dispersion, which characterises the interparticle interaction by its sign and its value. It is then shown how the NP/solvent parameters are obtained by Forced Rayleigh Scattering, through Soret coefficient (S_T) determinations as a function of temperature and NP's volume fraction. Finally, under-field properties are presented and discussed.

5.2 FF colloidal stabilization

What are the models at hand to describe the colloidal stabilization of ferrofluids based on NPs bearing a structural electrostatic charge?

5.2.1 Polar solvents

In polar solvents, the zero-field stabilization of a ferrofluid based on charged NPs is described, as in more usual colloidal dispersions, by the DLVO model, slightly modified to include the magnetic dipolar attraction in zero field [30]. This latter depends on the NP size and can be roughly seen as a stronger attraction, in terms of range and intensity, than van der Waals one. Each NP bears a structural charge Z_{st} of sign given by the NP coating, and smaller ions of opposite sign are necessarily present in the dispersion to ensure the electro-neutrality of the system, together with a number of other dissociated ionic species in the electrolyte. Thus, within the dispersion, a cloud of counterions surrounds each NP, which screens the electrostatic interparticle repulsion (see Fig. 5.1 - top of left box). The electrostatic Debye screening length κ^{-1} depends on the temperature, on the dielectric permittivity ϵ_r of the solvent (in water $\epsilon_r = 80$, in DMSO $\epsilon_r = 46$) and on the ionic strength associated to the concentration of free ions in the dispersion. For a monovalent electrolyte it reads:

$$\kappa^{-1} = \sqrt{\frac{\epsilon_0 \epsilon_r kT}{2e^2 c_s}} \tag{5.1}$$

where ϵ_0 is the vacuum permittivity, kT the Brownian energy, e the elementary charge and c_s the ionic concentration expressed in number of ions per cubic meter. If the Debye screening length κ^{-1} is too small with respect to the NP diameter (usually because of a too high ionic concentration and/or too large NP's diameter), a phase separation is observed which can be used for a size sorting of the NPs [20, 30, 31]. It leads to droplets of a highly concentrated phase (typically 30 vol%) surrounded by an extremely diluted phase, with a very low surface tension at the interface leading to spectacular under-field shape instabilities [32–34] (see Fig. 5.1 - right box). If the ionic concentration is increased an even more complete precipitation occurs between solid and liquid [20, 30, 31].

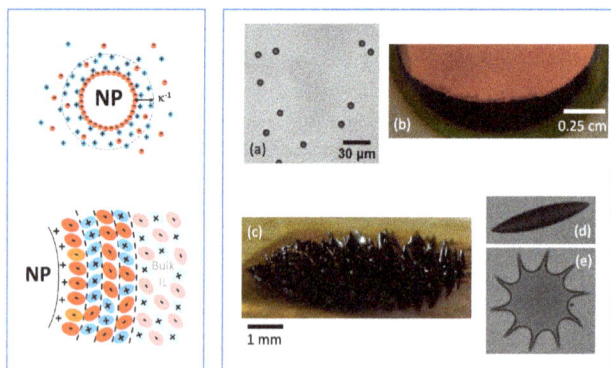

Figure 5.1: (Left box) - top - Schematic representation of a negatively charged NP in a polar solvent, the ionic Debye cloud surrounding the NP has a characteristic thickness κ^{-1}, - bottom - Schematic representation of the ionic layering around a positively charged NP in an Ionic Liquid and insuring its colloidal stability. (Right box) Illustration of phase separated FFs in water (a, d and e) and in EAN (b and c): (a) Zero-field observation of a phase separated sample with an optical microscope evidencing micro-droplets of concentrated phase; (b) Macroscopic observation after sedimentation in zero-field of the concentrated phase in an optical cell; (c) Under-field peak instability of the isolated concentrated phase of picture b; (d) Shape deformation and (e) instability under a rotating field, of a droplet similar to those of picture (a) [35, 36]; Full scale of pictures (d) and (e) is of the order of 100 microns.

5.2.2 Ionic Liquids

In Ionic Liquids, the colloidal stability cannot be described by DLVO theory anymore, the solvent is only constituted of ions and the ionic concentration is far too large. The colloidal stability of the charged NP is then ensured by a solvation layering of the IL (see Fig. 5.1 - bottom of left box), if the NP charge, the ions at the NP/IL interface and the size of IL ions are adapted to the IL self-organization [37,38]. MD simulations of ILs on flat surfaces [39] have evidenced the important role of the ratio $\kappa_{\text{ion}} = |\sigma/Q_{\text{ion}}^{\text{max}}|$ for obtaining an efficient ionic layering. In κ_{ion}, σ is the NP's structural charge density and $Q_{\text{ion}}^{\text{max}}$ the maximum charge density of a densely packed counterion monolayer. The optimum situation to produce an ionic layering perpendicular to the NP interface, over a large thickness corresponds to $\kappa_{\text{ion}} = 0.5$.

Also the NPs should not be too big to limit the dipole-dipole interaction. For example, in EAN (see [40]) with Li^+ ions at the interface, only the smallest NPs are dispersible in EAN. With larger NPs, a phase separation is observed in that case [41], similar to the one usually observed in water.

5.3 Ferrofluid synthesis

In this work, the magnetic NPs are all made of maghemite (γFe_2O_3). They are initially obtained in water by a coprecipitation process of $FeCl_2$ and $FeCl_3$ in strongly alkaline medium, which is known as Massart's method [42, 43]. At the end of the process, the NPs are dispersed in water with protonated hydroxyl surface groups and nitrate counterions at $pH \sim 2$. Their surface charge is then positive (with a superficial density of structural charge of the order of 2 elementary charges per nm^2 [44]), $pH \sim 7$ being their point of zero charge (PZC). The polydisperse NPs can then be size-sorted using the size-dependent colloid phase diagram as described for example in [43]. In order to produce in a reproducible and quantitative way, dispersions with a different coatings, different counterions, at different pH in water or in different media, the NPs are precipitated at their PZC. The NPs then wear no charge and the extra ions can be all washed out. The NPs can be subsequently recharged in a controlled way by using a chosen acid or base. It leads to dispersions of NPs with defined coating and counterions, either in water or DMSO, in particular with the sign and value of the structural surface charge monitored by the pH value and the nature of the used acid or base [45]. For example, negatively charged NPs in water at $pH \sim 7$ can be obtained by replacing protonated hydroxyl surface groups by citrate ones. NP's dispersions in EAN are obtained by mixing aqueous dispersions of citrate coated NPs with pure EAN, and then removing the water by a lyophilization process. In other ILs, aqueous or DMSO dispersions (with various coatings and counterions – (lithium Li^+, sodium Na^+, tetramethyl ammonium TMA^+, tetrabutyl ammonium $TBuA^+$) see [37, 38, 46]) are mixed to the chosen IL. Water or DMSO are then removed by lyophilization.

A NP's coating with other charged species is also used here for a comparison (see Table 5.1), namely the zwitterionic coupling agent 1-Methyl-3-(dodecyl-phosphonic acid) amidazolium bromide (CAS number [2230266-36-9]) $PAC_{12}MIM^{\pm}$ with Br^- counterions [46] and poly(acrylate-co-maleate) $PAAMA^-$ with NH_4^+ counterions [47].

What are the parameters with which it is possible to play to adapt the stability of the colloidal dispersions?

Table 5.1: Sample characteristics: Solvent, NP's coating, nature of the NP's counter-ions or of the superficial species, [c.i.] counter-ion's concentration, κ^{-1} electrostatic Debye length, d_{NP} volume averaged NP's diameter, A_2 second virial coefficient in the NP's dispersion at room temperature, ξ_{eff} NP's dynamic effective charge at room temperature and $\hat{S}_{NP}$ NP's Eastman entropy of transfer at room temperature - (*) 1-Butyl-(4-Sulfobutyl)-3-methylimidazolium ; (**) bistriflimide (***) triflate; (****) H3 series are based on ionic liquids IL3 to IL6 in Table 5.3.

Sample name	Solvent	Coating	Counter-ions (c.i.)	$[c.i.]_{free}$ (mol/L)	pH	κ^{-1} (nm)	d_{NP} (nm)	A_2	ξ_{eff}	$\hat{S}_{NP}/kT$ K^{-1}	ref
A1	water	citrate	Na$^+$	0.009	~7	3.3	11.4	+16	-35	-0.15	[48]
A2	water	citrate	Na$^+$	0.09	~7	1.3	11.4	+7.4	-35	-0.07	[48]
A3	water	citrate	Na$^+$	0.009	~7	3.2	10.8	+16	-35	-0.15	[48]
A4	water	citrate	Na$^+$	0.09	~7	1.3	10.8	+7.6	-35	-0.07	[48]
A5	water	citrate	Na$^+$	0.24	~7	0.9	10.8	+6.2	-35	-0.04	[48]
B1	water	citrate	Li$^+$	0.084	~7	1.0	8.5	+15.5	-28	-0.20	[49]
B2	water	citrate	Na$^+$	0.072	~7	1.1	8.5	+13	-30	-0.13	[49]
B3	water	citrate	TMA$^+$	0.081	~7	1.1	8.5	+10.5	-32	-0.02	[49]
B4	water	citrate	TBuA$^+$	0.093	~7	1.0	8.5	+14	-40	+0.09	[49]
B5	water	citrate	TBuA$^+$	0.070	~7	1.1	7.6	+13.5	-33	+0.20	[45, 50]
C1	water	H$^+$	NO$_3^-$	0.0091	~2	3.2	11.2	+15.6	+38	-0.45	[48]
C2	water	H$^+$	NO$_3^-$	0.0091	~2	3.2	10	+17.6	+30	-0.25	[48]
C3	water	H$^+$	NO$_3^-$	0.0091	~2	3.2	9.1	+19.6	+23	-0.02	[48]
C4	water	H$^+$	NO$_3^-$	0.0091	~2	3.2	8	+23.2	+16	+0.05	[48]
D1	water	H$^+$	TFSI$^-$	~0.01	~2	~3	9.6	-	-	-	[46]
D2	water	H$^+$	NO$_3^-$	0.021	1.7	2.07	10.6	+17	+30	-0.21	[45]
D3	water	H$^+$	ClO$_4^-$	0.015	1.8	2.45	10.6	+4.5	+39	-0.32	[45]
D4	water	H$^+$	ClO$_4^-$	0.010	2	3.0	9.0	+33	+65	-0.50	[51]
E1	water	PAC$_{12}$MIM‡	Br$^-$	-	~2		9.6	-	-	-	[46]
E2	water	PAAMA$^-$	NH$_4^-$	-	9.5		7.9	-	-	-	[47]
F1	DMSO	H$^+$	ClO$_4^-$	0.033	1.5	1.2	10.6	+10.5	+19	+0.45	[45]
F2	DMSO	H$^+$	ClO$_4^-$	0.010	2	2.3	9	+17	+43	+1.0	[50]
F3	DMSO	H$^+$	ClO$_4^-$	0.0014	2.8	6.0	7.6	+19.5	+15	+0.47	[45]
F4	DMSO	H$^+$	ClO$_4^-$	0.024	1.6	1.5	7.6	+12	+30	+1.9	[51]
F5	DMSO	H$^+$	ClO$_4^-$	0.013	1.9	2.1	6.7	+17.2	+27	+2.7	[21]
F6	DMSO	citrate	TBuA$^+$	0.077	~7	0.82	10.6	+14.5	-21	-0.39	[45]
G1	EAN	citrate	Li$^+$	-	~7 – 8		7.4	Att	-	-	[52]
G2	EAN	citrate	Na$^+$	-	~7 – 8		7.4	+4.6	+55	+6.0	[36]
H1	EMIM$^+$-TFSI$^-$	HSBMIM$^+$ (*)	TFSI$^-$ (**)	-	-	-	9.6	+7.4	-	-	[38, 53]
H2	EMIM$^+$-TFSI$^-$	HSBMIM$^+$ & HTfO (***)	TFSI$^-$ & TfO$^-$ (***)	-	-	-	9.6	Att.	-	-	[38]
H3 series	$\neq$ ILs (****)	HSBMIM$^+$ (*)	TFSI$^-$ (*)	-	-	-	9.6	Att	-	-	[46]

- The NP's diameter and the polydispersity of the size distribution are important parameters as they may drastically increase the NP/NP attraction. For the samples presented here, Table 5.1 shows the average NP's diameter $d_{\mathrm{NP}} = \sqrt[3]{\langle d^3 \rangle}$ over the whole distribution of NP's diameters d. These values are typical for stable colloids. Let us recall that for large NPs, the dipole-dipole magnetic interaction limits the colloidal stability.

- The NP's coating to adapt the structural charge of the NPs (in sign and value) to the nature of the solvent, for example the water pH [45], or to a particular IL [37].

- Nature and concentration of the species at the NP's interface: Counter-ions (necessarily present in the Debye cloud) in polar solvents [45, 49] or extra species that localize at the NP interface due to their charge/size/affinity for ILs [37, 38, 52].

5.4 Colloidal structure – NP/NP interaction

The colloidal stability can be checked by several techniques at the microscale: under-field diffraction of a light beam [44, 54] or under-field optical microscopy [52], to check the absence/ presence of concentrated droplets elongated along the applied field. Zero-field SAS on its part is very sensitive to the presence of NP's agglomerates [55]. At a given scattering vector Q, the scattered intensity $I(Q)$ is linked to the structure factor S(Q) via the form factor P(Q) in the expression:

$$I(Q, \phi) = (\Delta\rho)^2 \Phi V_{NP} P(Q) S(Q, \Phi) \qquad (5.2)$$

where $\Delta\rho$ is the contrast between NPs and solvent, Φ is the NPs' volume fraction, V_{NP} their volume, $P(Q)$ is the NPs' form factor and $S(Q)$ their structure factor. Measurements extrapolated at zero volume fraction enable to determine the experimental form factor, which corresponds to $S(Q) = 1$ in Equation 5.2.

Measurements of $I(Q)$ as a function of NP's volume fraction allows to distinguish between NPs in repulsive or in attractive interparticle interaction [19]. Indeed, if the form factor P(Q) of the NPs is determined, the value of the structure factor at zero scattering vector allows to discriminate between the two following situations:

- If $S(Q \to 0) < 1$ the interparticle interaction is repulsive and $S(Q \to 0)$ equals the compressibility χ of the NP system (see Fig. 5.2a and b),

- If $S(Q \to 0) > 1$, the interparticle interaction is attractive and $S(Q \to 0)$ equals the NP's number per aggregate [19] (if the inter-aggregate interaction is negligible); close to 1 a compressibility slightly larger than 1 is sometimes used (for an extension to slightly attractive systems [20]).

If the NP/NP interaction is repulsive, it thus allows to determine the compressibility χ of the NP system and also the most probable interparticle distance via Q^{max}, the scattering vector at the maximum of the structure factor [56]. We focus hereafter on systems with repulsive interaction, only a few slightly attractive systems are also referred to (see Table 1).

Figure 5.2 presents SANS and SAXS measurements obtained with aqueous dispersions based on NPs coated with citrate species with Na^+ counterions (samples A in Table 5.1 - data from [48, 56]) and SAXS measurements obtained with dispersions in EAN based on much smaller NPs (sample G2 in Table 5.1 - data from [36]) also coated with citrate species and some Na^+ counterions at the NP/IL interface.

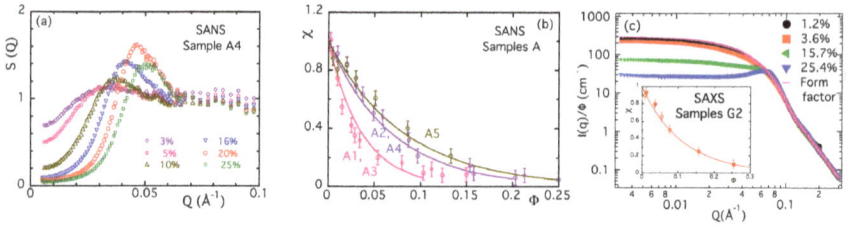

Figure 5.2: (a) SANS structure factor $S(Q)$ of colloidal dispersions in water at $pH \sim 7$ of NPs coated with citrate species at various NP volume fractions with a concentration of free Na^+ counterions of 0.09 mol/L (sample A4 from Table 5.1 - data from [56]); (b) Compressibility $\chi(\Phi)$ of the same NPs (samples A in Table 5.1 - data from [48]) at three different $[Na^+]$ values, full lines are fits of $\chi(\Phi)$ with the Carnahan-Starling formalism of Equation 5.4, the corresponding A_2 values being given in Table 5.1; (c) SAXS intensity $I(Q)$ scattered by colloidal dispersions in EAN, at various volume fractions Φ, with citrate-coated NPs and some Na^+ counterions at the NP/ IL interface; Inset: Associated $\chi(\Phi)$ determination for these dispersions, fitted with Equation 5.4 leading to $A_2 = 4.6$; (sample G2 in Table 5.1 - data from [36]).

The second coefficient A_2 of the Virial development of the osmotic pressure, Π, for a system based on nanoparticle of volume V_{NP}, is deduced from:

$$\chi = \frac{kT}{V_{NP}} \frac{\partial \Phi}{\partial \Pi} \quad \text{and if } \Phi \ll 1 \quad \chi \sim \frac{1}{1 + 2A_2\Phi}. \tag{5.3}$$

The right hand expression in Equation 5.3 is only valid for χ close to 1, and thus for $2|A_2|\Phi \leq 0.1$. Thus if measurements are performed at $\Phi = 1\%$, this expression only works for $-5 \leq A_2 \leq +5$. It allows anyway to discriminate between attractive ($A_2 < 0$) and repulsive ($A_2 > 0$) interaction.

In the case of interparticle repulsion, the compressibility can be modelled by the Carnahan-Starling formalism of effective hard spheres with an effective volume fraction Φ_{eff} [48,57,58] by:

$$\chi(\Phi) = \chi_{cs}(\Phi_{eff}) = \frac{(1 - \Phi_{eff})^4}{1 + 4\Phi_{eff} + 4\Phi_{eff}^2 - 4\Phi_{eff}^3 + \Phi_{eff}^4} \quad \text{with} \quad \Phi_{eff}/\Phi = A_2/4; \tag{5.4}$$

For hard spheres, $A_2 = 4$ and $\Phi_{eff} = \Phi$. This expression is valid up to the glass transition typically at $\Phi_{eff} \sim 40\%$. The fit of experimental determinations of $\chi(\Phi)$ by Equation 5.4 thus allows to determine A_2, even if the volume fraction is large, in this case of NP/NP repulsion. It is what is done in Fig. 5.2b and inset of Fig. 5.2c for different dispersions of NPs in repulsive interaction.

Table 5.1 gives some examples of A_2 values obtained for various samples used in the present work, where we mainly focus on samples with NP/NP repulsion. The range largely depends on the nature of the solvent.

- In water, depending on the NP diameter, the ionic strength and the nature of the counter-ion, A_2 can reach at room temperature values as large as 33 (sample D4 in Table 5.1 - data from [50]) or 22.6 in [20]. If the ionic strength is moderately large, A_2 values close to that of hard sphere can be obtained (see for example $A_2 = 4.4$

in [20]). If the ionic strength is too large, attractive situations are reached. Dispersions are still stable with small negative values A_2 values (See [20] and [45]), but if attraction is too large a phase separation is observed (see [30]).

- In DMSO, A_2 reaches a value of 17 in [50] and 19.5 is obtained in [45].

- In Ionic Liquids, comparatively slightly smaller A_2 values are obtained experimentally. In EAN, with small NPs ($d_{NP} = 7.4$ nm) and Na$^+$ counter-ions a value $A_2 = 4.6$ is obtained in [36], while with larger NPs ($d_{NP} \sim 11.5$ nm) a phase separation is observed in [41]. In contrast, with rather large NPs ($d_{NP} \sim 9.6$ nm) stable colloidal dispersions are obtained in EMIM$^+$-TFSI$^-$ with a still not so large $A_2 = 7.4$ (see [38,53]). Intermediate stable situations with small aggregates (associated to small attraction) are also frequently observed in Ionic Liquids (see in Table 5.1 and in [37,38,52]), which is very different from what is observed in water, where it usually produces a colloidal destabilization.

5.5 Forced Rayleigh Scattering - NP/solvent interaction

5.5.1 Forced Rayleigh Scattering

A forced Rayleigh scattering (FRS) device is used to determine the Soret coefficient S_T of the NPs in the colloidal dispersions. This device has been described in detail in [50,59]. Periodic spatial modulations of temperature are achieved by focusing a binary intensity modulation of light from a high power lamp in the solution, relying on the strong optical absorption of the NPs. By Soret effect, they induce periodic spatial modulations of NP's volume fraction. Both modulations are detected, thanks to the first order diffraction of a non absorbing laser beam (see [50,59] for details). In stationary conditions, local gradients of temperature T and of volume fraction Φ are related to S_T by:

$$\vec{\nabla}\Phi = -\Phi S_T \vec{\nabla} T \qquad (5.5)$$

where S_T can be either positive or negative. A positive S_T marks a thermodiffusion of the NPs towards cold regions and a negative Soret coefficient a NP's thermodiffusion towards hot regions.

5.5.2 Soret coefficient

In the present ionic dispersions, the charged NPs wear an effective dynamic charge ξ_{eff} (positive or negative), which can be measured by electrophoretic mobility coupled to DLS [45] (see Table 5.1), if the ionic concentration is not too large. By opposition in ILs, where the ionic concentration reaches values of the order of ~ 12 mol/L, ξ_{eff} is obtained by fitting $S_T(\Phi)$ [36]. According to Refs. [48–50] in water and DMSO, ξ_{eff} varies with temperature according to $\partial|\xi_{eff}|/\partial T > 0$, whatever be the sign of ξ_{eff}. It must also be noted that the direction of thermophoresis (or equivalently the sign of S_T) is not correlated to the sign of ξ_{eff}.

Moreover, in the ionic dispersions, the Soret effect is necessarily coupled to a thermoelectric effect. Namely a local electric field, $\vec{E} = S_e \vec{\nabla} T$, is induced in the ionic dispersion by temperature gradients [60,61], S_e being the Seebeck coefficient.

The measured Soret coefficient then results from a balance between:

(i) The NP/NP interaction; compressibility χ of the dispersion (see Equation 5.4 and Fig. 5.2) associated to the second virial coefficient A_2 (see Table 5.1) and the volume fraction Φ,

(ii) The NP/solvent interaction; the effective dynamic charge ξ_{eff} of the NPs and the NP's entropy of transfer $\hat{S}_{\text{NP}} = Q^*/T$, defined as in [62] with Q^* the NP's heat of transport, which characterizes the affinity of the NPs with the solvent (see Table 5.1),

(iii) The stationary Seebeck coefficient S_{e}^{St}, which is related to the concentration n_j and the entropy of transfer $\hat{S}_j = Q_j^*/T$ of all the j free charged species present in the dispersion - see Equation 5.7 below.

Indeed according to [48, 49][1] S_{T} can be written as:

$$S_{\text{T}} = \chi \left(\frac{\hat{S}_{\text{NP}}}{kT} - e\xi_{\text{eff}} \frac{S_{\text{e}}^{\text{St}}}{kT} \right). \tag{5.6}$$

For monovalent free ions $(+, -)$ in the NP's dispersion, the stationary Seebeck coefficient S_{e}^{St} can be written [63] as:

$$eS_{\text{e}}^{\text{St}} = \frac{n_+ \hat{S}_+ - n_- \hat{S}_- + Zn_{\text{NP}} \chi \hat{S}_{\text{NP}}}{n_+ + n_- + Zn_{\text{NP}} \chi \xi_{\text{eff}}} \tag{5.7}$$

where $\hat{S}_j$ and n_j are the entropy of transfer and the concentration of j species, respectively with $j = (+, -, \text{NP})$; Z is the (static) NP's effective charge given by the electro-neutrality of the dispersion (see [48, 49]). It is of the same order as ξ_{eff} and for sake of simplicity, it is frequently taken equal to ξ_{eff} in the fits of $S_{\text{T}}(\Phi)$ [48, 49].

5.6 Forced Rayleigh Scattering results

The Soret coefficient S_{T} is determined in the samples of Table 5.1 as a function of volume fraction Φ and temperature T.

5.6.1 Room temperature $S_{\text{T}}(\Phi)$ for FFs based on water and DMSO

The dependence of Soret coefficient S_{T} on various NP's parameters is probed in Figs. 5.3 and 5.4 as a function of the volume fraction at room temperature. The influence of the NP's diameter is presented in 5.3a for NPs with protonated hydroxyl surface species dispersed in water at $pH \sim 2$ with NO_3^- counter-ions [48]. S_{T} remains negative on the whole explored range, $|S_{\text{T}}|$ decreases with Φ and the larger d_{NP}, the larger $|S_{\text{T}}|$. In Fig. 5.3b, it is shown how, for citrate-coated NPs in water at $pH \sim 7$ [49], the nature of the counter-ions is able to change the sign of S_{T}. Indeed the Debye layer surrounds here the NPs, "dressing" them with the different cationic

---

[1]In a first approximation we do not consider in the modelling of the Soret coefficient (Equation 5.6) the term $\frac{1}{\Phi kT} \frac{\partial \Pi V_{\text{NP}}}{\partial T}$, which is small in these dispersions of charged NPs.

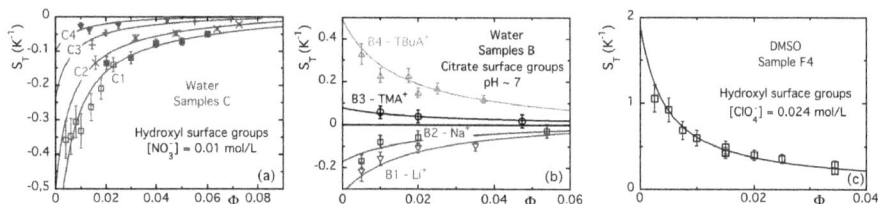

Figure 5.3: Φ-dependence of Soret coefficient for different colloidal dispersions of Table 5.1 in polar solvents. Full lines are fits of $S_T(\Phi)$ using Equation 5.6 with parameters of Table 5.1, as previously published in [48, 49, 51]; (a) Samples C from Table 5.1 in water with various NP's diameters - Note that open square symbols correspond to a sample based on mixed Mn-Zn ferrite NPs (from ref. [48]) with characteristics close to those of sample C1 in Table 5.1; (b) Samples B1 to B4 in water with different counter-ions; (c) Sample F4 in DMSO.

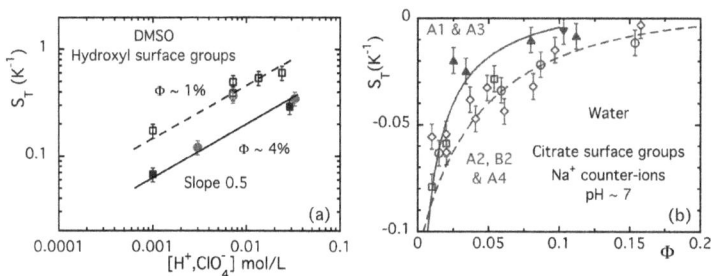

Figure 5.4: (a) Soret coefficient S_T as a function of the ionic concentration $[H^+, ClO_4^-]$ in DMSO dispersions based on NPs of sample F1 (squares) and of samples F3 and F4 (circles) from Table 5.1, at different ionic concentrations which are not all quoted in the Table, with a volume fraction $\Phi \sim 1\%$ (Open symbols) and $\Phi \sim 4\%$ (Full symbols). Full and dashed lines are guide for the eye of slope 0.5. (b) Φ-dependence of Soret coefficient S_T for aqueous samples based on citrate-coated NPs with Na^+ counter-ions at two different concentrations 0.009 mol/L (samples A1 and A3 in Table 5.1) and ~ 0.08 mol/L (samples A2 and A4 (at 0.090 mol/L) and B2 (at 0.072 mol/L) in Table 5.1). Full and dashed lines are respective fits of $S_T(\Phi)$ using Equation 5.6 with parameters of Table 5.1 for samples A, as previously published in [48].

counter-ions. Their nature changes S_T from negative with hydrophilic Li^+ and Na^+ cations to positive with archetypal hydrophobic TMA^+ and $TBuA^+$ cations [64].

If NPs are dispersed in DMSO, S_T is now positive whatever be the NP's concentration for the various NPs coating tested in ref. [45]. Figure 5.3c illustrates that point with NPs coated with protonated hydroxyl species and ClO_4^- counter-ions at various Φ's [51]. The other example here, is sample F6 in Table 5.1, corresponding to citrate-coated NPs and $TBuA^+$ counter-ions, for which $S_T = 0.13 \ K^{-1}$ have been determined at room temperature and $\Phi = 3.73\%$ in [45].

Both in water and DMSO, S_T depends on the concentration of free ions in the dispersion (the ionic strength), as illustrated by Fig. 5.4. In DMSO, at $\Phi \sim 1\%$ and $\sim 4\%$, S_T roughly increases as the square-root of the ionic concentration in the dispersion, thus roughly decreases as the inverse of the Debye length κ^{-1}. A similar trend is observed in water at $pH \sim 7$ and $\Phi \geq 2\%$ (see the Φ-dependence of S_T at two different ionic concentrations in Fig. 5.4b).

In conclusion, at room temperature, Soret coefficient S_T is often measured negative in water, but not always (see exceptions in Fig. 5.3b), as the sign of S_T depends on the nature of the counterions. In DMSO, Soret coefficient S_T is always observed positive at room temperature with NPs coated with protonated hydroxyl species and ClO_4^- counter-ions (samples F1 to F5 in Table 5.1) but also with citrate-coated NPs and TBuA$^+$ counter-ions (sample F6 - data from [65]). Whatever the colloidal characteristics of these dispersions (with repulsive interparticle interaction) and the sign of Soret coefficient, Figs. 5.3 and 5.4 both show $\partial |S_T|/\partial \Phi < 0$ at room temperature, never reaching the value $S_T = 0$ in the explored range of concentration.

5.6.2 T-dependence of S_T in FFs based on water and DMSO

By opposition to variations of NP's volume fraction, an increase of temperature above room temperature in water is able to change the sign of the Soret coefficient S_T in some aqueous dispersions, as attested by Figs. 5.5. It is true with NPs coated with protonated hydroxyl species with ClO_4^- counterions (Fig. 5.5a and c) and with citrate coated NPs with Na$^+$ counterions (Fig. 5.5b). It is also true with NP's coated with charged surface species PAC$_{12}$-MIM$^\pm$ (counterions Br$^-$ Sample E1) and PAAMA$^-$ (counterions NH$_4^+$ sample F2) - see Figs. 5.5a and b. Whatever the probed aqueous sample in these two figures (except for sample B4 based on citrate-coated NPs with rather hydrophobic TBuA$^+$ counterions), it is observed that $\partial S_T/\partial T > 0$.

Table 5.2 recapitulates the situation in water for the various couples coating/counterions explored here, giving the temperatures at which S_T changes its sign. Here, S_T cancels between 40°C and 65°C with four families of samples. According to the T-dependence of S_T in Fig. 5.5b, sample B1 would be expected to reach $S_T = 0$ at $T > 80$°C, while for samples D1 and B3 it would happen at $T < 20$°C. The more hydrophilic are the dressed NPs with their cloud of counter-ions, the higher is the temperature at which S_T cancels.

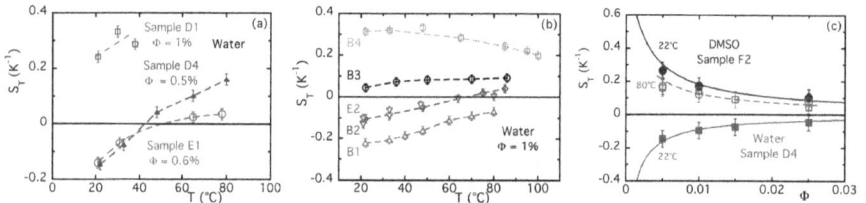

Figure 5.5: (a) and (b) Temperature-dependence of Soret coefficient S_T for various aqueous samples of Table 5.1 at $\Phi \leq 1\%$; Dashed lines are guides for the eye. (c) Φ-dependence of S_T for sample D4 in water (square symbols) and F2 in DMSO (circle symbols), both based on the same NPs and the same ClO_4^- counter-ions at the same concentration (see Table 5.1); Full symbols correspond to room temperature measurements while open symbols correspond to $T = 80$°C; Full lines are room temperature fits of $S_T(\Phi)$ using Equation 5.6 with parameters of Table 5.1, according to Ref. [51]; The same dashed line is a guide for the eye of the results for both samples at 80°C.

Table 5.2: Temperature at which S_T passes through zero for various aqueous NP's dispersions with different couples of NP's coating (first column) and NP's counter-ions (first line) for 7.9 nm $\leq d_{NP} \leq$ 9.6 nm.

	TFSI$^-$	TMA$^+$	ClO$_4^-$	Br$^-$	NH$_4^+$	Na$^+$	Li$^+$
H$^+$	< 20°C		40-45°C				
PAC$_{12}$-MIM$^\pm$				50-55°C			
PAAMA$^-$					~ 65°C		
Citrate^{3-}		< 20°C				~ 65°C	> 80°C

In DMSO, it is observed as for the aqueous sample B4, based on citrate-coated NPs with TBuA$^+$ counterions, that up to $T \sim 100°C$, $\partial S_T/\partial T < 0$ with S_T always positive.

Now if two analogous samples in water and DMSO with the same NPs are compared (D4 and F2), with the same coating by protonated hydroxyl species and the same counter-ions ClO_4^- at the same concentration, it is observed close to room temperature an opposite sign of S_T, as shown in Fig. 5.5c at 22°C. When T increases close to room T, $\partial S_T/\partial T$ are of opposite sign in the two solvents. S_T in water reverses its sign and at 80°C S_T(water) $\sim S_T$(DMSO) in the studied Φ-range[2]. Note that $|S_T|$ decreases with Φ in both solvents.

The negative values of S_T in water at room temperature are frequently ascribed to the existence of hydrogen bonds [66,67], which influence the NP/solvent interaction. These hydrogen bonds progressively disappear when temperature is increased. It is what happens in water, in Fig. 5.5c which shows that at 80°C, the influence of hydrogen bonds seems to have disappeared.

5.6.3 Soret coefficient in FFs based on Ionic Liquids

The Soret coefficient is now probed in FFs, which are based on various ILs (see Tables 5.1 and 5.3 with Ref. [36, 37, 53]). Figure 5.6a presents examples of the Φ-dependence of S_T at room temperature for two samples with interparticle repulsion. Figure 5.6b compares the T-dependence of S_T at $\Phi = 1\%$ for one sample with interparticle repulsion (sample H1) and four samples (H3 series) with slightly attractive interparticle interaction ($-7 \leq A_2 \leq -4$). All these samples are based on the same NPs, with the same coating and the same counterions. In these samples, it is only the IL's cation which has been modified (see Table 5.3), keeping the same IL's anion. In all these situations, we observe that $\partial S_T/\partial\Phi < 0$ and $\partial S_T/\partial T < 0$, as in DMSO with almost no influence of the IL's cation.

Whatever the samples which have been probed in Section 5.6, we always obtain $\partial|S_T|/\partial\Phi < 0$, while the T-dependence are different for samples in water on one side, and in DMSO and the various ILs, on the other side.

We always observe $\partial S_T/\partial T < 0$ with a positive S_T in the FF samples based on DMSO or on the various ILs probed here.

---

[2]The fit of $S_T(\Phi)$ with Equation 5.6 is only presented at 20°C, as the T-dependence of χ is not known.

Table 5.3: Various ILs on which are based the FFs probed by FRS in Fig. 5.6.

Sample	Name	IL
IL1	EAN	ethylammonium nitrate
IL2	$EMIM^+$-$TFSI^-$	1-ethyl 3-methylimidazolium bistriflimide
IL3	N_{114}-$TFSI^-$	N,N,N- triMethyl-N-Butyl Ammonium bistriflimide
IL4	$N_{112(301)}$-$TFSI^-$	N-Ethyl-N,N-diMethyl- N-(3- Methoxypropyl) Ammonium bistriflimide
IL5	Pyrr-$TFSI^-$	1-Methyl-1-(2-Methoxyethyl) Pyrrolidinium bistriflimide
IL6	Pi-$TFSI^-$	1-(2-Methoxyethyl)-1-Methyl Piperidinium bistriflimide

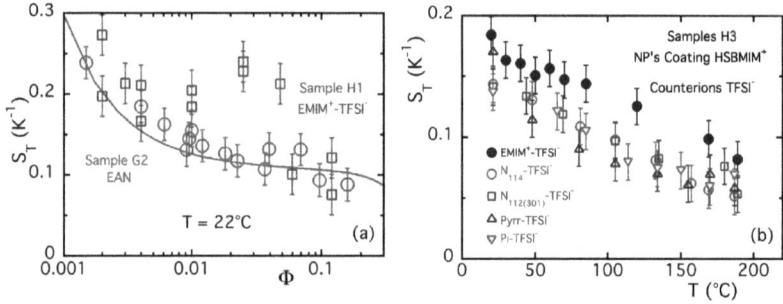

Figure 5.6: (a) Φ-dependence of Soret coefficient S_T for two NP's dispersions in room temperature ionic liquids (Open circles: sample G2 with citrate-coated NPs in EAN with Na^+ counter-ions [36]; Open squares: sample H1 with NPs coated with $HSBMIM^+$ and with $TFSI^-$ counter-ions in $EMIM^+$-$TFSI^-$ [37]); (b) T-dependence of S_T for dispersions at $\Phi \sim 1\%$ of NPs coated with $HSBMIM^+$ and with $TFSI^-$ counter-ions, in the room temperature ionic liquids IL2 to IL6 from Table 5.3 [37, 38] - Full symbols: sample H1; Open symbols: H3 series.

On the contrary in the FF samples based on water, S_T is either positive or negative, and in most of the case we obtain $\partial S_T / \partial T > 0$ due the strong T-dependence of H-bonds in water [66, 67]. The only exception is the sample with big hydrophobic counterions $TBuA^+$, which T-dependence is similar to that observed with the samples based on DMSO and on the various ILs.

5.7 Discussion: NP's entropy of transfer (NP/Solvent interaction) at room temperature in the probed FFs

The NP's entropy of transfer $\hat{S}_{NP}$ is determined by fitting $S_T(\Phi)$ with Equation 5.6 at constant T. The values obtained at room temperature are given in Table 5.1. Looking at all the samples in this table, the sign of $\hat{S}_{NP}$ can be either positive or negative, and it is not correlated to the sign of the NP's effective charge ξ_{eff}.

For FFs in water, $\hat{S}_{NP}$ is negative in most cases, except for samples B4, B5 and C4 in Table 5.1 for which $\hat{S}_{NP}$ is positive (B4, B5: citrate-coated NPs with hydrophobic counter-ions $TBuA^+$; C4: small NPs coated with protonated hydroxyl species). As shown in Fig. 5.7a, $\hat{S}_{NP}$ is, for citrate-coated NPs, an increasing function of $\hat{S}_+$ the entropy of transfer of the free counter-ions dressing the NPs in their surrounding Debye shell (see [49]). Moreover, Fig. 5.7a also shows that $\hat{S}_{NP}$ is an increasing

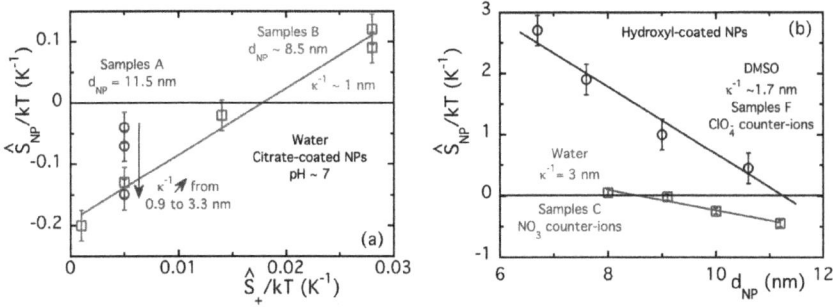

Figure 5.7: (a) Samples in water at $pH \sim 7$ with citrate-coated NPs at room temperature: Dependence of $\hat{S}_{NP}/kT$ on the entropy of transfer $\hat{S}_{+}/kT$ of the counter-ions dressing the NPs in their Debye shell; Open squares: samples B with various counter-ions - results from [49] at a Debye length $\kappa^{-1} \sim 1$ nm; Open circles: Samples A at various Na^{+} concentrations (thus various κ^{-1}) - results from [48]. (b) Samples with hydroxyl-coated NPs at room temperature: d_{NP}-dependence of $\hat{S}_{NP}/kT$ at roughly constant κ^{-1}; Open circles: Samples F in DMSO at $\kappa^{-1} \sim 1.7$ nm with ClO_4^- counterions; Open squares: Samples C in water at $\kappa^{-1} = 3$ nm with NO_3^- counter-ions - results from [48].

function of the concentration of these free counter-ions, thus a decreasing function of the electrostatic Debye length κ^{-1} (see Equation 5.1).

For NPs coated with protonated hydroxyl species, as illustrated in Fig. 5.7b, the NP's entropy of transfer $\hat{S}_{NP}$ decreases with d_{NP}, the NP diameter (at constant κ^{-1}) both in water with NO_3^- counter-ions [48] and in DMSO with ClO_4^- counter-ions. Moreover in DMSO with ClO_4^- counter-ions, the comparison between samples F3 and F4 with same d_{NP} (see Table 5.1) shows that here also $\hat{S}_{NP}$ decreases when κ^{-1} increases.

As described in [48–50], the NP's entropy of transfer $\hat{S}_{NP}$ results from the sum of two contributions:

$$\hat{S}_{NP} = \hat{S}_{NP}^{CM} + \hat{S}_{NP}^{Solv}. \tag{5.8}$$

The first contribution $\hat{S}_{NP}^{CM}$, associated to the Capacitor Model (see [68, 69]) is of pure electrostatic origin and can be calculated using [68,69] if, in addition to the NP's charateristics of Table 5.1, $\partial \xi_{eff}/\partial T$ is known (see [48]). It is always positive. The second contribution $\hat{S}_{NP}^{Solv}$ is a solvation one [70], usually negative. It is deduced from $\hat{S}_{NP} - \hat{S}_{NP}^{CM}$. The sign of $\hat{S}_{NP}$ results from a balance between these two contributions, which are often larger than $\hat{S}_{NP}$ itself. Let us quote the example of samples D4 and F2 at room T in Fig. 5.5c, same NPs with the same coating and same counter-ions (at the same concentration) in water and in DMSO [50]. The fits of $S_T(\Phi)$ give at room temperature:

- in water: $\hat{S}_{NP}^{CM}/kT \sim +3.0$ K^{-1}, $\hat{S}_{NP}^{Solv}/kT \sim -3.5$ K^{-1} and $\hat{S}_{NP}/kT \sim -0.5$ K^{-1};
- in DMSO: $\hat{S}_{NP}^{CM}/kT \sim +2.5$ K^{-1}, $\hat{S}_{NP}^{Solv}/kT \sim -1.5$ K^{-1} and $\hat{S}_{NP}/kT \sim +1.0$ K^{-1}.

In water, it has been shown in [48], that $\hat{S}_{NP}^{Solv}$ can be divided in two terms one coming from the NP's surface, which can be either sign, and the other one from the

Debye shell always negative:

$$\hat{S}_{NP}^{Solv} = \hat{S}_{NP,surf}^{Solv} + \hat{S}_{NP,shell}^{Solv}. \tag{5.9}$$

In [48], it is found both for NPs coated with protonated hydroxyl species, with NO_3^- counter-ions and for citrate-coated NPs with Na^+ counter-ions that:

$$\hat{S}_{NP,surf}^{Solv} \sim \xi_{eff}\hat{S}_{coat} \tag{5.10}$$

$\hat{S}_{coat}$ being the entropy of transfer of the ionic coating of the NPs, corresponding respectively to ions H^+ (for NPs coated with protonated hydroxyl species) and to acetate ions CH_3COO^- (for citrate-coated NPs) - see [48]. For its part, the shell contribution can be written as:

$$\hat{S}_{NP,shell}^{Solv} = -s^{Solv}\pi d_{NP}^2 \kappa^{-1} \quad \text{with} \quad s^{Solv} \sim 25 \; \mu eV.K^{-1}/nm^3. \tag{5.11}$$

s^{Solv} being the solvation entropy per unit volume of the aqueous ionic shell. To make a similar analysis in DMSO, it would be necessary to know $\partial \xi_{eff}/\partial T$ for all the different samples, and such measurements are not yet performed.

5.8 Under-field anisotropic properties of stable and monophasic magnetic colloids

Due to the rotational degree of freedom of the NPs in the fluid carrier, FFs are giant paramagnetic materials. They do not present under-field hysteresis of their magnetization M, which saturates in moderate magnetic fields H. At low NPs volume fraction, if the magnetic interparticle interactions are weak enough, the magnetization can be simply described by the Langevin model[3]:

$$M = \Phi m_s L(\xi) \quad \text{with} \quad \xi = \mu_0 m_s V_{NP} \frac{H}{kT} \quad \text{and} \quad L(\xi) = \cotan(\xi) - \xi^{-1} \tag{5.12}$$

m_s is the magnetization of the NP's material, ξ is the Langevin parameter, μ_0 the vacuum permeability and $L(\xi)$ the Langevin function. For describing more concentrated NP's dispersions, with interparticle magnetic dipolar interaction, we adopt here an effective field approach as in [49, 53, 71–73][4]:

$$M = \Phi m_s L(\xi_e) \quad \text{with} \quad \xi_e = \xi + \lambda\gamma L(\xi_e) \quad \text{and} \quad \gamma = \mu_0 m_s^2 V_{NP} \frac{\Phi}{kT} \tag{5.13}$$

λ is the effective field constant, taken here equal to 0.22 [72, 77] and γ the reduced (mean-field) parameter of dipolar interaction. On average, the anisotropic dipolar interaction is attractive.

In dispersions with a zero-field interparticle repulsion strong enough to ensure the under-field colloidal stability of the system and with γ high enough to be influent,

---

[3]We adopt here in this simple description a monodisperse distribution of NP's size.
[4]This model is valid if the dipolar interparticle interaction is not too large [56, 74], namely for $\gamma \leq 5$ this description is equivalent to those of [75] and [76].

SANS and SAXS scattering patterns of concentrated NP's dispersions can be experimentally anisotropic under an applied magnetic field [56,74,77–79]. This is due to an anisotropy of the structure factor $S(Q)$, explained by anisotropic fluctuations of concentration when a magnetic field field is applied [77,78]. In the same way, this leads to an under-field anisotropy of the NP's diffusion coefficient D_m [49,53,71,72,77]. Such an anisotropy is also observed for the Soret coefficient S_T [49,53].

Here, as illustrated in the inset of Fig. 5.8a, we limit ourselves to applied fields either parallel to the gradients of concentration $\vec{\nabla}\Phi$ and temperature $\vec{\nabla}T$ (the applied field is then denoted $\vec{H}_{||}$) or perpendicular to these gradients (the applied field is then denoted $\vec{H}_\perp$). The anisotropy of S_T is shown in Fig. 5.8b for sample A2 of Table 5.1 at room temperature and in Fig. 5.8c for sample H1 at different temperatures ranging between $T = 22°C$ and $188°C$.

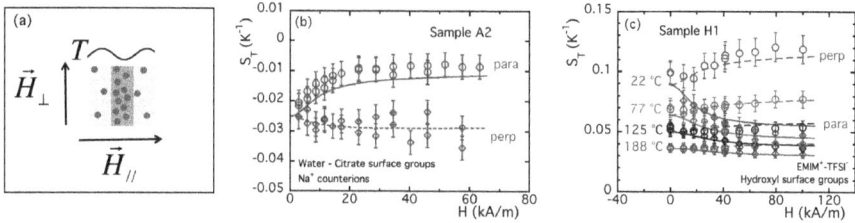

Figure 5.8: (a) Field directions $\vec{H}_{||}$ and $\vec{H}_\perp$ with respect to the gradients of concentration $\vec{\nabla}\Phi$ and temperature $\vec{\nabla}T$ in the FRS measurement. (b) Under-field anisotropy of S_T for the aqueous sample A2 from Table 5.1 at $\Phi = 5.9\%$ and room temperature (data and fit from [49]). (c) Under-field anisotropy of S_T for the sample H1 from Table 5.1 based on IL2, at $\Phi = 5.95\%$ and various temperatures (data and fit from [53]).

Under-field, the anisotropic gradients of Φ produce different conditions of continuity in Maxwell equations for a magnetic field either along or perpendicular to $\vec{\nabla}\Phi$. In the framework of the model of [71,72] where μ_{H} is the under-field contribution to the NP's chemical potential μ, it is shown that in Equation 5.6, the ratio $1/\chi$ should be replaced under-field by[5]:

$$\frac{1}{\chi} + \frac{\Phi}{kT}\left(\frac{\partial\mu_{\text{H}}}{\partial\Phi}\right)_{\text{H}} \quad \text{in } \vec{H}_\perp \text{ direction,} \tag{5.14}$$

and

$$\frac{1}{\chi} + \frac{\Phi}{kT}\left(\frac{\partial\mu_{\text{H}}}{\partial\Phi}\right)_{\text{H}} + \frac{\Phi}{kT\partial\Phi/\partial x}\left(\frac{\partial\mu_{\text{H}}}{\partial H}\right)_{\Phi} \quad \text{in } \vec{H}_{||} \text{ direction (denoted x direction).}$$

$$\tag{5.15}$$

In a mean-field approximation, the under-field chemical potential μ of the dispersion can be written as:

$$\mu = \mu^* + \mu_{\text{H}} \quad \text{with} \quad \mu_{\text{H}} = -kT\ln\left(\frac{\sinh\xi_e}{\xi_e}\right) \tag{5.16}$$

---

[5]This is true as well as in $S(Q=0)=\chi$, the low-Q value of the structure factor and in the diffusion coefficient, defined as $D_m = kT/\zeta\chi$ where ζ is the friction experienced by the NPs.

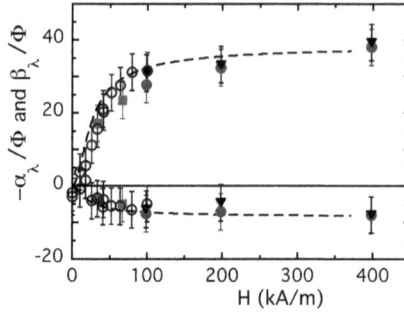

Figure 5.9: Magnetic field dependence of $-\alpha_\lambda/\Phi$ (bottom branch) at and β_λ/Φ (top branch) at room temperature deduced from the under-field anisotropy of the diffusion coefficient D_m determined by FRS measurements in [53] for sample H1 at $\Phi = 5.95\%$ (data: open circles, fits: dashed lines) and deduced from $S(Q=0)$ SAXS determinations in [79] for sample A4 (Full discs: $\Phi = 3.7\%$, full triangles: $\Phi = 9.7\%$) and sample A5 (Full squares: $\Phi = 13.3\%$).

μ^* is the zero-field chemical potential of the dispersion and μ_H its mean-field (negative and isotropic) under-field contribution. [49, 53, 71–73].

The term $-\alpha_\lambda = \frac{\Phi}{kT}\left(\frac{\partial\mu_H}{\partial\Phi}\right)_H$ is associated with the mean-field dipolar interaction, which is attractive on average. The term $\beta_\lambda = \frac{\Phi}{kT\partial\Phi/\partial x}\left(\frac{\partial\mu_H}{\partial H}\right)_\Phi$ is anisotropic. It is due to the discontinuity of $\vec{H}_\parallel$ in Maxwell equations across the spatial inhomogeneities of concentration (this term does not exist in $\vec{H}_\perp$ direction, the NP's concentration being homogeneous along the field in that case). They can be written as [49, 53, 71–73]:

$$\alpha_\lambda = \frac{\lambda\gamma L^2(\xi_e)}{1 - \lambda\gamma L'(\xi_e)} \quad \text{and} \quad \beta_\lambda = \frac{\gamma L^2(\xi_e)}{(1 - \lambda\gamma L'(\xi_e))(1 + (1 - \lambda)\gamma L'(\xi_e))}. \quad (5.17)$$

Figure 5.9 illustrates the variations of $-\alpha_\lambda/\Phi$ and β_λ/Φ as a function of H for several samples of Table 5.1 at room temperature with close diameter distributions. For the aqueous samples A4 and A5, α_λ/Φ and β_λ/Φ were deduced from underfield measurements of $S(Q = 0)$ at various concentrations in [79]. For sample H1 based on IL2 (see Table 5.3), α_λ/Φ and β_λ/Φ were deduced from under-field measurements of D_m at $\Phi = 5.9\%$ in [53]. The dashed lines correspond to the adjustments of α_λ/Φ and β_λ/Φ with Equation 5.17 for sample H1.

To describe the under-field anisotropy of S_T, in plus of α_λ and β_λ supplementary terms in Equation 5.6, two extra terms S_1 and S_2 should be added. They are related to derivatives of the NP's chemical potential with respect to temperature (see [49, 53, 73]). Their contribution is however shown to be small (see Discussion in [53]). Figure 5.8 shows the nice adjustments by the model, of the anisotropy of S_T at room temperature for the aqueous sample A2 presenting negative S_T values (Fig. 5.8b) and at different T's for the ferrofluid H1 based on the room temperature ionic liquid IL2, presenting positive S_T values (Fig. 5.8c). Note that as the dipolar parameter γ is a decreasing function of temperature, the anisotropy of S_T also decreases as T increases in sample H1 (see [53]).

5.9 Conclusion

Various ferrofluids (FFs) remaining stable and monophasic under an applied field and based on ionic nanoparticles (NPs) can be successfully obtained in different polar solvents and room temperature Ionic Liquids (ILs) with a repulsive interparticle interaction. This interparticle (NP/NP) interaction is quantified here at room temperature (and zero-field) via the determination of the second virial coefficient of the osmotic pressure A_2, by Small Angle Scattering of X-rays or of Neutrons as a function of the NP's volume fraction. The NP/solvent interaction is also quantified here at room temperature via the measurement and the analysis of the Soret coefficient at room temperature, as a function of the NP's volume fraction. Beside the effective dynamic charge ξ_{eff} of the NPs, the important parameter quantifying the NP/solvent interaction is the NP's entropy of transfer $\hat{S}_{\text{NP}}$, which results from a balance between a purely electrostatic term (Capacitor model) and a solvation term, which largely depends on the affinity between the NPs dressed with the counter-ions and the solvent.

For FFs based on water, $\hat{S}_{\text{NP}}$ is as S_{T} frequently negative, meaning that the NPs are thermophilic (but not always - hydrophobic counter-ions may render the NPs thermophobic [49]).

For FFs based on DMSO and EAN, $\hat{S}_{\text{NP}}$ is as S_{T} always positive and the NPs are thermophobic [36, 51] in all the cases investigated here. $\hat{S}_{\text{NP}}$ and S_{T} are larger than in water (even in absolute value), which suggests larger thermoelectric effects in these FFs [21, 36].

An investigation of the Soret coefficient as a function of temperature shows that for FFs in water, in most of the cases, S_{T} increases with T, most probably because of the progressive loss of H-bonds. On the contrary, for FFs in DMSO or in the various ILs investigated here, S_{T} decreases with T. Complementary measurements of the NP/NP interaction as a function of T becomes necessary to go further on and deduce the NP/solvent interaction as a function of T.

The under-field anisotropy of the Soret coefficient experimentally explored for some of the various synthesized FFs is successfully explained with the existing models whatever be the nature of the solvent carrier, the temperature and the sign of S_{T}.

Acknowledgments

We acknowledge here the numerous beam-time allocations at the facilities Orphée reactor in LLB-Gif-sur-Yvette-France and synchrotron SOLEIL-Gif-sur-Yvette-France, together with the precious help from the local contacts. We thank Dr Fabrice Cousin from LLB-CEN-Saclay-France and Dr Sawako Nakamae from SPEC-CEN-Saclay-France for their constant and precious collaboration. This work was supported by the French ANR TE-FLIC (Grant n° ANR-12-PRGE-0011-01), the European Union's Horizon 2020 research and innovation program under the grant n°731976 (MAGENTA) and the CAPES-COFECUB Ph 959/20 between France and Brazil.

References

[1] R. Rosensweig. Ferrohydrodynamics. Cambridge: Cambridge University Press (1985).

[2] B. Berkovski (ed.). Magnetic Fluids and Applications Handbook. Begell House Inc. Publ. New York (1996).

[3] E. Blums, A. Cēbers and M. Maiorov. Magnetic Liquids. W. de G. Gruyter, New York (1997).

[4] S. Odenbach (ed.). Ferrofluids: Magnetically Controllable Fluids and Their Applications. Berlin: Springer Verlag (2003).

[5] V. Socoloiuc, M. Avdeev, V. Kuncser, R. Turcu, E. Tombacz and L. Vekas. Ferrofluids and bio-ferrofluids: looking back and stepping forward. Nanoscale, to appear in (2022). Doi:10.1039/d1nr05841j.

[6] J. Philip and J. Kaskar. Optical properties and applications of ferrofluids—A review. J. of Nanofluids 1: 3–20 (2012).

[7] K. Raj and R. Moskowitz. A review of damping applications of ferrofluids. I.E.E.E. 16: 358–363 (1980).

[8] G. Béalle, R. D. Corato, J. Kolosnjaj-Tabi, V. Dupuis, O. Clément, F. Gazeau, C. Wilhelm and C. Menager. Ultra magnetic liposomes for MR imaging, targeting, and hyperthermia. Langmuir 28: 11834–11842 (2012).

[9] D. Mertz, O. Sandre and S. Begin-Colin. Drug releasing nanoplatforms activated by alternating magnetic fields. Biochimica et Biophysica Acta 1861(6): 1617–1641 (2017).

[10] V. Pilati, R. C. Gomes, G. Gomide, P. Coppola, F. G. Silva, F. L. Paula, R. Perzynski, G. Goya, R. Aquino and J. Depeyrot. Core/shell nanoparticles of non-stoichiometric zn-mn and zn-co ferrites as thermosensitive heat sources for magnetic fluid hyperthermia. J. Phys. Chem. C 122: 3028–3038 (2018).

[11] T. Krasia-Christoforou, V. Socoliuc, K. Knudsen, E. Tombácz, R. Turcu and L. Vékás. From single-core nanoparticles in ferrofluids to multi-core magnetic nanocomposites: Assembly strategies, structure, and magnetic behavior. Nanomaterials 10(11): 283–291 (2020).

[12] J. Wells, D. Ortega, U. Steinhoff, S. Dutz, E. Garaio, O. Sandre, E. Natividad, M. M. Cruz, F. Brero, P. Southern, Q. Pankhurst and S. Spassov. Challenges and recommendations for magnetic hyperthermia characterization measurements. Int. J. of Hyperthermia 38(1): 447–460 (2021).

[13] J. He and T. Tritt. Advances in thermoelectric materials research: Looking back and moving forward. Science 357: eaak9997 (2017).

[14] Z. He and P. Alexandridis. Ionic liquid and nanoparticle hybrid systems: Emerging applications. Adv. Colloid Interface Sci. 244: 54–70 (2017).

[15] T. J. Salez, B. T. Huang, M. Rietjens, M. Bonetti, C. Wiertel-Gasquet, M. Roger, C. L. Filomeno, E. Dubois, R. Perzynski and S. Nakamae. Can charged colloidal particles increase the thermoelectric energy conversion efficiency? Phys. Chem. Chem. Phys. 19: 9409–9416 (2017).

[16] A. Wurger. Thermoelectric ratchet effect for charge carriers with hopping dynamics. Phys. Rev. Lett. 126: 068 001 1–5 (2021).

[17] I. Nkurikiyimfura, Y. Wang and Z. Pan. Heat transfer enhancement by magnetic nanofluids—A review. Renew. Sustain. Energy Rev. 21: 548–561 (2013).

[18] N. Mereley (ed.). Magnetorheology: Advances and Applications. Cambridge, UK: R.S.C. Smart Materials (nov 2014).

[19] B. Frka-Petesic, E. Dubois, L. Almasy, V. Dupuis, F. Cousin and R. Perzynski. Structural probing of clusters and gels of self-aggregated magnetic nanoparticles. Magnetohydrodynamics 49: 0328–338 (2013).

[20] E. Dubois, R. Perzynski, F. Boué and V. Cabuil. Liquid-gas transitions in charged colloidal dispersions: Small angle neutron scattering coupled to phase diagrams of magnetic fluids. Langmuir 16: 5617–5625 (2000).

[21] B. Huang, M. Roger, M. Bonetti, T. J. Salez, C. Wiertel-Gasquet, E. Dubois, R. C. Gomes, G. Demouchy, G. Mériguet, V. Peyre, M. Kouyaté, C. L. Filomeno, J. Depeyrot, F. A. Tourinho, R. Perzynski and S. Nakamae. Thermoelectricity and thermodiffusion in charged colloids. J. Chem. Phys. 143: 054902 (2015).

[22] M. Dupont, D. MacFarlane and J. Pringle. Thermo-electrochemical cells for waste heat harvesting - progress and perspectives. Chem. Commun. 53: 6288–6302 (2017).

[23] A. Wurger. Thermopower of ionic conductors and ionic capacitors. Phys. Rev. Res. 2: 042030 (2020).

[24] M. Vasilakaki, I. Chikina, V. Shikin, N. Ntallis, D. Peddis, A. Varlamov and K. Trohidou. Towards high-performance electrochemical thermal energy harvester based on ferrofluids. Applied Materials Today 19: 100587 (2020).

[25] J. Duan, B. Yu, L. Huang, B. Hu, M. Xu, G. Feng and J. Zhou. Liquid-state thermocells: Opportunities and challenges for low-grade heat harvest. Joule 5: 283–291 (2021).

[26] X. Shi and J. He. Thermopower and harvesting heat. Science 371: 343–344 (2021).

[27] I. Plowas, J. Swiergiel and J. Jadzyn. Electrical conductivity in dimethyl sulfoxide + potassium iodide solutions at different concentrations and temperatures. J. Chem. Eng. Data 59(8): 2360–2366 (2014).

[28] J. Kiefer, K. Noack and B. Kirchner. Hydrogen bonding in mixtures of dimethyl sulfoxide and cosolvents. Current Physical Chemistry 1: 340–351 (2011).

[29] E. Fabre and S.M. Solel Murshed. J. Mater. Chem. A. 9: 15861–15879 (2021).

[30] F. Cousin, E. Dubois and V. Cabuil. Tuning the interactions of a magnetic colloidal suspension. Phys. Rev. E 68(2): 021405 (2003).

[31] J. Bacri, R. Perzynski, D. Salin, V. Cabuil and R. Massart. Phase diagram of an ionic magnetic colloid: Experimental study of the effect of ionic strength. J. of Colloid. Interface Sci. 132: 43–53 (1989).

[32] J.-C. Bacri and D. Salin. Instability of ferrofluid magnetic drops under magnetic fields. Phys. de Physique Lett. 43: L649–L654 (1982).

[33] J.-C. Bacri, A. Cebers and R. Perzynski. Adsorbed and near surface structure of ionic liquids at a solid interface. Phys. CRev. Lett. 72: 2705–2708 (1994).

[34] J. Erdmanis, G. Kitenbergs, R. Perzynski and A. Cebers. Magnetic microdroplet in rotating field: Numerical simulation and comparison with experiment. J. Fluid Mech. 821: 266–295 (2017).

[35] G. Kitenbergs. Hydrodynamic instabilities in microfluidic magnetic fluid flows. Ph.D. dissertation, Université Pierre et Marie Curie and University of Latvia (2015).

[36] K. Bhattacharya, M. Sarkar, T. J. Salez, S. Nakamae, G. Demouchy, F. Cousin, E. Dubois, L. Michot, R. Perzynski and V. Peyre. Structural, thermodiffusive and thermoelectric properties of maghemite nanoparticles dispersed in ethylammonium nitrate. Chem. Engineering 4(1): 5 1–25 (jan 2020).

[37] J. C. Riedl, M. A. Akhavan Kazemi, F. Cousin, E. Dubois, S. Fantini, S. Loïs, R. Perzynski and V. Peyre. Colloidal dispersions of oxide nanoparticles in ionic liquids: Elucidating the key parameters. Nanoscale Adv. 2: 1560–1572 (2020).

[38] J. C. Riedl, M. Sarkar, T. Fuzia, F. Cousin, J. Depeyrot, E. Dubois, G. Mériguet, R. Perzynski and V. Peyre. Design of concentrated colloidal dispersions of iron oxide nanoparticles in ionic liquids: Structure and thermal stability from 25 to 200°C. J. Coll. Int. Sci. 607: 584–594 (2022).

[39] V. Ivaništšev, S. O'Connor and M. Fedorov. Poly(a)morphic portrait of the electrical double layer in ionic liquids. Electrochem. Commun. 48: 61–64 (2014).

[40] M. Mamusa, J. Siriex-Plénet, F. Cousin, E. Dubois and V. Peyre. Tuning the colloidal stability in ionic liquids by controlling the nanoparticles/liquid interface. Soft Matter 10: 1097–1101 (2013).

[41] M. Mamusa, J. Sirieix-Plénet, R. Perzynski, F. Cousin, E. Dubois and V. Peyre. Concentrated assemblies of magnetic nanoparticles in ionic liquids. Faraday Discuss 181: 193–209 (2015).

[42] R. Massart. Préparation de ferrofluides aqueux en l'absence de surfactant; comportement en fonction du pH et de la nature des ions présents en solution. C. R. Acad. Sci. Paris 291(C1): 1–3 (1980).

[43] R. Massart, E. Dubois, V. Cabuil and E. Hasmonay. Preparation and properties of monodispersed magnetic fluids. J. Magn. Magn. Mat. 149: 1–5 (1995).

[44] J. Bacri, R. Perzynski, D. Salin, V. Cabuil and R. Massart. Ionic ferrofluids: A crossing of chemistry and physics. J. Magn. Magn. Mat. 85: 27–32 (1990).

[45] C. L. Filomeno, M. Kouyaté, V. Peyre, G. Demouchy, A. Campos, R. Perzynski, F. A. Tourinho and E. Dubois. Tuning the solid/liquid interface in ionic colloidal dispersions: Influence on their structure and thermodiffusive properties. J. Phys. Chem. C 121: 5539–5550 (2017).

[46] J. Riedl. Dispersions of nanomagnets in ionic liquids for thermoelectric applications. Ph.D. dissertation, Sorbonne Université (2020).

[47] E. Sani, M. R. Martina, T. J. Salez, S. Nakamae, E. Dubois and V. Peyre. Multifunctional magnetic nanocolloids for hybrid-solar-thermoelectric energy harvesting. Nanomat. 11: 10311–18 (2021).

[48] R. Cabreira-Gomes, A. da Silva, M. Kouyaté, G. Demouchy, G. Mériguet, R. Aquino, E. Dubois, S. Nakamae, M. Roger, J. Depeyrot and R. Perzynski. Thermodiffusion of repulsive charged nanoparticles—the interplay between single-particle and thermoelectric contributions. Phys. Chem. Chem. Phys. 20: 16402–16413 (2018).

[49] M. Kouyaté, C. Filomeno, G. Demouchy, G. Mériguet, S. Nakamae, V. Peyre, M. Roger, A. Cēbers, J. Depeyrot, E. Dubois and R. Perzynski. Thermodiffusion of citrate-coated γ-fe$_2$o$_3$ nanoparticles in aqueous dispersions with tuned counter-ions—anisotropy of the soret coefficient under magnetic field. Phys. Chem. Chem. Phys. 21: 1895–1903 (2019).

[50] C.L. Filomeno, M. Kouyaté, F. Cousin, G. Demouchy, E. Dubois, L. Michot, G. Mériguet, R. Perzynski, V. Peyre, J. Sirieix-Plenet and F.A. Tourinho. J. Magn. Magn. Mat. 431: 2–7 (2017).

[51] M. Sarkar, J. C. Riedl, G. Demouchy, F. Gélébart, G. Mériguet, V. Peyre, E. Dubois and R. Perzynski Inversion of thermodiffusive properties of ionic colloidal dispersions in water-dmso mixtures probed by forced rayleigh scattering. Eur. Phys. J. E 42: 979–2989 (2019).

[52] T. Salez, M. Kouyaté, C. L. Filomeno, M. Bonetti, M. Roger, G. Demouchy, E. Dubois, R. Perzynski, A. Cēbers and S. Nakamae. Magnetically enhancing the seebeck coefficient in ferrofluids. Nanoscale Adv. 1: 2979–2989 (2019).

[53] M. Mamusa, J. Sirieix-Plénet, F. Cousin, R. Perzynski, E. Dubois and V. Peyre. Microstructure of colloidal dispersions in the ionic liquid ethyl ammonium nitrate: Influence of the nature of nano particle's counterion. J. Phys.: Condens. Matter 26: 284113 (2014).

[54] T. Fiuza, M. Sarkar, J. C. Riedl, A. Cēbers, F. Cousin, G. Demouchy, J. Depeyrot, E. Dubois, F. Gélébart, G. Mériguet, R. Perzynski and V. Peyre. Thermodiffusion anisotropy under magnetic field in ionic liquid-based ferrofluids. Soft Matter 17: 4566–4577 (2021).

[55] O. Sandre, J. Browaeys, R. Perzynski, J. Bacri, V. Cabuil and R. Rosensweig. Assembly of microscopic highly magnetic droplets: Magnetic alignment versus viscous drag. Phys. Rev. E 59: 1736–1746 (1999).

[56] P. Rossmanith and W. Kohler. Polymer polydispersity analysis by thermal diffusion forced rayleigh scattering. Macromolecules 29: 3203 (1996).

[57] G. Mériguet, E. Wandersman, E. Dubois, A. Cēbers, J. de Andrade Gomes, G. Demouchy, J. Depeyrot, A. Robert and R. Perzynski. Magnetic fluids with tunable interparticle interaction: Monitoring the under-field local structure. Magnetohydrodynamic 48(2): 415–426 (2012).

[58] N. Carnahan and K. Starling. Thermodynamic properties of a rigid-sphere fluid. J. Chem. Phys. 53: 600 (1970).

[59] J. A. Barker and D. Henderson. Perturbation theory and equation of state for fluids. II a successful theory of liquids. J. Chem. Phys. 47: 4714–4721 (1967).

[60] G. Demouchy, A. Mezulis, A. Bée, D. Talbot, J. Bacri and A. Bourdon. Diffusion and thermodiffusion studies in ferrofluids with a new two-dimensional forced rayleigh-scattering technique. J. Phys. D: Appl. Phys. 37: 1417–1428 (2004).

[61] S. Putnam and D. Cahill. Transport of nanoscale latex spheres in a temperature gradient. Langmuir 21: 5317 (2005).

[62] C. Goupil, W. Seifert, K. Zabrocki, E. Muller and G. Snyder. Thermodynamics of thermoelectric phenomena and applications. Entropy 13(8): 1481–1517 (2011).

[63] J. Agar, C. Mou and J. Lin. Single-ion heat of transport in electrolyte solutions: A hydrodynamic theory. J. Phys. Chem. 93: 2079–2082 (1989).

[64] A. Majee and A. Würger. Collective thermoelectrophoresis of charged colloids. Phys. Rev. E 83: 061403 (2011).

[65] D. Bhowmik, N. Malikova, G. Mériguet, O. Bernard, J. Teixeira and P. Turq. Aqueous solutions of tetraalkylammonium halides: Ion hydration, dynamics and ion–ion interactions in light of steric effects. Phys. Chem. Chem. Phys. 16: 13447–13457 (2014).

[66] Y. Marcus. Effect of ions on the structure of water: Structure making and breaking. Chem. Comm. 109: 1346–1370 (2009).

[67] D. Niether and S. Wiegand. Thermophoresis of biological and biocompatible compounds in aqueous solution. J. Phys.: Condens. Matter 31: 503 003 1–25 (2019).

[68] M. Reichl, M. Herzog, A. Götz and D. Braun. Why charged molecules move across a temperature gradient: The role of electric field. Phys. Rev. Lett. 112: 198101 (2014).

[69] J. Dhont, S. Wiegand, S. Duhr and D. Braun. Thermodiffusion of charged colloids: Single particle diffusion. Langmuir 23: 1674 (2007).

[70] S. Duhr and D. Braun. Why molecules move along a temperature gradient. Proc. Natl. Acad. Sci. U.S.A. 103: 19678–19682 (2006).

[71] J.-C. Bacri, A. Cēbers, A. Bourdon, G. Demouchy, B. M. Heegaard and R. Perzynski. Forces rayleigh experiment in a magnetic fluid. Phys. Rev. Lett. 74: 5032–5035 (1995).

[72] J. C. Bacri, A. Cēbers, A. Bourdon, G. Demouchy, B. M. Heegaard, B. Kashevsky and R. Perzynski. Transient grating in a ferrofluid under magnetic field: Effect of magnetic interactions on the diffusion coefficient of translation. Phys. Rev. E 52: 3936–3942 (1995).

[73] T. Salez, S. Nakamae, R. Perzynski, G. Mériguet, A. Cēbers and M. Roger. Thermoelectricity and thermodiffusion in magnetic nanofluids: Entropic analysis. Entropy 20: 405 (2018).

[74] E. Wandersman, A. Cēbers, E. Dubois, G. Mériguet, A. Robert and R. Perzynski. The cage elasticity and under-field structure of concentrated magnetic colloids probed by small angle x-ray scattering. Soft Matter 9: 11480 (2013).

[75] K. I. Morozov and A. V. Lebedev. The effect of magneto-dipole interactions on the magnetization curves of ferrocolloids. J. Magn. Magn. Mat. 85: 51 (1990).

[76] A. O. Ivanov and O. B. Kuznetsova. Magnetic properties of dense ferrofluids: An influence of interparticle correlations. Phys. Rev. E 64: 041405 (2001).

[77] G. Mériguet, E. Dubois, M. Jardat, A. Bourdon, G. Demouchy, V. Dupuis, B. Farago, R. Perzynski and P. Turq. Understanding the structure and the dynamics of magnetic fluids: coupling of experiment and simulation. J. Phys. Cond. Matter 18: S2685–S2696 (2006).

[78] F. Gazeau, E. Dubois, J. Bacri, F. Boué, A. Ceners and R. Perzynski. Anisotropy of the structure factor of magnetic fluids under a field probed by small angle neutron scattering. Phys. Rev. E 65: 031 403 1–15 (2002).

[79] G. Mériguet, F. Cousin, E. Dubois, F. Boué, A. Cēbers, B. Farago and R. Perzynski. What tunes the structural anisotropy of magnetic fluids under a magnetic field? J. Phys. Chem. B 110: 4378–4386 (2006).

6

Potential Industrial Applications of Magnetic Fluid as Nanolubes

R.V. Upadhyay,[1,2,] Kinnari Parekh,[1] Kinajl Trivedi[2] and Anjana Kothari[1]*

6.1 Introduction

Theoretical studies on the influence of the properties of magnetic liquid lubricants in the hydrodynamic rolling system have been documented [1–12]. These studies have shown that the bearing lubricated with magnetic fluid enhances the load-carrying capacity, damping and stiffness and reduces the friction force and side leakages compared to the conventional lubricating oil. Deysarkar et al. [13] experimentally investigated the tribological properties of the commercial magnetic fluid (Ferrofluid Corporation, Catalog No. EMG905, 2.5% by weight of magnetic nanoparticle size 40 nm) in the presence and absence of a magnetic field using four ball testers. The magnetic field was applied by a permanent magnet placed around the ball pot assembly. In the absence of a magnetic field, the friction coefficient and wear scar diameter were found as 0.11 and 0.52 mm, respectively. The anti-wear property is reinforced by 10% by applying the magnetic field, while the anti-friction property stays the same. However, the rationale behind this improvement was not explicitly explained.

[1] Dr. K C Patel R & D Centre, Charotar University of Science & Technology, CHARUSAT Campus, Changa- 388 421, Anand, India.

[2] P D Patel Institute of Applied Sciences, Charotar University of Science & Technology (CHARUSAT), Changa-388 421, Anand, India.

* Corresponding author: rvu.as@charusat.ac.in

Li-Jun et al. [14] studied the tribological properties of turbine oil 46 with lubricant additive as 20 nm sized $Mn_{0.78}Zn_{0.22}Fe_2O_4$ magnetic nanoparticle with varying particle concentration from 0.5 to 6 wt%. The turbine oil doped with the optimum concentration (6 wt%) showed the best tribological properties among the tested oil samples and obtained enhancement in anti-wear by 25% and anti-friction by 27% compared to the base oil. Its improvement in tribological properties is attributed to forming protective films of magnetic nanoparticles. The influence of the magnetic field on tribological properties was investigated using stainless steel and a soft iron runner with 10 holes of 6 mm diameter each. The permanent magnet, NdFeB, is built into the holes. The applied magnetic field was 12, 22, 36 or 45 mT. The anti-wear properties of the optimized magnetic field (22 mT) are 7.8% higher than the zero field value. The growth in magnetic fields increases the friction coefficient while the wear scar's diameter remains nearly constant. The adsorption of magnetic nanoparticles on the worn surface and the formation of a protective film can be a possible mechanism for improving the diameter of the wear scar and the coefficient of friction. The same is attributed to energy dispersive spectroscopy analysis, which shows the presence of the Zn element on the spent surface. With an increasing field, the viscosity of fluid grows and particles get aggregated; as a result, the shear stress between the rubbing surfaces increases, resulting in a higher value of the coefficient of friction without influencing the wear scar diameter [15].

Huang et al. [16] deliberated the tribological properties of Fe_3O_4 magnetic nanoparticles covered with secondary surfactant (T161) at different concentrations (1–7% by weight) dispersed in synthetic oil (50 mPa·s). The particle size determined by the transmission electron microscope was 14.2 nm. Their results indicate that the optimum concentration of these particles was 4 wt%, at which the anti-friction was enhanced by 31% and anti-wear by 28% compared to the base oil. The improvement observed was explained by the argument that nanoparticles occupy the space between the steel balls and eliminate contact. The surface adhesion of the particles has been confirmed by energy dispersion spectroscopy. Thus, the lubrication mechanism was attributed to rolling friction and protective film forming.

Huang et al. [17] examined the magnetic fluid's static support capability and tribological properties under the external magnetic field. The test was carried out using a universal mechanical device made of aluminum plates and cylinders. The cylinder had eight holes (4.0 mm 4.2 mm), and each hole was filled with NdFeB permanent magnets (4.0 mm 4.0 mm). The Fe_3O_4-based magnetic fluid with 0–8.52% by volume was dispersed in synthetic oil of alpha-olefin hydrocarbons. The result showed that the magnetic fluid in the presence of a magnetic field has a higher holding capacity than the carrier liquid. Its support capacity expands with increasing volume percentage or magnetization of magnetic dispersal. It is due to the force supported by the magnetic fluid being a product of the fluid's magnetization and the gradient field's intensity (Equation 6.1).

$$F_m = \mu_0 M\nabla H \qquad (6.1)$$

where, μ_0 is the permeability of the free space, M is the magnetization of magnetic fluid and ∇H is the magnetic field gradient. Thus, as the fluid's volume fraction increases, the fluid's magnetization expands, leading to an escalated magnetic fluid's holding capacity. The results indicate that the coefficient of friction and wear was minimal at 4.83% by volume. At this volume fraction, magnetic nanoparticles act like a roller element between the two friction surfaces. At a higher concentration, coagulation of the particles takes place, which reduces the ball rolling effect. The result is an increase in the coefficient of friction and the size of the wear scar.

In order to reduce the size of the bulk magnet, Shen et al. [18] introduced a novel concept of magnetic surface texture using magnetic fluid. Magnetic surface textures were produced on 316 stainless steel surfaces through photolithography, electrolytic machining, plating and polishing. These micro-dimples have a diameter of 500 μm, a depth of 45 μm, a surface density of 5% and a magnetic film of 25 μm thickness arranged in a square lattice. The Fe_3O_4 magnetic nanoparticles synthesized in diester with 4.83% of the volume have a saturation magnetization of 15.9 kA/m. The tribological properties of the magnetic surface texture lubricated with magnetic fluid were studied using the disk on a disk rig with varying speed (0 to 0.18 m/s) and load (0 to 0.25 MPa). Both disks were constructed of 316 stainless steel. The result showed that the magnetic surface after being magnetized has better lubricating properties compared to non-magnetized surfaces. In the case of a non-magnetized surface, the centrifugal force can dissipate the ferrofluid, thus weakening the lubricating effect. While on the surface, the magnetic force can attract the magnetic fluid to the edge of the contact area of the micro magnets and restrict the spread of magnetic fluid from the friction surface.

Andablo-Reyes et al. [19] studied the use of magnetic fluid to control the lack of lubricating oil. The test was conducted using a mini traction machine with a variable magnetic field (25, 50 and 150 mT). The magnetic field is obtained by placing constant circular magnets under the contact. Magnetic fluid obtained from FerroTec, Co. has a volume fraction of 5.5 and a saturation magnetization of 24.3 kA/m. The experiment was carried out in starvation mode with speed, viscosity, load and magnetic field variations. The result showed that the magnetic fluid deprivation was negligible because the magnetic fluid was confined to the contact area under a magnetic field.

Huang et al. [20] studied the influence of the Tribo-tester material, the surface ratio and the thickness of the magnetic film. They selected 45# steel materials for the lower disk and fabricated it with magnets by varying the area ratio (0, 5, 10, 15 and 20%), magnetic film thickness (0, 10, 20, 30 and 50 μm) and the upper plate of 316 stainless steel material. The computational simulation showed that the magnetic flux was concentrated at the edge of the magnetic film and that the flux overlaps increased according to the surface ratio. The result showed that an area ratio of 5% with a magnetic film thickness of 30 μm has the best friction and wear reduction because the maximum magnetic flux intensity is similar for every area ratio at every micro magnet boundary. Thus, with the increasing value of the surface ratio, the flow line overlaps, weakening the field's intensity. From Equation (6.1), it could also be indicated that the magnetic force induced on the ferrofluid decreases. The different substrate materials change the performance of friction. The hydrodynamic region extended at the lower sliding speed using 45 #. In another work, they showed the effect of changing the diameter of the micro magnet from 500 μm to 300 μm, fixing area density by 10% and varying magnetic film thickness (20, 60 and 80 μm). The tribo test material was 316 stainless steel. They also looked at the angle of contact of the magnetic liquid. The results showed that the angle of contact increases with the thickness of the film as the magnetic field prevents the magnetic fluid from extending away from the surface. The lowest coefficient of friction and wear was observed for a magnetic lattice film having a film thickness of 80 μm relative to the normal surface. It is due to the strength of the magnetic field area of the sample having grown with increasing film thickness [21].

Gao et al. [22] assessed the tribological properties of various forms of Fe_3O_4 nanoparticles (hexagonal, octahedral and irregular) added to a base oil (40 motor oils) as lubricating additives. The mean particle size was 46 nm for hexagonal particles, 49 nm for octahedral particles and 50 nm for irregularly shaped particles. The hexagon-shaped Fe_3O_4 (1.5% by weight) showed a 58% and 14% reduction in friction coefficient and the wear scar's diameter compared to the base oil, respectively. The lubrication improvement mechanism was better with hexagonal particles because the hexagonal particles have a flaky structure and a higher contact area than the other particle morphology. The hexagon-shaped particle forms a protective layer on the worn surface. At the same time, the other particle morphology exhibited thick, deep furrows.

Zhou et al. [23] studied the tribological properties of Fe_3O_4 (10 nm) magnetic nanoparticles dispersed in liquid paraffin oil using a ball on a flat reciprocating tribotester. The test was performed with different loads (10, 30 and 50 N) and concentrations of magnetic nanoparticles (0, 0.5, 1 and 2 g/L). The mean coefficient of friction and the wear volume decrease by 25 and 64%, respectively, at a magnetic nanoparticle concentration of 2 g/L and a load of 10 N. They explained that nanomagnetic particles form a protective film on the antagonistic surface that reduces friction and wear.

In this arrayed magnetic film, the maximum sliding speed of 0.18 m/s and the load of 0.25 MPa is very low considering the actual condition. However, earlier literature had shown that lubrication is ineffective at this speed and load. Chen et al. [24] studied the tribological properties using the ring-on-disk tribotester, where the ring and the disk are in aluminum and 316 stainless steel, respectively. Whereas the magnetic material chosen was nanomagnetic particles CoNi. The test was performed at various film depths/thicknesses (0, 20, 50, 80 μm) and surface densities (0, 5, 10, 15%) of the micro-magnet. The magnetic film with ferrofluid enhanced the tribological properties compared to the normal surface. The optimized film thickness was 80 μm with an area density of 15%. The magnetic strength to keep the magnetic fluid at the contact area grows due to the increased intensity of the magnetic field. The magnetic liquid also showed a better lubricating property, even at a higher load of 1.73 MPa and at a speed of 1320 mm/s.

Zeqi et al. [25] investigated the tribological properties of various tri cresyl phosphate (TCP) anti-wear additives (0–2.5%) dispersed in 150SN mineral base oil under magnetic field conditions. The constant magnetic field (10 mT) was applied by wrapping the coil around the ball pot. The friction coefficient and the wear scar's diameter were 0.11 and 0.61 mm, respectively, in the absence of a magnetic field for a TCP concentration of 2%. The diameter of the wear scar is reduced by 3%

and the coefficient of friction increases by 18% in the presence of a magnetic field with respect to the absence of a magnetic field. The mechanism is explained using X-ray Photoelectron Spectroscopy that showed the presence of tribofilm on the worn surfaces, mainly composed of ferric-containing compounds such as Fe2O3, Fe3O4 and FePO4 on the ball lubricated with the TCP contained base oil. The diameter of the wear scar was reduced due to the formation of tribofilm but increased the shear stress between tribopair due to the escalated coefficient of friction.

From literature mentioned above, the following points are concluded. (i) With the particle size increasing from 14 nm to 40 nm, the friction coefficients expand from 0.063 to 0.110 and the wear scar diameter remains the same (0.53 mm and 0.52 mm). (ii) Having a different particle morphology, spherical and hexagonal, the coefficient of friction decreases from 0.063 to 0.04. In contrast, the diameter of the wear scar increased from 0.53 mm to 0.70 mm. (iii) With different types of particulate matter (Fe_3O_4 and $MnZnFe_2O_4$), the diameter of the wear scar decreases (0.53 mm to 0.45 mm) and the coefficient of friction increases from 0.063 to 0.07. (iv) In the presence of a magnetic field, the friction coefficient and the wear scar's diameter depend on the orientation and force of the magnetic field. The microarray results better than the bulk magnet because of field orientation and distribution.

It can be concluded from the literature reviewed that there is a need to study different types of materials, higher magnetic fields, different carriers and also other tribo-testers. More effort is still required to provide benchmarks for the future design of the magnetic fluid as a lubricant. Next the experimental section is briefly described, which flows from the tribology properties of base oils and magnetic fluids.

6.2 Experimental

Quantifying the tribological properties for different applications requires the knowledge of magnetic particles, the physical properties of the carrier in which these are dispersed, physic-chemical properties of magnetic fluids and finally, the tribology behavior of a magnetic fluid. All these lead to defining a good magnetic fluid as a lubricant. They will be briefly covered next.

6.2.1 Synthesis of magnetic nanoparticles and their characterization

This is the backbone of nanomagnetic particle lubricant applications. The control of size and size distribution is crucial as it decides the end properties of magnetic fluids. Among all available techniques to synthesize nanoparticles, the chemical co-precipitation approach is considered a better way to synthesize Fe_3O_4 nanoparticles [26]. This technique has the advantage of varying the size and shape of magnetic nanoparticles and the process is scalable and relatively fast [27–31]. The structural characterization of magnetic nanoparticles was performed using Bruker X-ray Powder Diffractometer (D2 PHASER), where CuKα ($\lambda = 0.15414$ nm) radiation is used as a source of X-ray.

Figure 6.1 shows the X-ray Diffraction pattern of Fe_3O_4 magnetic nanoparticles. A series of characteristic peaks are indexed and these planes correspond to a face-centered cubic unit cell with a spinel structure of Fe_3O_4 (JCPDS 19-629). The average

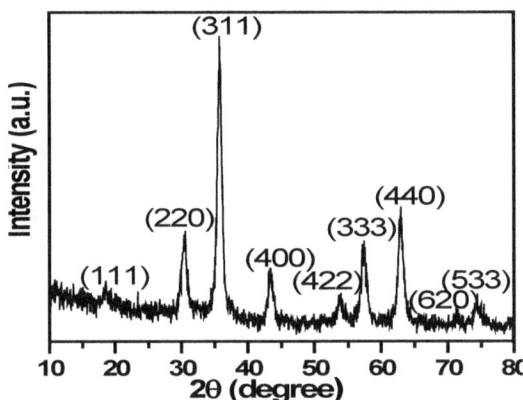

Figure 6.1. X-ray Diffraction patterns of Fe_3O_4 nanoparticles. The number in brackets shows the Miller plane for each peak.

crystallite size (D_x) was calculated using Scherrer's formula [32] and was found to be 11.7 ± 0.5 nm. The lattice parameter (a) calculated for all planes is 0.8369 ± 0.0001 nm, which is close to that of the bulk Fe_3O_4 at 0.8369 nm [33].

6.2.2 Magnetic fluid synthesis and particle morphology

The present work uses two different types of oils based on their wide engineering applications: synthetic oil (alpha-olefin hydrocarbon oil, G0) and high viscosity mineral oil (HV oil). The physical properties of the carrier liquids are given in Table 6.1. The magnetite particles were coated with oleic acid and dispersed in carrier fluids, as shown in Table 6.1. To stabilize the particles, a secondary surfactant was used in small quantities and the stability of the fluid was verified. Thus obtained fluids are stable against gravitational force. The volume fraction of nanoparticles varied as per the requirements.

Transmission Electron Microscope (TEM) model JEOL, JEM 2100, is used to visualize the morphology of the magnetic nanoparticles. The TEM result (Fig. 6.2) shows that particles are nearly spherical.

The equivalent diameter of the particles was measured from images at different areas of the grid and fitted to the log-normal size distribution function. The magnetic nanoparticles' average particle size is 12.1 ± 0.7 nm. The size obtained from TEM is

Table 6.1. Physical properties of carrier fluids.

Physical Properties	Synthetic oil (G0)	High viscosity mineral oil (HV oil)	Low viscosity mineral oil (LV oil)
ρ @ 303 K (± 2 kg/m^3)	865	890	878
η @ 313 K (± 1 mPa.s)	254	219	10.3
Flashpoint (K)	523	503	422
Pour point (K)	< 233	< 263	< 228

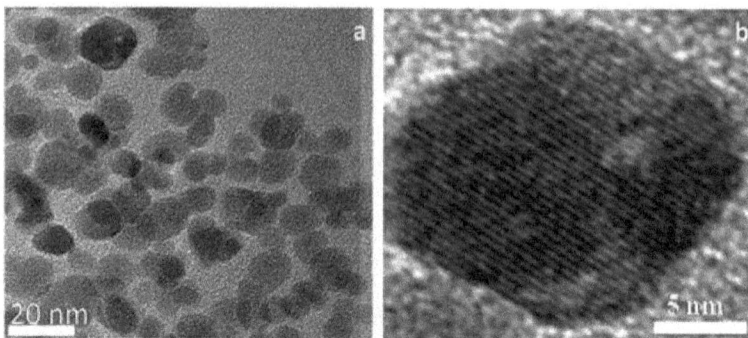

Figure 6.2. (a) TEM image and (b) HRTEM image for synthesized magnetite particles.

slightly higher than that obtained from XRD. As the sub-nanometer scale magnetic nanoparticles are almost single crystals; their physical and crystallite size can be nearly the same. The crystalline nature of the sample is confirmed by the high-resolution TEM image of magnetic nanoparticles (Fig. 6.2 (b)). The HRTEM image demonstrates well-developed lattice fringes.

6.2.3 Tribology tester

Tribological properties, namely anti-friction, anti-wear and weld load of all samples, are evaluated using a four-ball test machine (TE-800-FBT, MAGNUM engineering, Bangalore, India). The experimental set-up of the four-ball test machine is shown in Fig. 6.3, along with a schematic of the ball pot assembly. In a four-ball tester, three balls are kept stationary and the fourth ball rotates. The geometry of the four balls is similar to the tetrahedral shape. The test balls are made up of chrome alloy steel, according to AISI E-52100, with a diameter of 12.7 mm, containing extra polish grade 25 and hardness between 60–65 HRC (Rockwell C Hardness).

The friction and wear properties are studied as per ASTM D 4172 protocol. In this, the rotating speed is fixed at 1200 ± 20 rpm and the applied load is 392 N. Data are recorded for 60 min at 348 K. The extreme pressure was studied at 1770 ± 20 rpm for 10 s at 303 K and the load is gradually increased until the balls get welded. Before each test, new steel balls were ultra-sonicated for 5 min in acetone and wiped using a fresh and lint-free industrial wipe. The steel balls were placed into the ball pot assembly by lifting a ball race and were tightened using a torque wrench (68 N-m) to prevent the motion of three balls during the experiment. The ball pot assembly was filled with test oil. To ensure voids, the ball pot assembly was filled with nearly 3 mm oil layer oil that was kept above the ball surface. The Coefficient Of Friction (COF) was measured for all samples. Each measurement was repeated three times and the average values were reported.

The Wear Scar Diameter (WSD) was measured using an optical microscope with 8X magnification with 0.01 mm resolution and using two perpendicular lines, the average wear scar diameter was calculated. The standard deviation is calculated for the three tests and is plotted as an error bar in the experimental data.

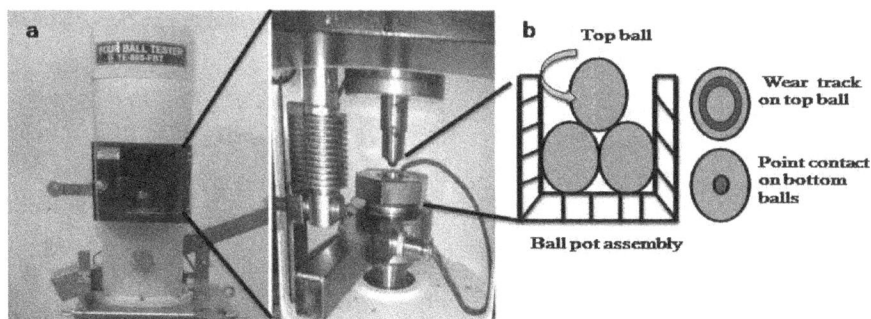

Figure 6.3. (a) Four ball tester and (b) Schematic of the ball pot assembly.

6.3 Tribology of Base oil

In order to understand the tribological properties of nano lubricants, it is essential to understand the properties of the basic oils used to synthesize magnetic fluids. Various types of base oils are available in the market for different applications. However, choosing an oil that fits a particular application is complex. The tribology properties of the base oil used to synthesize magnetic fluids are focused here.

The G0 oil stands for synthetic oil, while HV oil is a type of mineral oil. Due to the change in their chemical structure, both fluids exhibit different behavior. The observed behavior shown in Fig. 6.4a can be explained by dividing it into three regions: static, running in and steady state. The first region shows that COF increases to 0.072 within 8 s, which is attributed to the static friction offered by G0 and HV oil to achieve the speed of 1200 rpm. In the second region, COF increased from 0.072 to 0.090 for G0 and HV oil. The observed fluctuation in COF around an average value is attributed to the changes in oil film thickness. It can be considered a mixed lubrication region where the load is carried by oil film and asperities. In this region, asperities are deformed elastically. In the third region, COF increases because of a decrease in oil film thickness. At the localized point contact, oil undergoes high pressure due to applied load and high friction. It results in a temperature rise, thus the reduction in viscosity eventually reduces the film thickness. Hence, mixed lubrication converts into boundary lubrication and asperities deform elastically and plastically [34]. An intense shoot-out at 103 s in HV oil indicates the making and breaking of film. Similar phenomena were also observed earlier. The average value of COF obtained for G0 and HV oil is 0.105 ± 0.005 and 0.091 ± 0.004, respectively. The COF of G0 is higher than HV oil because the viscosity of the G0 is 1.16 times higher than HV oil. It leads to increased shear stress between two rubbing surfaces.

Figure 6.4b shows the optical image of the ball after the test. The average wear scar diameter was calculated using two mutually perpendicular lines and the standard deviation was calculated for WSD on nine balls from three tests. The average WSD of G0 oil is 0.545 ± 0.027 mm, while HV oil is measured to be 0.865 ± 0.042 mm. The WSD is 58% higher for HV oil compared to G0. The ragged edges with parallel grooves were observed on the worn surface lubricated with G0 oil. While HV oil shows deep parallel grooves, as seen in Fig. 6.4b.

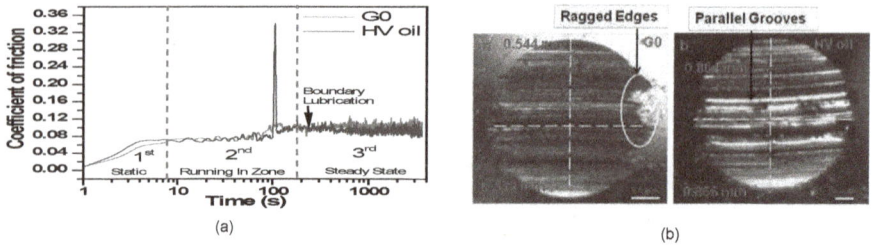

Figure 6.4. Co-efficient of friction (a) and wear scar diameter (b) of two base oils used for magnetic fluids.

The morphology of surfaces, specifically worn scars resulting from shearing of the surfaces, was examined to understand the lubrication mechanism. In the present work, the Scanning Electron Microscope (SEM) of JEOL, JSM 6010LA, was used for obtaining surface images of worn balls. Similarly, the detailed study of the worn ball surface was carried out by Atomic Force Microscope (AFM) model Nanosurf easy scan 2 basic AFM, Switzerland.

The surface lubricated with the G0 oil, SEM and AFM images are presented in Figs. 6.5a and 6.5c. The worn surface shows cracks and pits on the surface of a ball due to surface fatigue. Parallel grooves and worn debris are also observed. It indicates adhesive and fatigue wear. The AFM image of G0 also shows worn debris adhering to the surface. This is due to the material transfer during the adhesive wear. The surface roughness of the worn surface is found to be 796 nm^2.

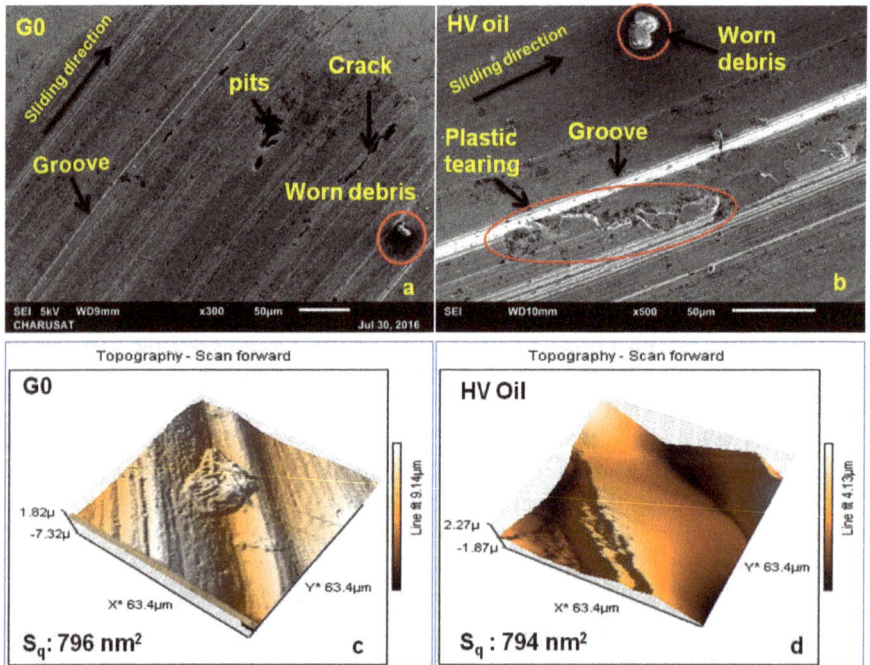

Figure 6.5. Scanning Electron Microscope (SEM) and Atomic Force Microscope (AFM) 3D images of the worn ball surface lubricated with synthetic oil (G0), (a) and (c); and high viscosity mineral oil (HV oil), (b) and (d), respectively.

SEM and AFM images are presented in Figs. 6.5b and 6.5d for the surface lubricated with the HV oil. The SEM image shows the worn debris, parallel grooves and plastic tearing of the surface, which results from momentary making of bonds between rubbing surfaces. The AFM image also shows a broad peak on the 3D topography surface. The grooves are very broad compared to the scanning area of the AFM image and can be seen as a flat surface with a surface roughness (794 nm^2). It may be ascribed to the occurrence of abrasive wear. The observed wear mechanism is different in both oils. A homogeneous structure consists of synthetic oil, while mineral oil has a heterogeneous molecular structure [35]. Hence metal-to-metal contact may occur more in HV oil than in G0 oil. It can also be seen as spikes in COF for HV oil, as shown in Fig. 6.4. Plastic tearing and debris on the worn surface (Fig. 6.5) are also present. Hence it is concluded that the observed tribological properties of these oils are due to their different molecular structures.

6.3.1 Physical properties of Magnetic nanoparticles dispersion in G0 and HV oils

The magnetic fluid is synthesized in G0 and HV oil. The wt% of magnetic particles was kept at 10%. These are designated as G10 and HV10. Several dilutions (G0 to G10: G series and HV oil to HV10: HV series) were prepared using this parent fluid. The sample codes with the weight fraction of magnetic nanoparticles (MNP) are given in Table 6.2 and Table 6.3.

Table 6.2. Physical and magnetic properties of various diluted fluids in G0 oil.

Sample code	wt% of MNP	ρ @ 303 K ($\pm$ 2 kg/m^3)	η @ 313 K ($\pm$ 1 mPa.s)	M_s ($\pm$ 0.1 kA/m)	χ_i
G0 oil	0	865	254	0	0
G2	2	847	257	1.0	0.042
G4	4	870	258	2.2	0.078
G6	6	882	262	3.1	0.115
G8	8	904	260	4.3	0.144
G10	10	916	261	5.8	0.213

Table 6.3. Physical and magnetic properties of various diluted fluids in HV oil.

Sample code	wt% of MNP	ρ @ 303 K ($\pm$ 2 kg/m^3)	η @ 313 K ($\pm$ 1 mPa.s)	M_s ($\pm$ 0.1 kA/m)	χ_i
HV oil	0	890	219	0	0
HV2	2	897	228	1.2	0.066
HV4	4	923	239	2.5	0.141
HV6	6	948	251	3.5	0.194
HV8	8	959	263	4.4	0.244
HV10	10	970	276	5.4	0.299

In Tables 6.2 and 6.3, the magnetic properties were obtained from magnetization (M) versus field (H) measured using Vibrating Sample Magnetometer. The saturation magnetization (M_s) was calculated using the standard method of plotting the M versus 1/H curve and extrapolating the value to H tends to infinity. The initial susceptibility (χ_i) was calculated from the low field slope of the M versus H curve. Both parameters show a linear increase with the growth in magnetic particle concentration. It is valid for both oil dispersions.

Rheological properties for all the diluted fluids were carried out using Anton-Paar Rheometer (MCR301). All the fluids exhibit Newtonian behavior and the viscosity obtained at 313 K is presented in Tables 6.2 and 6.3. The viscosity increases with an escalation in nanoparticle concentration.

6.4 Tribology of magnetic fluids

Figure 6.6 (a,b) shows the variation in friction co-efficiency over time for the G0 and HV series for the variation in the concentration of nanoparticles.

Figure 6.6 shows that with the increase in the concentration of nanoparticles, the friction decreases. However, the behavior in the first static area is different when the base oil is changed. The reduction in COF in the magnetic liquid may be explained as follows. The result of TEM shows (Fig. 6.2a) that magnetic nanoparticles are almost spherical; these particles enter between gliding surfaces and act as a rolling element. Therefore, the sliding friction is transformed into a sliding and rolling mixture [36–38]. Hence, it will separate the asperities and keep the system in a mixed lubrication region. The variation of COF for the concentration of nanoparticles is illustrated in Fig. 6.7a for the purpose of comparison .

The COF value decreases and stays constant in both cases, but the HV oil shows a lower COF value than G0 oil. What is observed is explained below. The viscosity of the G-series increases from 257 mPa·s to 261 mPa·s as the concentration of nanoparticles increases. This small change in viscosity will not affect the shear stress between two friction surfaces, so the COF stays nearly constant with the concentration of nanoparticles after 2% by weight. However, in the HV series, the viscosity increases by 228 mPa·s to 276 mPa·s with concentrations of nanoparticles. This 21% increase in viscosity affects the stress between the surfaces, which reduces the COF by up to 6% by weight, but with further increase in particle weight fraction, the COF begins to increase. A rise in shear stress leads to an increase in COF. COF decreases by 56 and 45% for G4 and HV6 with G0 and HV oil, respectively. The surface analysis of the balls was performed to understand the wear mechanism and the extent of the damage.

Figure 6.7b shows the change in wear scar diameter with nanoparticle concentrations for G0 and HV-based oils. In contrast to COF, the wear scar diameter of HV oil is higher than that of G0 oil. With the addition of magnetic nanoparticles in the base, oil WSD reduces even with a small concentration of 2% weight for both fluids. WSD is reduced by 30% in the G series until G4 and remains almost constant for higher concentrations. In the HV series, WSD decreased by 46 for 6% by weight of magnetic nanoparticles compared to HV oil and at higher concentrations, the WSD values overlap with those of the G series. The reduction in WSD shows that

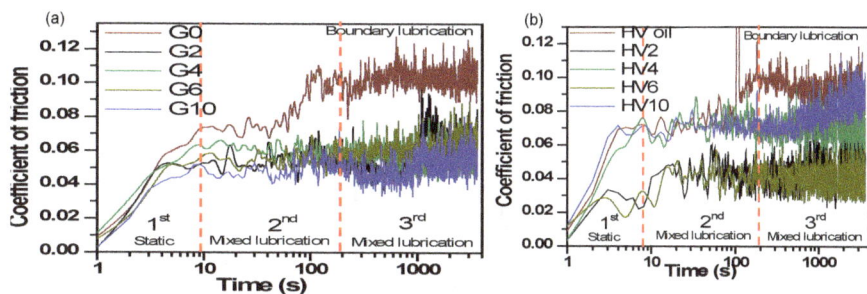

Figure 6.6. Coefficient of friction as a function of time for various weight % of magnetic nanoparticles dispersed in (a) G0 and (b) HV oils.

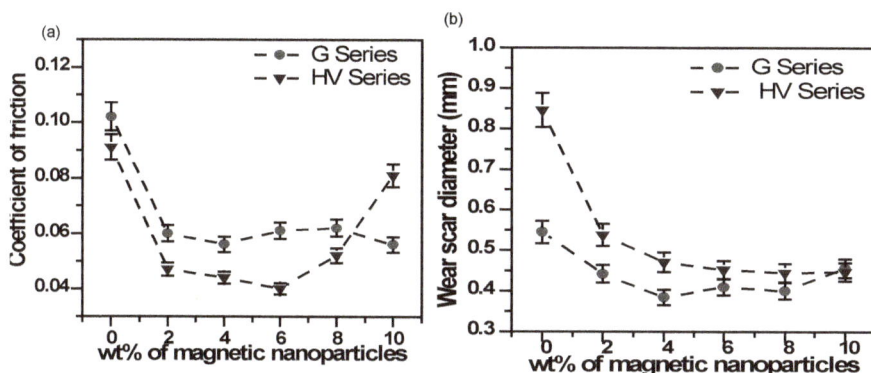

Figure 6.7. Variation of (a) COF and (b) wear scar diameter with MNP concentration.

particles act as fillers between the asperities due to their small size and adherence to the metal surface [39, 40], which eliminates the contact between the rubbing surfaces and reduces wear scar diameter. The detailed explanation of the observed effect is described in greater detail in [41 Kinjal thesis/paper].

A qualitative analysis of the worn surface was carried out using a Scanning Electron Microscope and Atomic Force Microscope for typical concentrations of nanoparticles, i.e., G4 and G10 and HV6 and HV10. Both samples are selected based on the observed COF and WSD results. Figure 6.8 shows the SEM images of the balls.

The smoother surface is observed for G4 (Fig. 6.8(a)), whereas parallel grooves have been observed over a growing concentration of nanoparticles (G6, Fig. 6.8(b)). The furrows and scratches were also observed in 10% weight (Fig. 6.8(c)). The observed scratches in G10 (10 wt%) with increased WSD value could be correlated with the argument that on increasing magnetic nanoparticles, the distance between the particles decreases and under shear forms an aggregation. The worn surface lubricated by HV oil has parallel grooves and plastic deformation becomes smoother for HV4 (Fig. 6.8(d)). The smoothest surface was found in HV6 (Fig. 6.8(e)). Moreover, at higher particle concentrations, more scratches on the worn surface (Fig. 6.8(f)) are displayed. These HV series SEM images correlate with the worn surface's optical images.

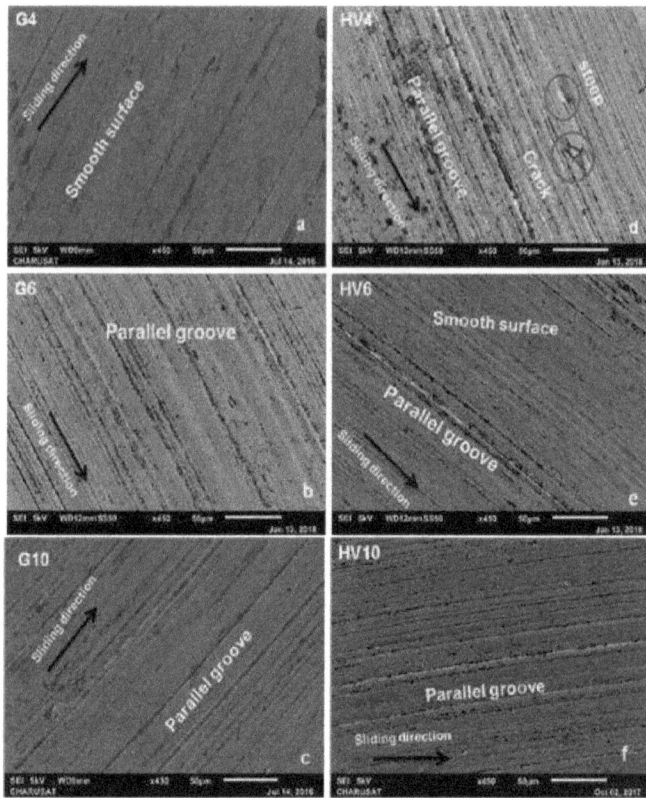

Figure 6.8. SEM images of balls after the tribology experiments.

Figure 6.9 illustrates three-dimensional AFM images of the worn surface lubricated with (a) G4, (b) G6, (c) G10, (d) HV4, (e) HV6 and (f) HV10. AFM images of base oils were presented earlier in Figs. 6.5c and 6.5d. Abrasion of the adhesive was observed for G0, which is not present when adding magnetic nanoparticles. Minor scratches with a shallow peak and valley were observed for G4 (Fig. 6.9(a)), which changes to a shallow ploughing track for G6 (Fig. 6.9(b)) and with increasing concentration (G10), it becomes deeper and broader (Fig. 6.9(c)). In HV oil, the ploughing track is extensive and looks like band scratches with a flat surface. In the case of HV4, there are still wide tracks, but instead of a flat strip, a shallow ploughing trail was observed (Fig. 6.9(d)). In addition, the increasing concentration of the magnetic nanoparticle track becomes narrow (Fig. 6.9(e)) and, for HV10, resembles small piles and depths (Fig. 6.9(f)). Similar results were observed in SEM and optical images of the used surface with added nanoparticles.

The results show that the optimum concentration of magnetic particles in G0 oil and HV oil is 4% by weight and 6% by weight, respectively. Both SEM and AFM images also confirmed these results. Therefore, this optimal nanoparticle concentration in two different fluids is possible for better tribological properties. Other than the diameter of the wear scar and COF is an extreme pressure (welding load). It is significant when applying magnetic fluid as a lubricant. There are additives

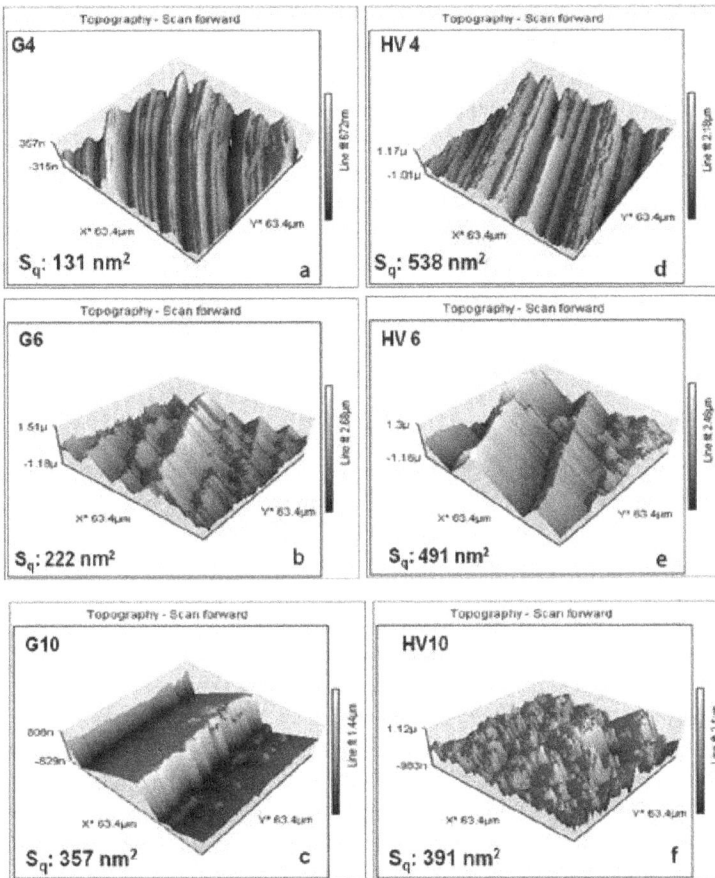

Figure 6.9. AFM images of the worn surface of the balls for low concentration of MNP.

for improving the welding load, the diameter of the wear scar and the COF of the lubrication oil. These additives are described as performance enhancers. Next the influence of one such performance enhancer additive for improving weld load and, in order of preference, this supplement combined to the weld load properties of the magnetic fluid lubricant is shown.

6.4.1 Performance-enhancing additives and their influence on lubrication properties.

To increase the weld load of conventional oil, Performance-Enhancing Additives (PEA) were combined such as molybdenum disulfide, graphite, sulfurized olefins, dialkyl dithiocarbamate, boron, complex ester, chlorine, phosphorus, sulfur and nanoparticles [42–45]. The performance-enhancing additive reacts with metallic surfaces while operating at high temperatures. In high-pressure environments, avoid adhesive wear by forming a protective coating that prevents metallic surfaces from welding together [46–48]. Adding 2% by weight of cadmium diisopropyl dithiophosphate (CODP, oil additive/PEA) increases the welding load almost twice [49]. The MoS$_2$ as an additive [50] increased weld load by 30 kg (12% compared to base oil, even though the wt% of the particle varies from 0.1 wt to 1 wt%. The 1 wt% of Cu nanoparticles (5 nm) increased the weld load of liquid paraffin from 120 kg to 308 kg, which is nearly the same as the addition of 2 wt% of CdDDP [51]. At the same time, CCTO and ZnO did not significantly increase the basic oil soldering load [52]. These studies have shown that the PEA dramatically contributes to the weld load's properties.

Furthermore, our recent study focuses on the sequence in which this PEA is added. Here the influence of PEA in sample G4 is demonstrated. The design protocol for the weld load is ASTM compliant. First the welding charge for pure oil G0 and fluid G4 were measured. The resultant values are shown in Table 6.4.

The weld load on the addition of magnetic nanoparticles was increased by 25%. It can be explained as follows when magnetic nanoparticles are dispersed into the synthetic oil, magnetic nanoparticles work as a spacer between tribopairs and also get mended on the surface (as evident from SEM and AFM images). Therefore, it prevents two surfaces from bonding, increasing the weld load. However, the required industrial value is more than 250 kg. This can be improved through PEA. However, choosing the EPA is very important. In this case, the magnetic particles are coated with some fatty acids and dispersed using a secondary surfactant in the G0 oil. This oil is synthetic, so the PEA should be compatible with the oil and particle surfaces. Besides, it must have a greater affinity towards the steel balls.

Complex ester is selected as a performance-enhancing additive (PEA) as it has a higher affinity to the ball surface. At the outset, different volumetric fractions of PEA were added to the G0 oil and the welding load was recorded. It was found that the maximum welding load (200 kg) was reached when a volumetric fraction of PEA was 5%. Thus, the 5% volumetric fraction was fixed during the entire experiment. This value is well under the required value. Therefore, to study the influence of the magnetic liquid synthesized with the PEA, two samples (G4-A and G4-B) were prepared. In the first sample, PEA was added to the synthesized magnetic fluid having secondary amide surfactant (G4+PEA), while in the latter (G4-B), a mixture of G0 and PEA magnetic nanoparticles was added and magnetic fluid was synthesized by adding secondary amide. A detailed study was carried out on these two samples to verify the influence of the order in which the PEA was added on the lubrication properties.

Table 6.4. Weld load parameters for G0 and G4 fluids.

Sample code	ρ @ 303 K (± 2 kg/m³)	η @ 313 K (± 1 mPa·s)	Coefficient of friction	Wear scar diameter (mm)	Weld load (kg)
G0	865	254	0.102 ± 0.005	0.545 ± 0.027	160
G4	870	258	0.056 ± 0.003	0.384 ± 0.025	200

6.4.1.1 Coefficient of friction, wear scar size and worn surface analysis

The frictional coefficient for all samples is shown in Fig. 6.10. The nature of the graph shows a decreasing trend. The coefficient of friction decreases with the addition of PEA in G0 by 37% compared to G0, which is greater than G4. The COF value for

G4-A is inside the error bar with G4. The G4-B COF is slightly lower than the G4. The marginal decrease indicates the influence of the order with the addition on COF.

Figure 6.11 presents the optical images of the worn surfaces. The oil with PEA shows deep parallel scratches compared to pure oil without PEA (Fig. 6.11a,b). The smooth surface was observed for G4, while parallel grooves were observed for G4-A (Fig. 6.11c,d). The worn surface seemed smoother for G4-B than for all other samples (Fig. 6.11). The observed results validate the argument that the sequence of PEA addition is crucial in the final results. There is a marginal reduction in the wear scar diameter in G4-B relative to G4-A.

SEM and AFM were recorded for all ball surfaces treated with different liquids. Figures 6.12 and 6.13 depict both SEM and AFM. The worn surface of G0 and G0 PEA shows worn debris and parallel grooves on the surface seen from the Scanning Electron Microscopy (SEM) images (Figs. 6.12a and b). The surface observed under SEM exhibits a smoother surface for G4 and G4-B compared to G0 (Fig. 6.12c,d). Thus, PEA has less influence over the smoothing of the surface.

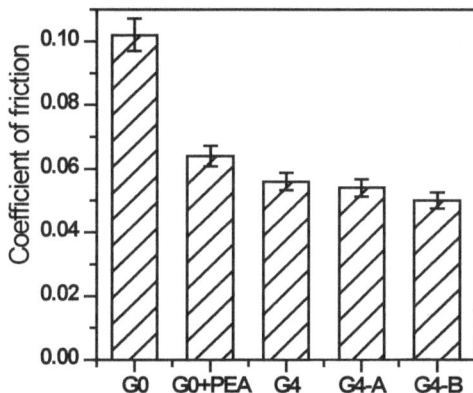

Figure 6.10. COF for different samples.

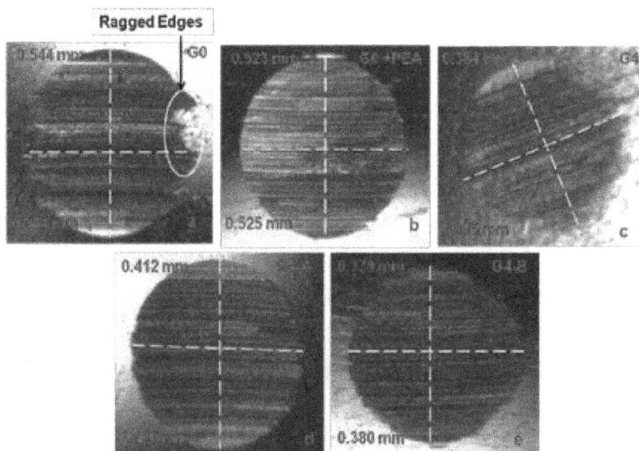

Figure 6.11. Wear scar diameter of (a) G0, (b) G0+PEA, (c) G4, (d) G4-A and (e) G4-B.

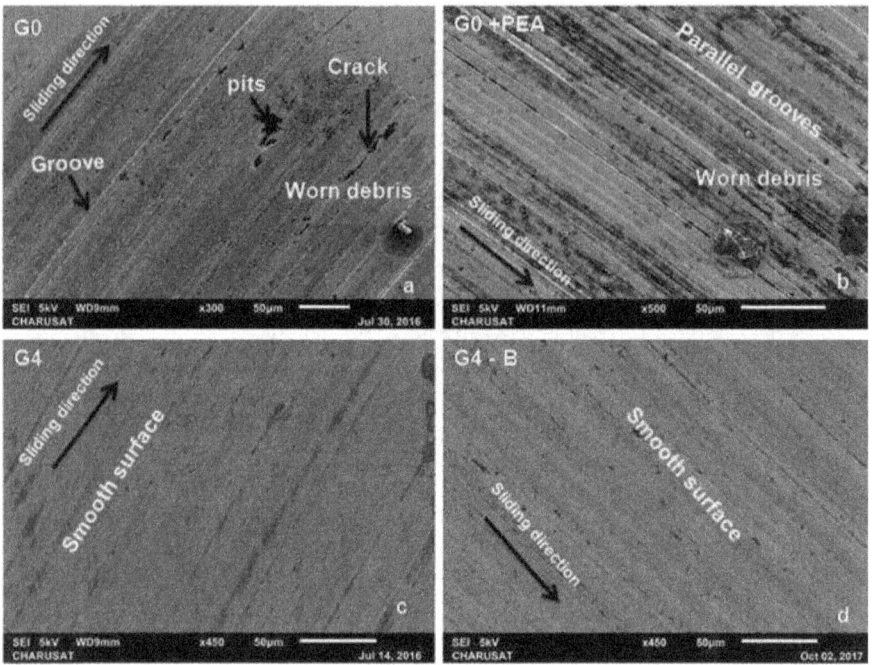

Figure 6.12. SEM images (a) G0, (b) G0+PEA, (c) G4 and (d) G4-B.

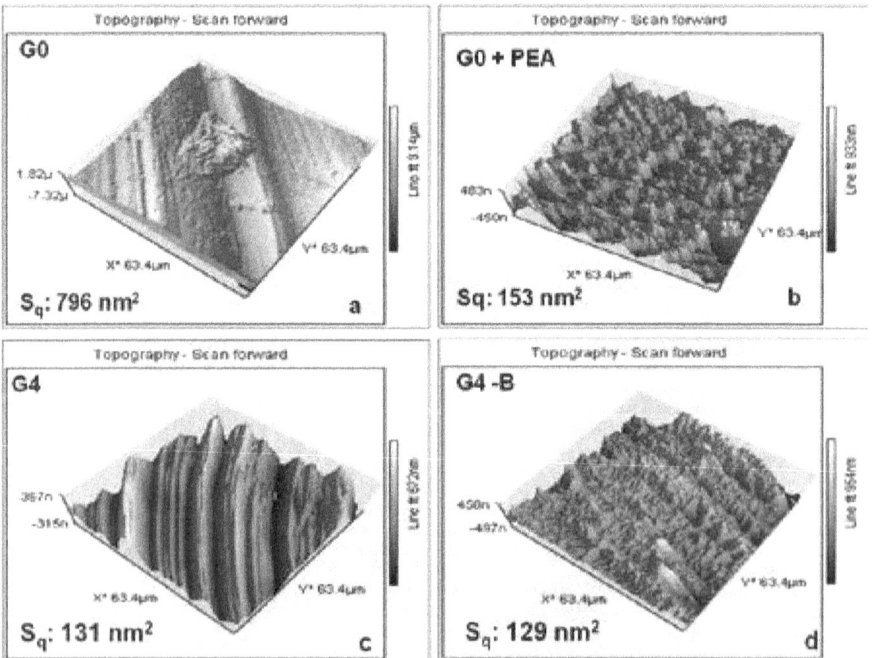

Figure 6.13. AFM images. (a) G0, (b) G0+PEA, (c) G4 and (d) G4-B.

The atomic force microscope (AFM) also shows adherence and deep gouges in G0 (Fig. 6.13a). The small peak and valleys are seen for G0 PEA (Fig. 6.13b). The surface roughness gets a minimum for G0 PEA. For G4 and G4-B (Fig. 6.13 c,d), the surface roughness decreased by almost 14% compared to G0+PEA.

This observed effect may be explained in the following way. In sample G0, metal-to-metal contact occurs, which is indicated by the presence of worn debris adhering to the ball surface. It means that they have a higher COF and WSD value. Adding PEA to G0 helps reduce COF because of its physical attachment to the metallic surface. It helps reduce shear stress and surface roughness by 80% compared to G0, but fails to prevent wear and converts adherence to abrasive wear. When magnetic nanoparticles are dispersed in synthetic oil, they act as a gap between tribopairs, reducing friction, while wear decreases as it is repaired on the surface. This explanation can be checked if this effect is noticeable in the welding load.

6.4.1.2 Weld load for different fluids with PEA

With the addition of PEA to G0, the improved welding load of G0 is the same as with the addition of magnetic nanoparticles (Fig. 6.14). Thus, one can achieve the value of the weld load of the base with the addition of magnetic nanoparticles and PEA. With the combination of magnetic nanoparticles and PEA, when PEA is added to the synthesized magnetic fluid (G4), the welding charge increases by 58% for G4-A compared to G4. There is a drastic increase in weld load of 300% for G4-B compared to G4. The process of improving the welding load value is included in the diagram in Fig. 6.15 [53, 54].

For G4-A, PEA is added to the synthesized magnetic liquid. In this way, PEA creates a link with secondary amide, G0 and some remain free in the system. PEA has the greatest affinity to the available amide on the surface of magnetic nanoparticles relative to the ball's surface. It renders the magnetic nanoparticles to attract the ball's surface and PEA stays away from it (Fig. 6.15a). Thus, the weld load value is set by nanoparticles. Therefore, there is a minor increase in the weld load compared to G4. However, in G4-B, PEA was added first in G0, which modified the base oil and nanoreactor. The particle is contained within this G0 PEA nanoreactor. PEA has a stronger affinity to the spherical surface than the G0. It may allow the PEA molecules to form a protective film over the friction surface, as shown in Fig. 6.15b.

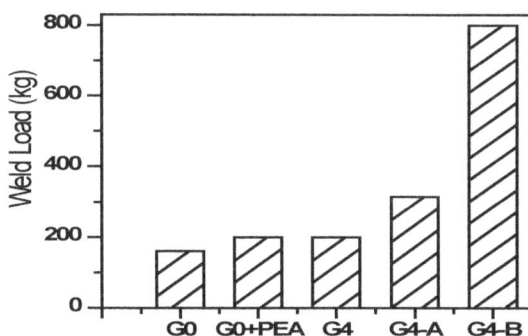

Figure 6.14. Welding load of the dispersed base oil with nanomagnetic particles and a performance-enhancing additive (PEA).

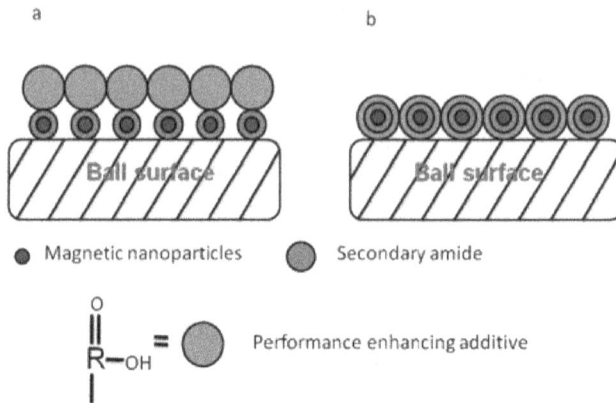

Figure 6.15. Schematic of the PEA on spherical surfaces, (a) the PEA remains on the surface of the encased secondary particles and (b) the particle is housed in this G0 PEA nanoreactor.

Under these conditions, the magnetic particle acts like a spacer and may support a higher load. It will not allow contact between the friction surfaces. However more details and understanding are still under investigation. It demonstrates that the order in which the PEA is added affects the weld load properties.

6.5 Conclusion

This study demonstrates that magnetic fluid can be used as a lubricant, but the properties of each component need to be understood in detail. The size and distribution of nanomagnetic particles also affect lubrication properties. Moreover, the order in which different additives are included in the preparation of the magnetic fluid-based lubricant influences the welding load, the COF and the diameter of the wear scar. Currently, there is very limited research in this area. It is subject to scientific understanding, but the actual application needs a large-scale production of magnetic fluid with desired lubrication properties. The future use of magnetic fluid lubricant is one area worth exploring.

Acknowledgments

The authors are grateful to the Department of Science & Technology (DST), Govt. Of India, New Delhi and Charotar University of Science & Technology for the financial support.

References

[1] C. Q. Chi, Z. S. Wang and P. Z. Zhao. Research on a new type of ferrofluid-lubricated journal bearing. J. Magn. Magn. Mater. 85: 257–260 (1990).

[2] T. A. Osman, G. S. Nada and Z. S. Safar. Static and dynamic characteristics of magnetized journal bearings lubricated with ferrofluid. Tribol. Int. 34: 369–380 (2001).

[3] G. S. Nada and T. A. Osman. Static performance of finite hydrodynamic journal bearings lubricated by magnetic fluids with couple stresses. Tribol. lett. 27: 261–268 (2007).

[4] A. A. Elsharkawy and S. F. Alyaqout. Optimum shape design for surface of a porous slider bearing lubricated with couple stress fluid. Lubr. Sci. 21: 1–12 (2009).

[5] P. Sinha, P. Chandra and D. Kumar. Ferrofluid lubrication of cylindrical rollers with cavitation. Acta Mechanica 98: 27–38 (1993).

[6] R. C. Shah and M. V. Bhat. Ferrofluid squeeze film between curved annular plates including rotation of magnetic particles. J. Eng. Math. 51: 317–324 (2005).

[7] R. C. Shah and K. S. Parikh. Comparative study of ferrofluid lubricated various designed slider bearings considering rotation of magnetic particles and squeeze velocity. Int. J. Theor. Math. Phys. 4: 63–72 (2014).

[8] J. R. Lin, R. F. Lu, M. C. Lin and P. Y. Wang. Squeeze film characteristics of parallel circular disks lubricated by ferrofluids with non-Newtonian couple stresses. Tribol. Int. 61: 56–61 (2013).

[9] R. C. Shah and M. V. Bhat. Ferrofluid lubrication of a parallel plate squeeze film bearing. Theor. Appl. Mech. 30: 221–240 (2003).

[10] R. C. Shah and M. V. Bhat. Lubrication of porous parallel plate slider bearing with slip velocity, material parameter and magnetic fluid. Ind. Lubr. Tribol. 57: 103–106 (2005).

[11] N. D. Patel and G. Deheri. A ferrofluid lubrication of a rough, porous inclined slider bearing with slip velocity. J. Mech. Eng. Technol. 4: 259–268 (2012).

[12] R. C. Shah and M. V. Bhat. Ferrofluid squeeze film in a long journal bearing. Tribol. Int. 37: 441–446 (2004).

[13] A. K. Deysarkar and B. H. Clampitt. Evaluation of ferrofluid as lubricants. J. Synth. Lubr. 5: 105–114 (1988).

[14] W. Li-jun, G. Chu-wen and R. Yamane. Experimental research on tribological properties of $Mn_{0.78}Zn_{0.22}Fe_2O_4$ magnetic fluids. J. Tribol. 130: 031801-1-5 (2008).

[15] W. Li-jun, G. Chu-wen, Y. Ryuichiro and W. Yue. Tribological properties of Mn–Zn–Fe magnetic fluids under magnetic field. Tribol. Int. 42: 792–797 (2009).

[16] W. Huang, X. Wang, G. Ma and C. Shen. Study on the synthesis and tribological property of Fe_3O_4 based magnetic fluids. Tribol. Lett. 33: 187–192 (2009).

[17] W. Huang, C. Shen and X. Wang. Study on Static Supporting capacity and tribolgical performance of ferrofluid.Tribol. Trans. 52: 717–723 (2009).

[18] C. Shen, W. Huang, G. Ma and X. Wang. A novel surface texture for magnetic fluid lubrication. Surf. Coat. Technol. 204: 433–439 (2009).

[19] E. Andablo-Reyes, J. de Vicente, R. Hidalgo-Alvarez, C. Myant, T. Reddyhoff and H. A. Spikes. Soft Elasto-Hydrodynamic Lubrication. Tribol. Lett. 39: 109–114 (2010).

[20] W. Huang, W. B. Wu and X. L. Wang. Tribological properties of magnetic surface lubricated by ferrofluid. Eur. Phy. J. Appl. Phys. 59: 31301–31308 (2012).

[21] W. Huang, S. Liao and X. Wang. Wettability and friction coefficient of micro-magnet arrayed surface. Appl. Surf. Sci. 258: 3062–3067 (2012).

[22] C. Gao, Y. Wang, D. Hu, Z. Pan and L. Xiang. Tribological properties of magnetite nanoparticles with various morphologies as lubricating additives. J. Nanopart. Res. 15: 1502-1-10 (2013).

[23] G. Zhou, Y. Zhu, X. Wang, M. Xia, Y. Zhang and H. Ding. Sliding tribological properties of 0.45% carbon steel lubricated with Fe_3O_4 magnetic nanoparticle additives in baseoil. Wear 301: 753–757 (2013).

[24] W. Chen, W. Huang and X. Wang. Effects of magnetic arrayed films on lubrication transition properties of magnetic fluid. Tribol. Int. 72: 172–178 (2014).

[25] J. Zeqi, F. Jianhua, C. Boshui, W. Jiang, W. Jiu and Z. Zhe. Improvement of magnetic field on tribological properties of lubricating oils with zinc butyloctyldithiophosphate. China Petroleum Processing & Petrochemical Technology 18: 92–98 (2016).

[26] G. M. Sutariya, R. V. Upadhyay and R. V. Mehta. Preparation and properties of stable magnetic fluid using Mn substituted ferrite particles. J. Colloid Inter. Sci. 155: 152–155 (1993).

[27] G. Cao. Nanostructures & nanomaterials: synthesis, properties & applications. Imperial college press (2004).

[28] M. Aliofkhazraei (ed.). Handbook of nanoparticles. Cham: Springer International Publishing (2016).

[29] C. Daraio and S. Jin. Synthesis and patterning methods for nanostructures useful for biological applications. Nanotechnology for Biology and Medicine. Springer, pp. 27–44 (2012).

[30] H. S. Nalwa. Encyclopedia of Nanoscience and Nanotechnology. American Scientific Publishers (2004).

[31] M. Singh, S. Manikandan and A. K. Kumaraguru. Nanoparticles: a new technology with wide applications. Res. J. Nanosci. Nanotech. 1: 1–11 (2011).

[32] B. D. Cullity. Elements of X-ray diffraction. Addison- Wesley (1956).

[33] J. Smit and H. P. J. Wijn. Ferrites. Wiley (1959).

[34] H. G. Phakatkar and R. R. Ghorpade. Tribology. Nirali Prakashan (2012).

[35] G. W. Stachowiak, A. W. Batchelor and T. A. Stolarski. Engineering Tribology Elsevier (1994).

[36] X. Tao, Z. Jiazheng and X. Kang. The ball-bearing effect of diamond nanoparticles as an oil additive. J. Phys. D: Appl. Phys. 29: 2932–2937 (1996).

[37] T. Xu, J. Zhao, K. Xu and Q. Xue. Study on the tribological properties of ultradispersed diamond containing soot as an oil additive. Tribol. Trans. 40: 178–182 (1997).

[38] T. Xu, J. Zhao, K. Xu and Q. Xue. Study on the tribological properties of ultra-dispersed diamond containing soot as an oil additive. Tribol. Trans. 40: 178–182 (1997).

[39] M. Kalin, J. Kogovsek and M. Remskar. Mechanisms and improvements in the friction and wear behavior using MoS$_2$ nanotubes as potential oil additives. Wear 280: 36–45 (2012).

[40] T. Sui, B. Song, F. Zhang and Q. Yang. Effect of particle size and ligand on the tribological properties of amino functionalized hairy silica nanoparticles as an additive to poly-alphaolefin. J. Nanomater. 2: 427–436 (2015).

[41] Kinjal Trivedi. Tribological Properties of Magnetic Fluids. PhD Thesis, Charotar University of Science & Technology (2018).

[42] L. Pena-Paras, J. Taha-Tijerina, A. Garcia, D. Maldonado, J. A. González, D. Molina, E. Palacios and P. Cantu. Antiwear and extreme pressure properties of nanofluids for industrial applications. Tribol. Trans. 57: 1072–1076 (2014).

[43] W. Dai, B. Kheireddin, H. Gao and H. Liang. Roles of nanoparticles in oil lubrication. Tribol. Int. 102: 88–98 (2016).

[44] M. Gulzar, H. H. Masjuki, M. A. Kalam, M. Varman, N. W. M. Zulkifli, R. A. Mufti and R. Zahid. Tribological performance of nanoparticles as lubricating oil additives. J. Nanopart. Res. 18: 223–248 (2016).

[45] H. Spikes. Friction modifier additives. Tribol. Lett. 60: 1–26 (2015).

[46] W. J. Bartz. Lubricants and the environment. Tribol. Int. 31: 35–47 (1998).

[47] E. R. Booser. Tribology data handbook: an excellent friction, lubrication, and wear resource. CRC Press (1997).

[48] E. R. Booser. CRC Handbook of Lubrication and Tribology, Volume III: Monitoring, Materials, Synthetic Lubricants, and Applications. CRC Press (1993).

[49] H. Jianqiang, Z. Huanqin, W. Li, W. Xianyong, J. Feng and Z. Zhiming. Study on tribological properties and action mechanism of organic cadmium compound in lubricants. Wear 259: 519–523 (2005).

[50] V. Srinivas, C. K. R. Rao, M. Abyudaya and E. S. Jyothi. Extreme pressure properties of 600N base oil dispersed with molybdenum disulphide nanoparticles. Univ. J. Mech. Eng. 2: 220–225 (2014).

[51] B. Li, X. Wang, W. Liu and Q. Xue. Tribochemistry and anti-wear mechanism of organic–inorganic nanoparticles as lubricant additives. Tribol. Lett. 22: 79–84 (2006).

[52] R. N. Gupta and A. P. Harsha. Synthesis, characterization, and tribological studies of Calcium–Copper–Titanate nanoparticles as a biolubricant additive. J. Tribol. 139: 021801-1-11 (2017).

[53] R. E. Rosensweig. Ferrohydrodynamics. Dover Publication (1997).

[54] L. R. Rudnick. Lubricant additives: chemistry and applications. CRC press (2017).

7

Novel Magnetic Fluid Polishing with the Control of Force and Distribution of Non-magnetic Abrasives

Noritsugu Umehara

◇◇◇

7.1 Introduction

Magnetic fluid has attractive properties that can be applied to machines and devices. Magnetic fluid seal [1], magneto hydrostatic separation and dampers [3] are successful examples that have been invented in the past. However, magnetic fluid has not been well applied in manufacturing.

Many issues can not also be overcome by using traditional grinding and polishing methods for brittle and hard materials such as ceramics and aspherical lenses with complicated shapes.

The starting point of the idea to overcome various issues in manufacturing was given from the ferrohydrodynamics theory of R.E. Rosensweig [4, 5]. It explains that the buoyant force F acts on the non-magnetic body in the magnetic fluid under the magnetic field as shown in Fig. 7.1. According to this principle, non-magnetic abrasives can be dispersed at a certain position in the fluid under a specially designed magnetic field.

Figure 7.2 shows one example where abrasives are at the bottom of the case in Fig. 7.2(a) and float in the middle in Fig. 7.2(b) [6]. If a workpiece is submerged and rotated in the floating abrasive layer, as shown in Fig. 7.2(b). The surface should be finished with free abrasives.

Dept. of Micro-nano Mechanical Science and Engineering, Graduate School of Engineering, Nagoya University, Furo-cho, Chikusa-ku, Nagoya, 464-8603, Japan.
Email: ume@mech.nagoya-u.ac.jp

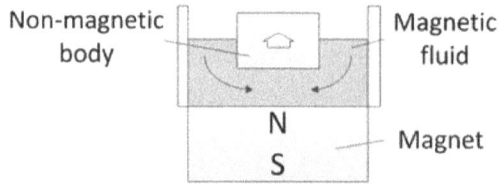

Figure 7.1. Buoyant force acting on a non-magnetic body in magnetic fluid under a magnetic field.

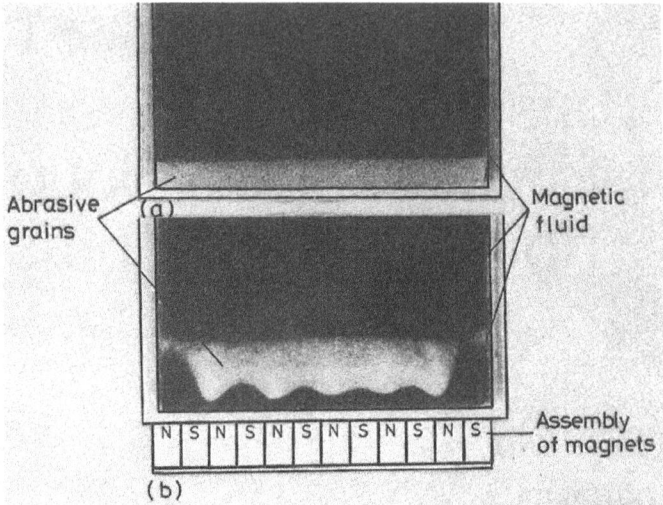

Figure 7.2. Floatation of abrasives in a magnetic fluid under a magnetic field.

Free non-magnetic abrasive in polishing liquid can also be controlled based on research [7, 8].

Non-magnetic particles in the magnetized magnetic fluid are subject to an attracted force for two particles aligned along magnetic field lines and the repulsive force for two particles aligned perpendicular to magnetic field lines, as shown in Fig. 7.3 based on this research, it can be considered that microparticles in the magnetic fluid can be dispersed uniformly by the magnetic field if the direction of applying the magnetic field is perpendicular to the mating surfaces which hold non-magnetic microparticles between them as shown in Fig. 7.4 [9].

In this chapter, the principle and importance of magnetic fluid are introduced not only for achieving excellent polishing, but also the method of design for optimum achievement of the magnetic field.

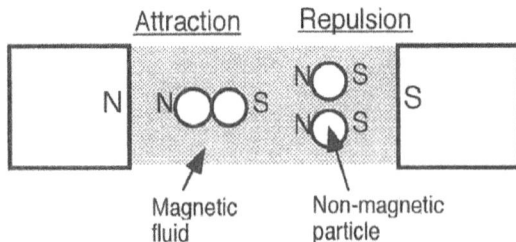

Figure 7.3. Interaction between non-magnetic particles in magnetized magnetic fluid.

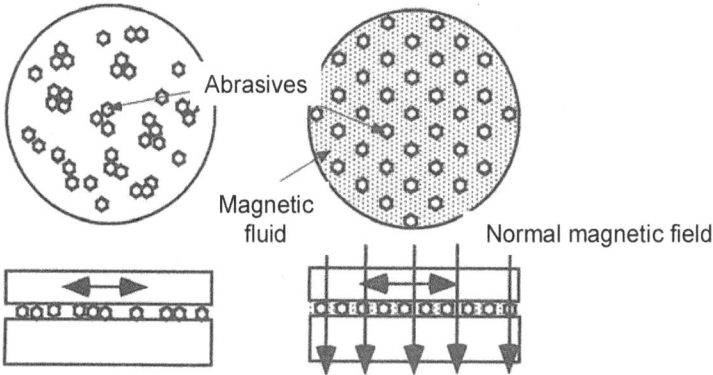

Figure 7.4. Uniform distribution of abrasives in magnetic fluid [9].

7.2 Principal of magnetic fluid grinding

According to this principle, non-magnetic abrasive grains can be dispersed at a certain position in a magnetic fluid under a specially designed magnetic field. Figure 7.5 shows an example of abrasive grains floating at a certain height. If a workpiece is submerged and rotated in the layer of abrasive grains, as shown in Fig. 7.6(a), the surface is ground by free abrasive grains. However, the removal rate is very low and the control of shape is poor since the total buoyant force of abrasive grains is too small to accomplish large removal rates. If a float is introduced to this system, as shown in Fig. 7.6(b), larger grinding pressure can be produced since a large buoyant force near the magnet pole surface is transmitted to the grinding surface of a workpiece.

Figure 7.7 shows an example of grinding pressure P as a function of distance h of an abrasive grain layer or a float from the magnet. Solid lines in this figure show the theoretical values of grinding pressure calculated using the following equation developed by Rosensweig [5],

$$F_b = -\iint_S \left(\mu_0 \frac{Mn^2}{2} + \int_0^H MdH \right) n \bullet dS \tag{7.1}$$

where F_b is a buoyant force of the non-magnetic body, s area of the non-magnetic body, μ_0 permeability of free space, M magnetization of magnetic fluid, Mn normal component of M to the non-magnetic body, H the strength of the magnetic field and n normal unit vector to the non-magnetic body.

The lower theoretical and experimental values show the grinding pressure caused by abrasive grains only, and the higher values show the pressure caused by abrasive grains and a float. The float has a square shape (38 mm × 20 mm) and its thickness is 1 mm. It has a step of 1 mm thickness and a contact area of 30 mm² with the workpiece. The volume concentration of abrasive grains is 30 vol%. A water-based magnetic fluid was used.

From Fig. 7.7, it is clear that the estimated grinding pressure agrees reasonably well with the measured grinding pressure. The grinding pressure with a float at

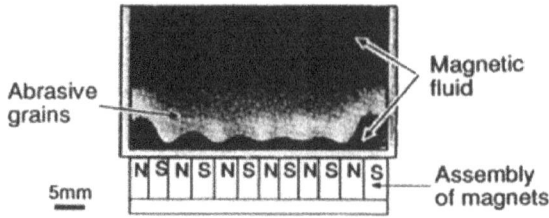

Figure 7.5. Floating of abrasive grains in a magnetic fluid under the action of a magnetic field.

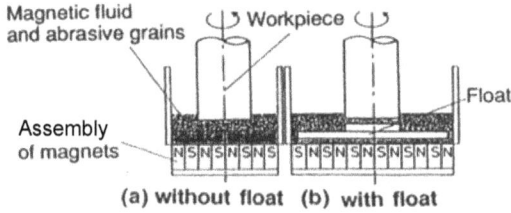

Figure 7.6. The principle of magnetic fluid grinding with and without a float.

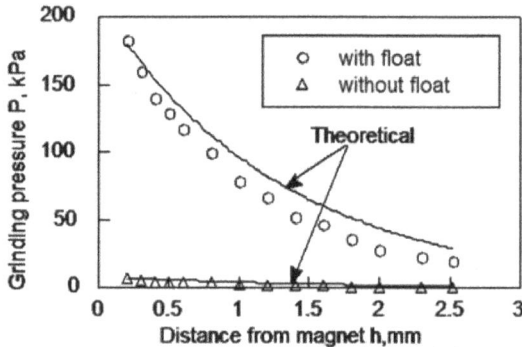

Figure 7.7. Grinding pressure P as a function of distances h of a floating abrasive grain layer and a float from the magnet.

$h = 1$ mm is also 20 times larger than that without a float. It is evident from these results that a float can easily produce large grinding pressure [10].

The contact stiffness, defined as the grinding load divided by the elastic displacement of the contact surface, can be calculated from Fig. 7.7 to be 3.5×10^3 N/m at $h = 0.6$ mm. In contrast, the contact stiffness of a standard grinding wheel of 5 mm width and polyurethane polisher of 1 mm are $5–50 \times 10^6$ N/m² and $9–12 \times 10^3$ N/m, respectively. These results show that the contact stiffness in magnetic fluid grinding is smaller than that of the grinding wheel or the polyurethane polisher. Therefore, it is considered that such low contact stiffness in loading with a float can prevent the workpiece surface from severe damage or generation of cracks in the finishing of ceramics.

The salient features of magnetic fluid grinding are the following:

- Vibration and impact produced between the workpiece and the tool at high grinding speeds can be reduced by the float flexibly supported by the magnetic fluid. Hence the system can operate at high speeds (more than 10,000 rpm) to accomplish high removal rates.

- If the workpiece is placed in the floating abrasive grain layer, abrasive grains in a magnetic fluid can be supplied to the workpiece surface continuously from the outside.

- The wear of abrasive grains is decreased by the cooling effect of the magnetic fluid. As a result, abrasive grains have a longer life.

- All sharp edges of abrasive grains come into contact with the surface of the workpiece in grinding. Thus, one abrasive grain can work more effectively than that fixed in a standard hard grinding wheel.

7.3 Design of the apparatus for optimal grinding load and stiffness

7.3.1 Magnetic field

Childs and Yoon have calculated a float's magnetic field and magnetic buoyant force using a finite difference method [11]. However, in their method, the physical meaning of each effect of magnetic fluid properties, magnet properties and apparatus geometry on the magnetic field and magnetic buoyant force of a float could not be understood easily. So, in this study, a float's magnetic field and magnetic buoyant force are theoretically analyzed.

In order to obtain a large buoyant force of the float as a grinding load, a large magnetic field gradient and large magnetic field are necessary. The most simple magnetic assembly for satisfying such conditions is the assembly of magnets with opposing polarity in adjacent magnets, as shown in Fig. 7.8. Next, the magnetic field above such a magnet assembly is analyzed. Since the ordinary magnetic fluid is a nonconductor, the current density is zero in a magnetic fluid and a potential function ϕ_1 above the magnet exists. Such a potential function above a magnet should satisfy the following Laplace's Equation.

$$\nabla^2 \phi_1 = 0 \qquad (7.2)$$

Other possible functions ϕ_2 should be given in the magnet by the following Laplace's Equation.

$$\nabla^2 \phi_2 = 0 \qquad (7.3)$$

Trial solutions for each equation may be written as follows;

$$\phi_1 = \left(be^{\frac{\pi}{a}z} + ce^{-\frac{\pi}{a}z} \right) \sin\left(\frac{\pi}{a}x \right) \qquad (7.4)$$

$$\phi_2 = \left(de^{\frac{\pi}{a}z} + fe^{-\frac{\pi}{a}z} \right) \sin\left(\frac{\pi}{a}x \right) \qquad (7.5)$$

where a is the width of the magnet.

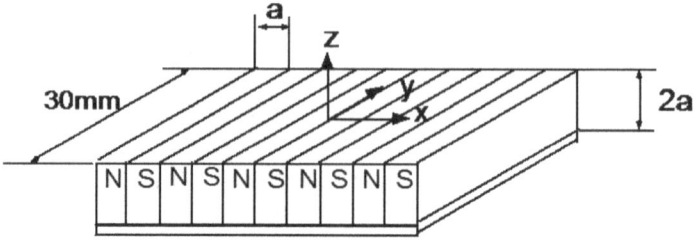

Figure 7.8. The size and shape of the permanent magnet assembly.

The boundary conditions are given as follows,

1. $$z = 0 \quad H_s = \frac{\pi}{2} B_r \sin\left(\frac{\pi}{a} x\right)$$ (7.6.a)

2. $$z = +\infty \rightarrow \phi_1 = 0, \; z = -\infty \rightarrow \phi_2 = 0$$ (7.6.b), (7.6.c)

3. $$\phi_1 = \phi_2 \text{ at } z = 0$$ (7.6.d)

4. $$(B_1 - B_2) \cdot n = 0$$ (7.6.e)

where Br is the residual flux density of the magnet.

Using these boundary conditions (7.6.a)–(7.6.e), both potential functions can be written as follows:

$$\phi_1 = -\frac{a}{4} B_r e^{-\frac{\pi}{a} z} \sin\left(\frac{\pi}{a} x\right)$$ (7.7)

$$\phi_2 = -\frac{a}{4} B_r e^{\frac{\pi}{a} z} \sin\left(\frac{\pi}{a} x\right)$$ (7.8)

The two components of the magnetic field intensity can be written as follows:

$$Hz = \frac{\partial \phi_1}{\partial z} = \frac{\pi}{4} B_r e^{-\frac{\pi}{a} z} \sin\left(\frac{\pi}{a} x\right)$$ (7.9)

$$Hx = \frac{\partial \phi_1}{\partial x} = -\frac{\pi}{4} B_r e^{-\frac{\pi}{a} z} \cos\left(\frac{\pi}{a} x\right)$$ (7.10)

7.3.2 *Magnetic buoyant force of a float (Grinding load)*

The magnetic buoyant force F_b of a non-magnetic body can be calculated using Equation (7.11). Langevin's function can give Magnetization M of magnetic fluid [4]. It means M is nonlinearly related to the magnetic field. However, to calculate F_b analytically, it was assumed that magnetization of the magnetic fluid is constant at the saturated magnetization *Ms* of the magnetic fluid. It is reasonable since the magnetic field close to the magnet is sufficiently strong for the saturation of magnetic fluid.

The following equations can give magnetic buoyant force F_b;

$$F_b = -\int\int_s (\mu_0 MsH)n \bullet ds \qquad (7.11)$$

$$F_b = \mu_0 MsA\left(\frac{\pi}{4}B_r\right)e^{-\frac{\pi}{a}h}\left(1 - e^{-\frac{\pi}{a}t}\right) \qquad (7.12)$$

where Ms is the saturated magnetization of the magnetic fluid, A area of a float, and t thickness of the float

From Equation (7.12), the effect of saturated magnetization Ms of the magnetic fluid, area of the float A, residual flux density B_r of the magnet, the width of magnets 'a' and thickness of a float t on the buoyant force of the float can be found. Equation (7.10) is very convenient for the design of magnetic fluid grinding apparatus. A comparison between calculated values using Equation (7.12) and the measured values is shown in Fig. 7.9. The calculated values agree well with the measured values over 1.5 mm of distance h. Under 1.5 mm of distance h, measured values are larger than calculated values on magnets. Magnetic saturation Ms of magnetic fluid depends on the density of the magnetic particles in a magnetic fluid.

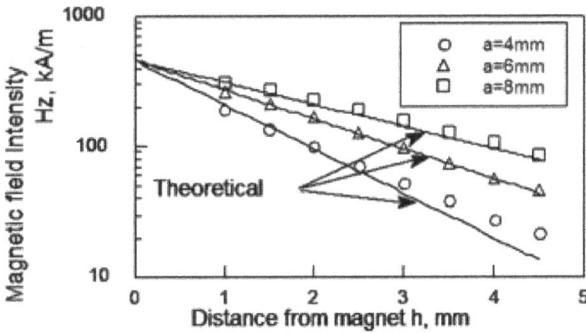

Figure 7.9. Variations of magnetic field intensity Hz with distance from magnet h as a function of magnet width a [12].

7.3.3 Supporting stiffness of the float (Grinding stiffness)

Supporting stiffness of a float can be obtained from Eq. (7.13) as follows [12],

$$\frac{\partial F_b}{\partial h} = \mu_0 MsA\left(\frac{\pi}{4}B_r\right)\left(-\frac{\pi}{a}\right)e^{-\frac{\pi}{a}h}\left(1 - e^{-\frac{\pi}{a}t}\right) \qquad (7.13)$$

From this equation, it can be seen that the supporting stiffness of a float is proportional to the buoyant force. The gradient of the supporting stiffness of the float to the buoyant force also depends on the magnet width. It means a change in magnet width can only change the supporting stiffness of the float under the same grinding load.

7.4 Application of magnetic fluid for manufacturing

7.4.1 Magnetic fluid grinding for Silicon nitride balls

Figure 7.10 shows a schematic diagram of the grinding apparatus for finishing ceramic balls used for ceramic ball-bearing application [6]. Magnetic fluid, float, abrasive grains, as-sintered ceramic balls and permanent magnets are arranged as shown in Fig. 7.10. A float, abrasive grains and ceramic balls are all non-magnetic materials floated in the magnetic fluid by magnetic buoyant force. Balls are rotated along the guide ring's inner wall by the driving shaft. They are ground by the abrasive grains in the magnetic fluid.

The removal rate of the silicon nitride balls (made by pressureless sintering) increases with the grinding load and the rotational speed of the driving shaft. The maximum removal rate of silicon nitride balls was 12.4 µm/min with SiC abrasive grains. This removal rate is about 40 times larger than the traditional V-groove lapping method. Surface roughness became less with the decrease in the mean grain size of the abrasive grains. The minimum surface roughness was 0.1 µm Rmax.

Figure 7.11 shows the sphericity and the grinding time for cases with and without a float. In Fig. 7.11, sphericity is given by the difference between the ball's maximum and minimum diameters, and L is the mean grinding load. In the case without a float, the sphericity of the balls increases gradually with grinding time. In contrast,

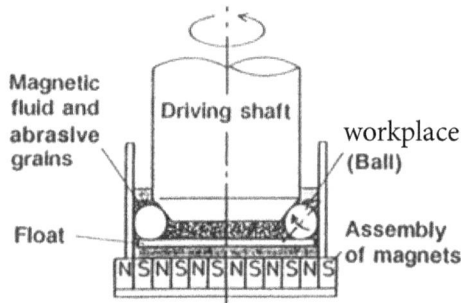

Figure 7.10. Schematic diagram of the magnetic fluid grinding apparatus for finishing advanced ceramic balls [6].

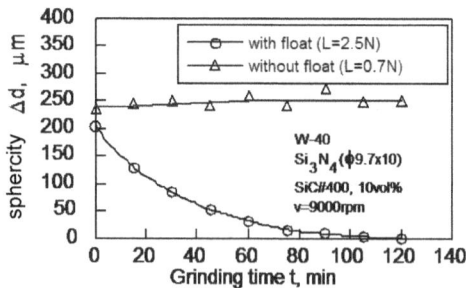

Figure 7.11. The relationship between grinding time and the sphericity of silicon nitride balls in magnetic fluid grinding with or without a float [6].

Table 7.1. Surface roughness and sphericity of an as-sintered ball and a ground ball of silicon nitride under optimum grinding conditions.

	As-sintered	Ground
Optical view of Si₃N₄ balls		
Ball diameter	7.7mm	7.1mm
Roughness	10μm Rmax	0.1μm Rmax
Sphericity	500μm	0.14μm
Grinding time : 180min		

the sphericity of a ball with a float decreases rapidly with grinding time and reaches 2 μm after 120 min of grinding. It is clear from this that a float is indispensable for finishing balls by this method.

Table 7.1 shows grinding properties under optimum grinding conditions. The sphericity of the as-sintered balls was reduced from 500 μm to 0.14 μm and its original rough surface looks shiny after 180 min of grinding.

7.4.2 Micro local area polishing using the magnetic fluid distribution of free abrasives

In order to polish the aspherical glass lens in a proper shape and smooth surface, a local area of 1 mm is needed to be polished uniformly. For polishing such a tiny area, the moving stroke of the tip tool is also tiny less than 1 mm. Hence the distribution of free abrasives on the tip tool should be uniform.

Based on research [7, 8], non-magnetic particles in the magnetized magnetic fluid are subject to attract forces for two particles aligned along magnetic field lines and the repulsive force for two particles aligned perpendicular to magnetic field lines, as shown in Fig. 7.3 based on this research, it can be considered that microparticles in the magnetic fluid can be dispersed uniformly by the magnetic field if the direction of applying the magnetic field is perpendicular to mating surfaces which are holding non-magnetic microparticles between them as shown in Fig. 7.4. Therefore we have tried to fabricate a local area polishing tool with permanent magnet and magnetic field as shown in Fig. 7.12 [10].

The magnetic metal ball is used as a polisher with a permanent magnet. Tiny drops of magnetic fluid, including non-magnetic abrasives, were supplied between the magnetic ball and a glass disk workpiece. When the normal magnetic field is applied to the magnetic fluid, as shown in Fig.7.12, non-magnetic abrasives should be dispersed uniformly, as shown in Fig. 7.4. When the magnetic ball is moved in a circle whose diameter is 100 μm by a Piezo-electric actuator, abrasives can do local area polishing in a local minute area.

Figure 7.13 shows cross-sectional profiles of the polishing area without and with a normal magnetic field. If there is no normal magnetic field, the depth of the polishing scar does not increase with polishing time. On the other hand, the normal magnetic field can increase the polishing depth which is proportional to the polishing time.

Figure 7.14 also shows the optical image and cross-sectional profiles without and with a normal magnetic field. It can be seen from Fig. 7.14(b) that normal magnetic fields improve the dispersal of abrasives and polish uniformly. On the other hand, agglomerated abrasives make many deep scratches if there is no normal magnetic field, as shown in Fig. 7.14(a).

From Figs. 7.13 and 7.14, it appears that the magnetic field can control the distribution of non-magnetic abrasives and polish the local minute area uniformly in free abrasive polishing.

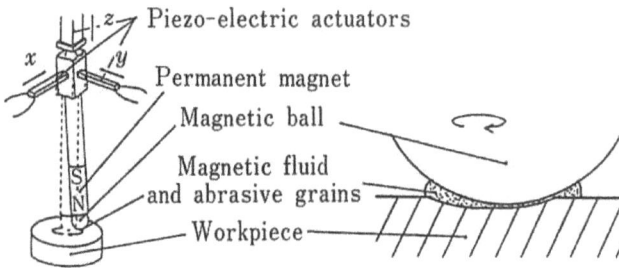

Figure 7.12. Schematic of local minute area polishing method with magnetic fluid.

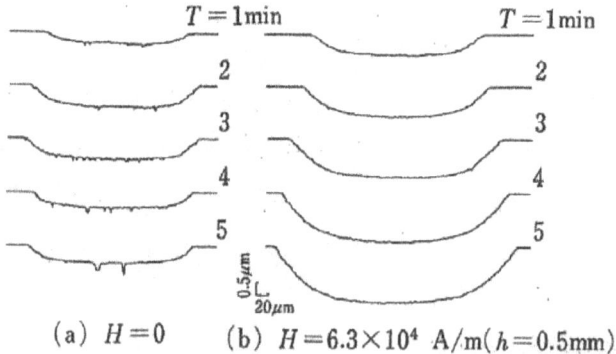

(a) $H=0$ (b) $H=6.3\times10^4$ A/m($h=0.5$mm)

Figure 7.13. Cross-sectional profiles of polishing a scar without and with normal magnetic field.

7.4.3 *Magnetic Intelligent Compound (MAGIC) grinding wheel*

Using magnetic fluid in grinding, we can control the distribution and apply the force of abrasives in a magnetic liquid by a magnetic field. We can make a novel grinding tool if the magnetic fluid can be solidified. To check if this concept is adequate, the

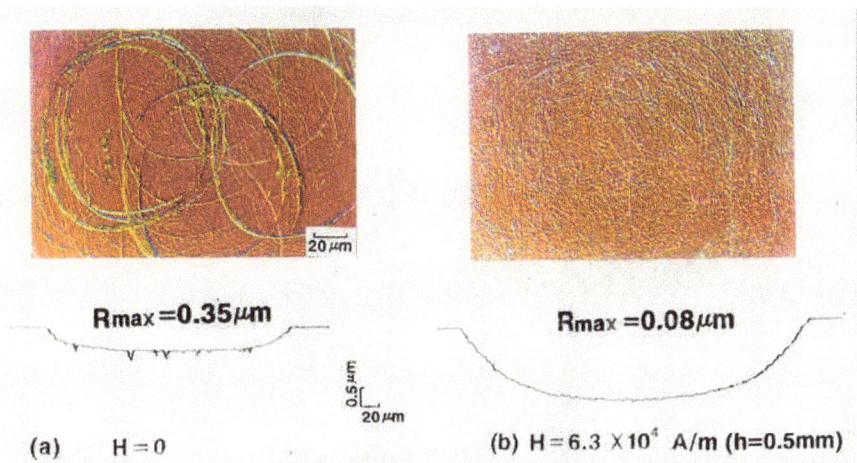

(a)   H = 0   (b) H = 6.3 × 10⁴ A/m (h=0.5mm)

Figure 7.14. Optical microscope images of a polishing scar on the glass workpiece without and with a normal magnetic field.

"Magnetic Fluid Frozen Grinding Wheel" was investigated, made, and used in a cold room after applying a magnetic field during freezing. As a result, it was found that the frozen tool provided enough removal rate and surface roughness for polishing the inner surface of the mold [13, 14]. However, the tool has a disadvantage in that this could be used in a cold room. To spread this new polishing tool, the tool should be made and used at room temperature. Therefore we made a new magnetic fluid whose base liquid can be solidified at room temperature. We named this liquid "Magnetic Intelligent Compound". This liquid is a mixture of the base polymer and magnetic particles. Many abrasives such as Alumina or Diamond are added to make a polishing tool with this liquid. After heating the mixture, including abrasives for melting the base polymer, the mixture was solidified by cooling then adequately applied to a magnetic field. After cooling well, a new polishing tool whose abrasives are adeptly arranged for excellent polishing properties could be obtained [15, 16].

Figure 7.15 shows the process of making a polishing tool of magnetic, intelligent composite liquid with abrasives. In Fig. 7.15(a), the magnetic, intelligent composite liquid was melted at a high temperature and mixed with abrasives during melting. In Fig. 7.15(b), the melted magnetic, intelligent composite liquid was poured into the mold by applying a magnetic field. The magnetic field was applied successively during solidification with cooling, as shown in Fig. 7.15(c). After solidification of the magnetic, intelligent composite liquid, the inner surface of the mold is polished by the relative oscillation motion, as shown in Fig. 7.15(d).

The advantages of this new method can be considered as the following:

- Mold surface can be finished quickly because abrasives were well-distributed without forming agglomerated clusters.
- Bonding materials are so soft that deep scratch marks could not be generated with the finishing process.

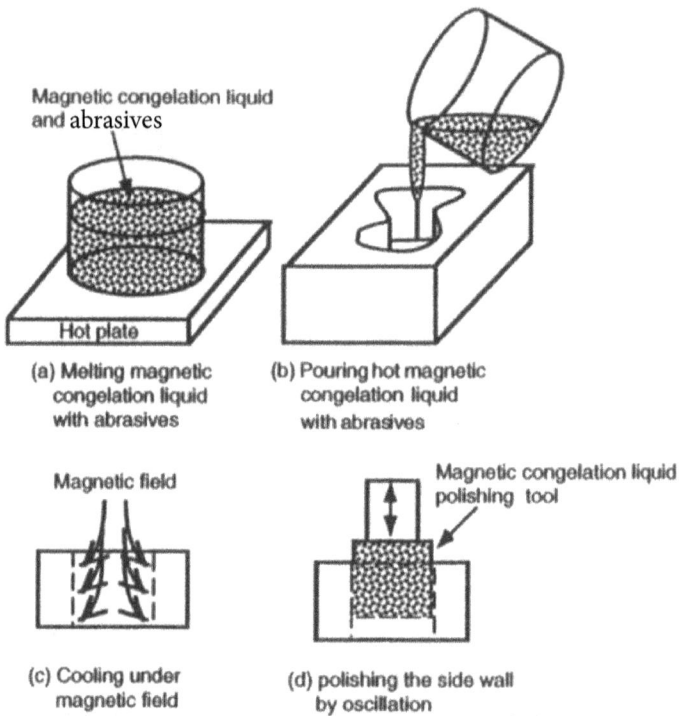

Figure 7.15. Preparation process of the new polishing tool with magnetic intelligent compound.

- Dressing this tool can be done easily, because bonding materials can be solved easily by the organic liquid.
- This pellet can be recycled after wearing out of the pellet. The magnetic intelligent composite liquids and abrasives can be used many times.
- By pouring magnetic, intelligent composite liquid, a complicated-shaped surface could be finished with oscillation.
- Sputtering bond materials can be collected easily because a magnetic field can collect them.

In order to confirm the well-arranged abrasives in the magnetic, intelligent composite grinding wheel, abrasives in magnetic, intelligent composite liquid were observed with an optical microscope. Magnetic, intelligent composite liquid with an abrasive is poured into the gap between two glass plates. The magnetic field was applied perpendicular and parallel to the glass with a permanent magnet during the cooling process.

Figure 7.16 shows the observation results with an optical microscope of the distribution of abrasives applied magnetic field in the parallel direction to the glass plates. It can be seen that magnetic particles in a magnetic, intelligent compound made some clusters, and abrasives also made a cluster in the direction of the magnetic field in the remaining area. It can also be seen that large magnetic field strength provided a slim cluster.

Fundamental polishing properties of the magnetic, intelligent compound grinding wheel were investigated. The grinding wheel was pressed and sledded against the workpiece.

Figure 7.17 shows the variation of surface roughness with polishing time under the different magnetic fields in the making process. It can be seen that large magnetic field strength provided small surface roughness. Based on the observation results of the distribution of abrasives, it can be seen that uniformly distributed abrasives provided small surface roughness because it prevented the agglomeration of abrasives. An oscillating motion was applied to the polishing tool in the present polishing apparatus. Therefore, if abrasives are distributed locally, not uniformly,

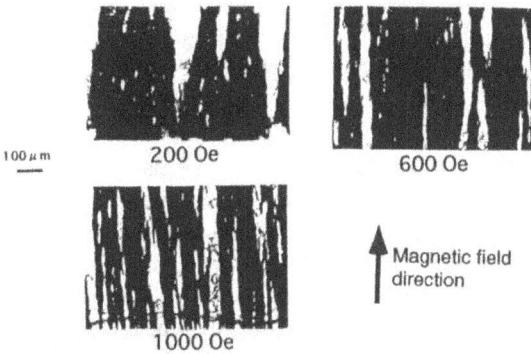

Figure 7.16. The surface roughness variation with polishing time under the different magnetic fields in the making process.

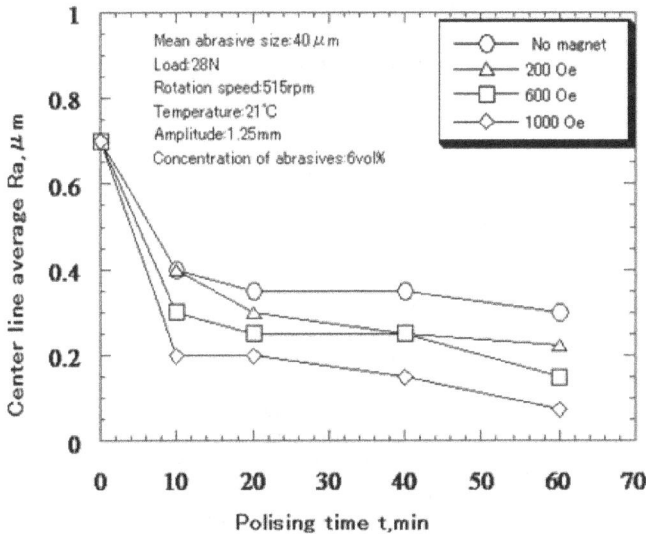

Figure 7.17. Observation results with an optical microscope of the distribution of abrasives applied magnetic field in the parallel direction to the glass plates.

the whole of the surface can not be polished uniformly. As a result, deep scratches will be generated without a magnetic field. These effects of the magnetic field on polishing properties were also confirmed for polishing the WC mold surface.

7.5 Summary

Magnetic fluid (Ferrofluid) can apply magnetic force to non-magnetic floats and non-magnetic abrasives. The magnetic force can be a nonlinear polishing force for grinding and polishing. The nonlinear polishing force has many advantages in silicon nitride ball polishing because the nonlinear polishing force could avoid a fracture of brittle workpiece materials such as Silicon Nitride ceramics.

Magnetic fluid could also cause diverse non-magnetic abrasives uniformly. The identical distributed abrasives showed uniform polishing which can be especially useful for local minute area polishing in a glass lens.

In order to polish the inner surface of a mold with a complicated shape, a magnetic intelligent compound (MAGIC) grinding wheel was developed. This liquid is a mixture of the base polymer and magnetic particles. Many abrasives such as Alumina or Diamond are added for polishing tools with this liquid. After heating the mixture, including abrasives for melting the base polymer, the mixture was solidified by cooling with applying an adequate magnetic field according to the shape of workpiece. After cooling well, a new polishing tool whose abrasives are well arranged for excellent polishing properties was obtained. This method can use this tool again and again without generating any waste. This procedure also recognized the automation polishing process of the surface of the mold.

References

[1]   R. Moskowitz. Dynamic Sealing with Magnetic Fluids. A S L E Transactions 18(2): 135–143 (1975).
[2]   R. E. Rosensweig. Material separation using ferromagnetic liquid techniques. Google Patents (1969).
[3]   R. E. Coulombre, H. d'Auriol, L. Schnee, R. E. Rosensweig and R. Kaise. Goddard Space Flight Center, Greebelt, Maryland, in Rep. No. NASA-9432 (1967).
[4]   R. Rosensweig. Ferrohydrodynamics Cambridge University Press Cambridge. New York, Melbourne (1985).
[5]   R. E. Rosensweig. Fluidmagnetic buoyancy. AIAA Journal 4(10): 1751–1758 (1966).
[6]   N. Umehara and K. Kato. Principles of magnetic fluid grinding of ceramic balls Applied Electromagnetics in Materials 1: 37–43 (1990).
[7]   A. Skjeltorp. Monodisperse particles and ferrofluids: a fruit-fly model system. Journal of Magnetism and Magnetic Materials 65(2-3): 195–203 (1987).
[8]   T. Fujita and M. Mamiya. Interaction forces between nonmagnetic particles in the magnetized magnetic fluid. Journal of Magnetism and Magnetic Materials 65(2-3): 207–210 (1987).
[9]   N. Umehara et al. Magnetic dispersion of microparticles using magnetic fluid—application to texturing process for magnetic rigid disk. CIRP Annals 46(1): 155–158 (1997).
[10] N. Umehara and S. Kalpakjian. Magnetic fluid grinding–a new technique for finishing advanced ceramics. CIRP Annals 43(1): 185–188 (1994).
[11] T. Childs and H. Yoon. Magnetic fluid grinding cell design. CIRP Annals 41(1): 343–346 (1992).

[12]   N. Umehara et al. Micro surface polishing using magnetic fluid in local area. Journal of the Japan Society for Precision Engineering 60(11): 1606–1610 (1994).

[13]   N. Umehara and M. Kawauchi. Fundamental polishing properties of frozen magnetic fluid grinding. Journal of Magnetism and Magnetic Materials 201(1): 364–367 (1999).

[14]   K. Kato, N. Umehara and M. Suzuki. A study of hardness of the frozen magnetic fluid grinding wheel. Journal of Magnetism and Magnetic Materials 201(1): 376–379 (1999).

[15]   N. Umehara. MAGIC polishing. Journal of Magnetism and Magnetic Materials 252: 341–343 (2002).

[16]   S. Hagiwara et al. Proposal for die polishing using a new bonding abrasive type grinding stone: development of MAGIC grinding stone. Machining Science and Technology 7(2): 267–279 (2003).

8

Ferrofluid Composites
Synthesis and Characterization

Komal Jain,[1,] Prashant Kumar,[1,2,4] Arjun Singh,[1,2,3]*
Ajay Shankar[5] and R.P. Pant[1,2,]*

8.1 Introduction

Composites generally are a mixture of two or more substances whose segregation into its components is impossible without destroying any components. Some well-known composites are vulcanized rubber, cement, paint, ferrofibers, glass wool, etc. These magnetic composites are of significant interest due to their vast applications, from oil spillage cleaning to EMI shields, coolants to sealants, MRI contrast to hyperthermia treatment and many more [1–4]. These extensive applications make magnetic composites a vital material for commercial purposes. In all these composites, one thing common about magnetic components is that they are generally made of micron or nano-sized magnetic nanoparticles combined with polymers/fluid/CNTs or a combination of all [5, 6]. Some of the most frequent polymers or organic compounds used in these composites are polyvinyl alcohol (PVA), polyaniline, chitosan, CNTs, 2-acetoacetoxyethylmethacrylate (AEMA), etc. Magnetic components are mainly confined to ferrite particles (Fe_3O_4), mixed ferrites ($A_xB_{1-x}Fe_2O_4$, $A_xFe_{3-x}O_4$), iron particles, etc. Generally, these polymer macromolecules get adsorbed on magnetic particles or combined using surfactants to form magnetic composites or magnetopolymers.

[1] Indian Reference Materials Division, CSIR-National Physical Laboratory, New Delhi-110012, India.
[2] Academy of Scientific and Innovative Research (AcSIR), Ghaziabad-201002, India.
[3] Department of Physics, Indian Institute of Technology, Jammu-181221, India.
[4] School of Science, RMIT University, Melbourne, VIC 3000, Australia.
[5] Department of Chemistry, Indira Gandhi National Tribal University, Madhya Pradesh 484887, India.
* Corresponding authors: komaljain90@gmail.com, rppant@nplindia.org

8.2 Synthesis of magnetic fluid composites

Magnetic composites can be synthesized by mixing magnetic particles in polymeric or hydrogel solutions. But the question with simple mixing comes from the fact that magnetic particles tend to agglomerate due to magnetic interactions. So mixing the homogenous distribution of magnetic particles becomes difficult even after long hours because of the varying sizes of these magnetic agglomerates. It is easily tackled by using magnetic fluids with desired magnetic nanoparticles in the same base fluids instead of as such particles. Magnetic fluids are homogenous dispersion of nanomagnetic particles with the most negligible aggregation and this help in forming uniform magnetic composites in different matrixes with comparatively much less mixing time. The synthesis or manufacturing process of these composites can be divided into two categories: (a) mixing components magnetic and polymer components and (b) processing magnetopolymer to get fibers, sheets, etc.

The first category can further be divided into two stages which can be worked on separately. In the first stage, one must process the polymer or Base Matrix Material (BMM). For example, mixing of polymer, such as PVA in water, functionalization of polymer/BMM. In the second stage, synthesize the magnetic fluid with desired stoichiometric magnetic nanoparticles in a compatible carrier medium. To shed light on this point, one can take the functionalization of MWCNTs as an example. CNTs can be functionalized using various methods depending on the desired functional group needed to be attached to the CNT walls or inside. For example, CNTs developed using the CVD process. These CNTs will contain impurities inherent to the synthesis process and need to be purified. It can be done by treating CNTs with acid for a couple of hours to obtain purified CNTs ready to be used for functionalization [7, 8]. In this example, CNTs are functionalized with the –COOH group using nitric acid (HNO_3) treatment. The detailed schematic is shown in Fig. 8.1.

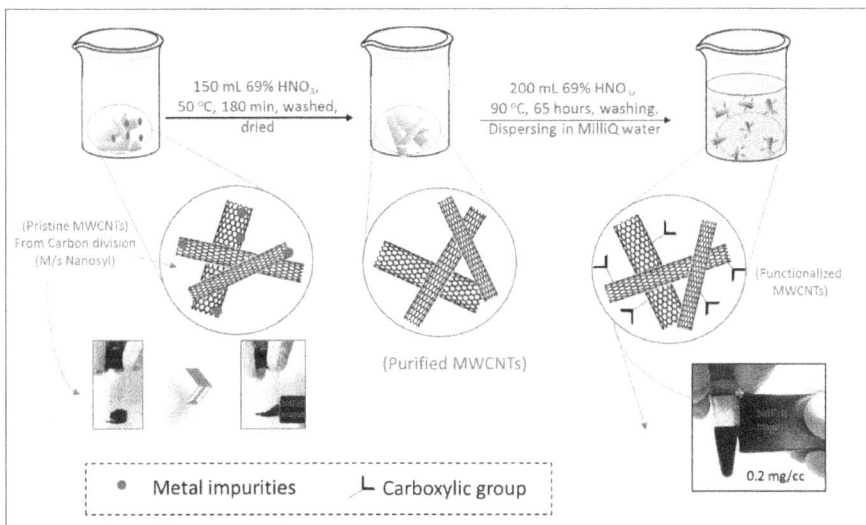

Figure 8.1. A schematic diagram showing details for the preparation of MWCNT dispersion.

Simultaneously, magnetic water fluid with Fe_3O_4 nanoparticles can be prepared using co-precipitation, hydrothermal, etc. Magnetic fluid can be made using the co-precipitation method in the current example. For this purpose, ferrous and ferric salts are taken in a 2:1 molar ratio and oleic acid as a surfactant. After that, magnetic nanoparticles are co-precipitated using an ammonia solution. The residue is washed, dried and dispersed in a sodium oleate solution. Here oleic acid acts as a primary surfactant and sodium oleate as a secondary surfactant. With both components being prepared, a simple mixing of functionalized MWCNTs and magnetic fluid gives the desired MWNCT-Fe_3O_4 composite in water. Similarly, simple mixing of PVA solution and magnetic fluid gives ferrogel/magnetopolymer.

Under the second category, ferrogel or MWCNT-Fe3O4 composite produced in category (a) is further processed to give magnetic films, sheets or fibers. For example, the ferrogel prepared above can produce ferrofibers using the electrospinning technique. In an electrospinning setup, polymer/hydrogel is pumped through a stainless steel needle at a fixed rate using a syringe pump (Fig. 8.2). After that, a potential difference is maintained between the needle and collector drum rotating at a fixed rpm, about 2000 rpm. Now depending on the collection procedure, threads of ferrofibers or woven sheet of fibers are obtained.

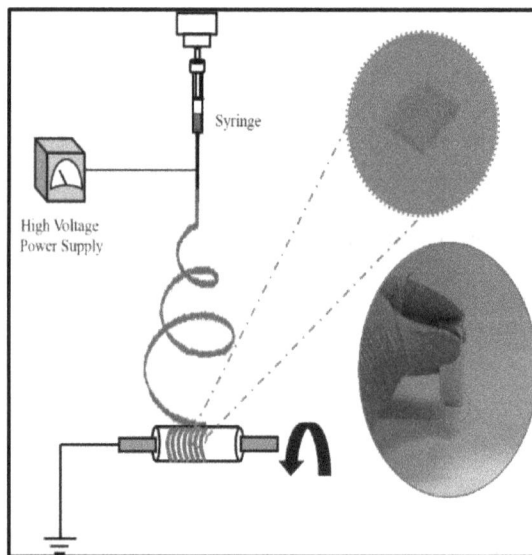

Figure 8.2. Schematic of electro spin set up for making ferrofiber.

8.3 Physical properties of magnetic composites

The developed magnetic composites need to be tested for their characteristics. It requires a detailed study of the structural, polymer matrix, morphology, magnetization, viscoelasticity, chemical bonding (confirming if the desired functional group has been attached) and other properties relevant to the final product specification and development process.

8.3.1 *Structural and morphological testing*

Before getting the final product, one must confirm that the intermediary components, magnetic fluid, functionalized CNTs or polymer, are as per requirement. The easiest way of doing this is by using the X-ray diffraction technique. It is well known that every material in the world can be divided into two categories based on the crystallite structure: (i) crystalline and (ii) amorphous. There exist 14 basic crystal structures and all materials are made of them or their combination. All these materials structures have to follow the well-known Bragg's law of X-ray diffraction given by the Equation (8.1):

$$2d sin\theta = n\lambda \tag{8.1}$$

where d is interplanar spacing, θ is diffraction angle and λ is X-ray wavelength. The X-ray diffraction pattern is the characteristic fingerprint of the crystal structure. Most magnetic nanoparticles comprising magnetic fluid are ferrites with spinel structures corresponding to the Fd-3m space group. The phase of material or any impurities can be established by matching the X-ray diffraction pattern of developed magnetic nanoparticles with the corresponding structure diffraction pattern. Further, the crystallite size of the material can also be determined using the Debye–Scherrer Equation given as [9]:

$$d = \frac{k\lambda}{\beta cos\theta} \tag{8.2}$$

where k: constant is dependent on the shape of the particle (k = 0.9 for spherical), λ: X-ray wavelength, β: Full With Half Maximum (FWHM) of the peak, θ: diffraction angle. The case of particles are straight forward and the crystallite size can be calculated easily by simply substituting values. But in the case of CNTs, few things may change. For example, 'd' is replaced with L_c, called average coherence length along the c-axis (c-axis means along the length of CNTs). In the current example, Fig. 8.3 shows the X-ray diffraction pattern of CNTs at various stages. Besides calculating coherence length by comparing the diffraction patterns at various stages of the functionalization process, i.e., pristine or as such CNTs, purified CNTs (PCNTs) and functionalized CNTs (FCNTs), it is observed that as the acid treatment proceeds some damage had occurred to the outer graphite layers progressively. This is deduced based on the broadening of FWHM and vanishing of a small peak at 43.6°. The extent of damage can be determined by calculating a mean number of graphite walls (N) using the Equations (8.10, 8.11):

$$N = \frac{L_c}{d_{(002)}} \tag{8.3}$$

where d (002) is interplanar spacing corresponding to (002) plane represented by a peak of 25.3°. The values of L_c and N are tabulated in Table 8.1. The progressive decrease in N values indicates damage and functionalization at the outer graphite layers. But the drop in the mean number of the graphite layer is small at each acid treatment stage, indicating the functionalization process is acceptable for product development.

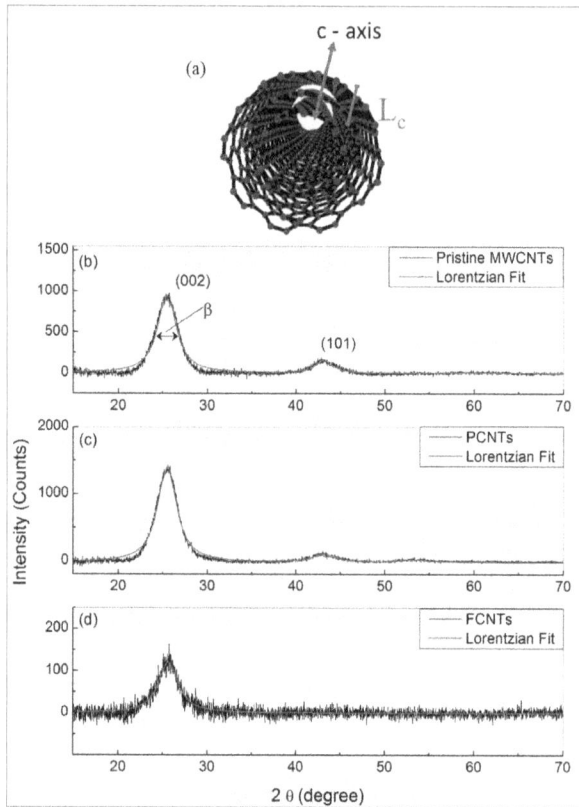

Figure 8.3. (a) A MWCNT shows its c-axis and average coherence length. XRD pattern of (b) pristine, (c) purified and (d) functionalized MWCNTs with Lorentzian fits.

Table 8.1. The effect of acid treatment on crystallinity at various stages of acid treatment.

Sample	Average coherence length, L_c (nm)	Mean number of graphitic walls, N
Pristine MWCNT	3.40 ± 0.05	9.7
Purified MWCNT	3.29 ± 0.02	9.4
Functionalized MWCNT	3.14 ± 0.06	9.0

A great deal of information can be derived using the X-ray diffraction technique when no actual sample image can be seen. The same and other information is expected when we visualize the sample at the nanoscale. For this purpose, electron microscopy can be used. Figure 8.4 shows Transmission Electron Microscopy (TEM) images of MWCNTs-Fe_3O_4 hybrid nanosuspension. From the images, it is clear that CNTs have a diameter in the range of $14 - 18$ nm and a length of $1.7 - 2.5$ μm. After subtracting the coherence length calculated earlier from the diameter, one gets the inner diameter of CNTs in the range of $11 - 15$ nm. The TEM images show that magnetic nanoparticles are random in shape with an average particle size of 6.3 nm.

Moreover, many nanoparticles are present in free states and a few are adsorbed on the surfaces of CNTs with no particle inside the CNTs. It leads to discarding

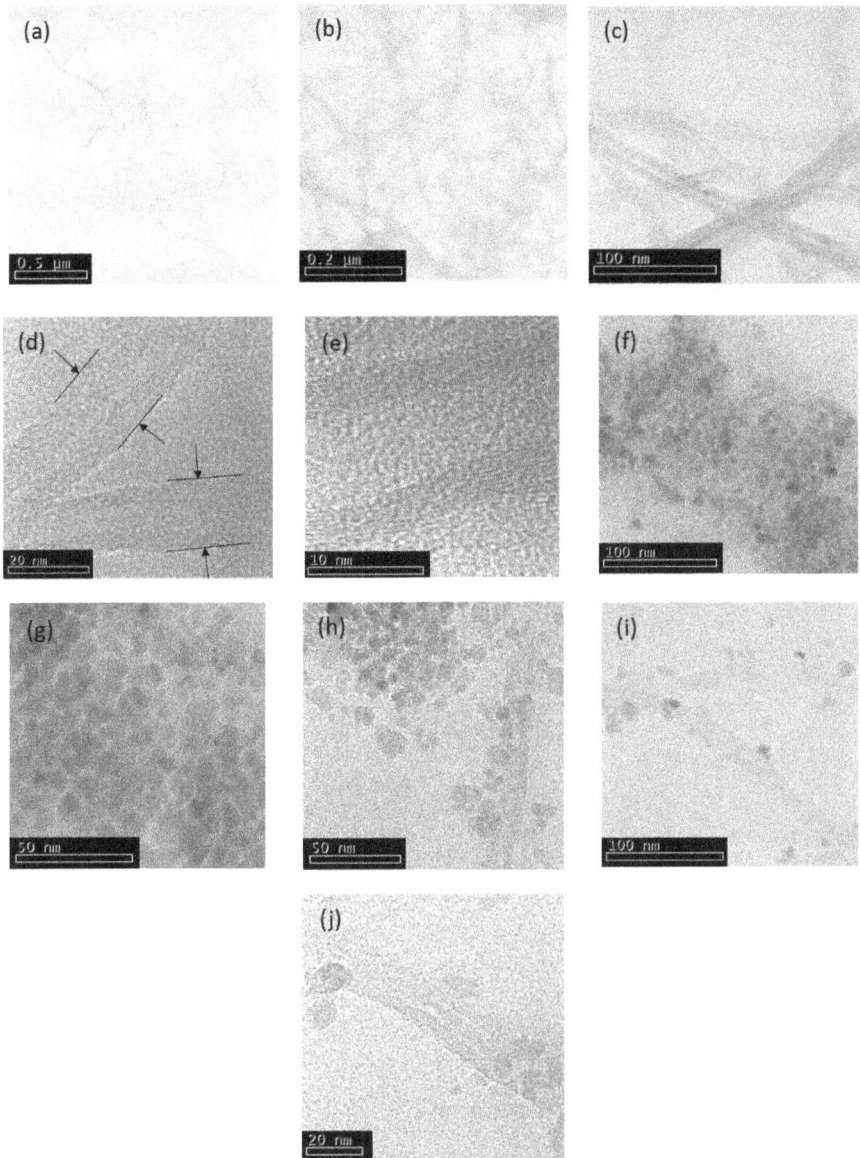

Figure 8.4. (a), (b), (c), (d), (e) TEM images of FCNTs at different magnifications, (f), (g) Fe$_3$O$_4$ nanoparticles, (h), (i) and (j) Fe$_3$O$_4$/MWCNTs nanocomposites obtained after drying hybrid fluid [6].

any capillary forces as particle size and CNTs diameters match, but none of the particles do [12, 13]. From this, it can be inferred that nanoparticles are adsorbed on the CNT's surface via ion-dipole interaction. Figure 8.5 shows Scanning Electron Microscopy (SEM) images of FCNTs and FCNTs/Fe$_3$O$_4$ particles dried in a magnetic field (H = 3 kG) mixed in a 1:2 ratio. Figure 8.5 (c) depicts CNTs bundled together with some surface exfoliation. The bundling is attributed to hydrogen bonding between carboxylic groups of different nanotubes. Exfoliation is expected due to

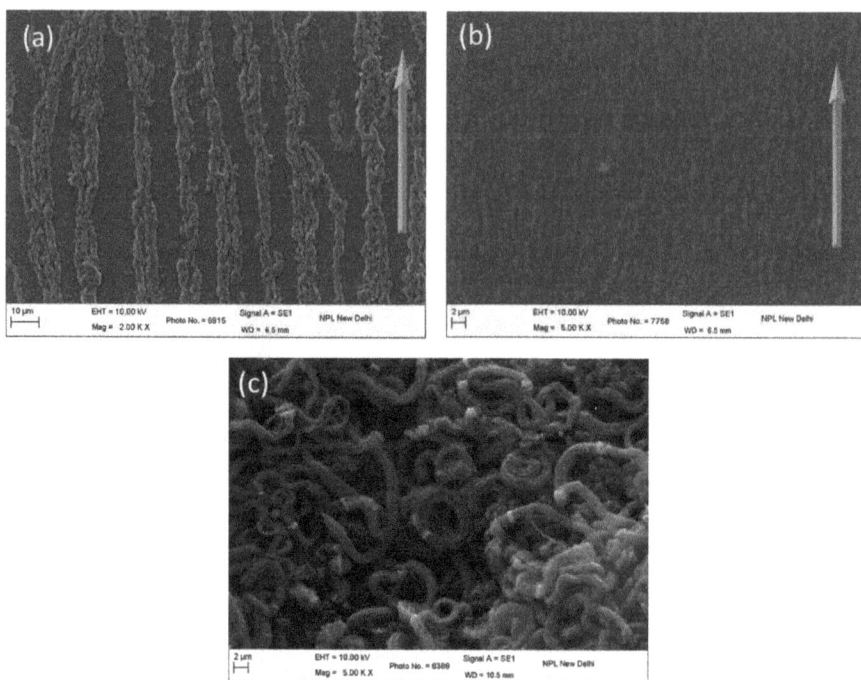

Figure 8.5. SEM images of Fe₃O₄/FCNTs nanocomposites obtained after drying hybrid fluid having different v/v ratios. (a) 1:30 WFF/WCNT; H = 3 kG, (b) 1:10 WFF/WCNT; H = 3 kG. The arrows correspond to the direction of the applied magnetic field (H = 3 kG), (c) bundles of CNTs after the purification process [6].

acid treatment, as found from the decrease in average wall number calculated from XRD. The SEM image of FCNT/Fe₃O₄ reveals the clear response of the FCNT/Fe₃O₄ system to the magnetic field via fibrous structures. The thickness of these fibrous structures varies greatly depending on FCNT:Fe₃O₄ ratio [6].

Figure 8.6 (a) shows an optical microscopy image of ferrofibers after 5 min of electrospinning. The figure shows fibers forming in the preferred direction without any globule formation. To ensure no globules, it has to be made sure that no extra material comes out of the syringe, which depends on the viscosity of the ferrogel formed by mixing magnetic fluid and polymer solution. SEM images (Fig. 8.6 (b)) of ferrofibers confirm the alignment of ferrofibers after electrospinning for 10 hr. The point to note here is that this alignment had been attained without assistance from external magnetic fields in contrast to previously reported ones. It could be achieved only after understanding the material's nature and synthesis process correctly. The sodium oleate increases the adsorption of PVA on the surface of Fe3O4 nanoparticles and also induces a negative charge on the surface. It leads to an increased charge in ferrogel, thus higher electrostatic forces between the collector and ferrogel threads coming out of charged syringe needle [5]. Further, the high electrostatic forces keep individual threads repealed while depositing on the collector aligning threads in parallel. At the same time, hydrogen bonding between carboxylic group (−COOH) of the surfactant, hydroxyl group (−OH) of PVA and water form a homogenized structure of ferrogel which after being electrospun gives oriented/aligned ferrofibers

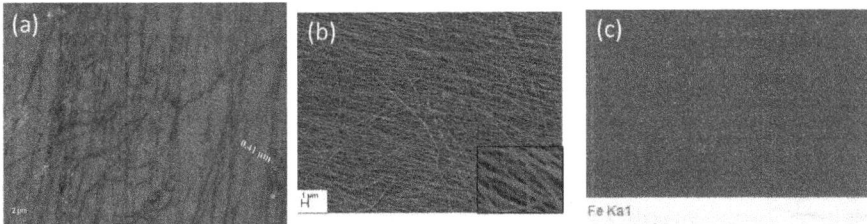

Figure 8.6. (a) Optical micrographs of ferrofibers after 5 min, (b) SEM micrograph after 10 hr of electrospinning, (c) Fe distribution in ferrofibers [5].

sheet with aggregated free uniformly distributed Fe_3O_4 nanoparticles. This is confirmed in Fig. 8.6 (c) that show uniform Fe distribution in the spatial elemental mapping of fibers. The diameter of ferrofibers, magnetization, viscosity and other properties greatly depend on the nanoparticle concentrations.

8.3.2 Magnetic studies of composites

To understand the magnetic properties of ferrofluid-based composites, the ferrofibers developed earlier are considered here as an example. These fibers show high magnetization and anisotropy. Figure 8.7 (a) shows the M-H loop of ferrofibers was made using 5, 10 and 15% w/w ferrofluid in PVA solution using the electrospinning technique described earlier. All samples show a superparamagnetic nature with saturation magnetization coercivity and retentivity in Table 8.2.

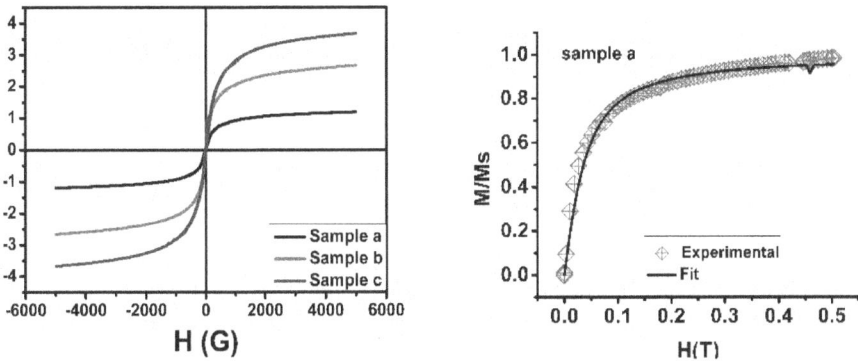

Figure 8.7. (a) M-H loop shows superparamagnetic nature of ferrofibers, (b) Langevin fit of ferrofibers sample a [5].

Table 8.2. Langevin fitting parameters [5].

| Sample | Experimental | | | Fitted Parameters | | | | | | |
	M_s (emu/g)	H_c (G)	M_R (G)	M_s^f (emu/g)	M_d (G)	D (nm)	σ_D	$\emptyset$	n'	r (nm)
a	1.18	6.14	1.8	1.21	384	10.9	0.32	0.018	2.59×10^{22}	33.8
b	2.68	7.4	7.2	2.72	384	10.9	0.32	0.042	6.04×10^{22}	25.5
c	3.66	8.98	8.4	3.72	384	10.9	0.32	0.058	8.35×10^{22}	22.8

Due to the superparamagnetic nature of the sample, the virgin curve in the 1st quadrant of the M-H loop can be fitted using the Langevin function, which describes the paramagnetic systems (Fig. 8.7(b)). In the case of superparamagnetic nanoparticles having a distribution, the Langevin function is given as [14]:

$$M = \int_0^{-\infty} L(\alpha)f(D)dD - \chi_i H \tag{8.4}$$

where L (α): Langevin function

$$L(\alpha) = M_s^f(\coth(\alpha) - \frac{1}{\alpha}) \text{ with } \alpha = \frac{M_d H((\frac{1}{6})\pi D^3}{kT}; \tag{8.5}$$

$$\text{and } M_S^f = \emptyset M_d \tag{8.6}$$

and f (D): log-normal size distribution of magnetic nanoparticles.

$$f(D) = \frac{1}{\sqrt{2\pi}\,\sigma_D\,D} e^{\{-\frac{\ln(\frac{D}{D_o})^3}{2\sigma_D^2}\}} \tag{8.7}$$

where σ_D: standard deviation; H: applied field; k: Boltzmann constant; T: temperature; M_s^f: fluid magnetization; Ø: solid volume fraction of magnetic nanoparticles in ferrofibers; M_d: domain magnetization; D: particle diameter; D_o: median diameter and $\chi_i H$: diamagnetic contribution from surfactant and PVA.

Using the information obtained from the fit, the particle concentration can be found using:

$$\emptyset = n'(\frac{1}{6}\pi D^3) \tag{8.8}$$

Further, inter-particle distance can be calculated using $(1/n')^{1/3}$, assuming uniform particle distribution. The parameters obtained from the best fit of Equation (8.4) are given in Table 8.2 with respective particle concentrations and volume fractions for each sample.

The spin dynamics and anisotropic effects of the ferrofibers can be understood using ferromagnetic resonance (FMR), which utilizes microwave absorption and the Zeeman effect to shed light on the spin relaxation mechanism. Figure 8.8 (a) shows the microwave resonance spectra in the plane (‖) and out of the plane (⊥) for two of the above-mentioned samples at 300K. A broad resonance signal indicates the ferromagnetic nature of ferrofibers. A shift in the resonance field (H_{res}) with an angle confirms the orientation of magnetic domains in ferrofibers. Further, in-plane configuration, a short relaxation behavior at $H_r = \sim 3380$ G (inset Fig. 8.8 (a)) confirms the superparamagnetic nature of Fe_3O_4 nanoparticles in the magnetic fluid. However, the absence of the same signal in out-of-plane (⊥) geometry indicates magnetic anisotropy due to the alignment of the domains in ferrofibers. In the out-of-plane configuration, the relaxation of superparamagnetic particles is strongly coupled with fibers. It occurs via Brownian relaxation dominantly, merging a small peak in the broad signal. In-plane geometry, the relaxation mechanism is mainly via short Neel relaxation showing superparamagnetic signals of Fe_3O_4 nanoparticles embedded in the polymer matrix. The angular dependence of resonance field (H_{res}) for both

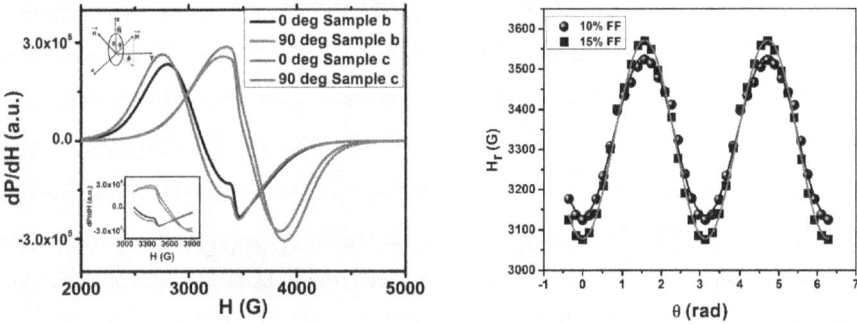

Figure 8.8. (a) FMR spectra at 0° and 90°, inset shows the appearance of the shoulder at 3380 G due to the superparamagnetic nature of Fe_3O_4 nanoparticles. (b) shows the angle-dependent microwave resonance at room temperature showing uniaxial anisotropy [5].

samples 'b' and 'c' is fitted using a simple equation (Equation 8.9) as shown in Fig. 8.8(b) [5, 15].

$$\frac{H_{res}(\Theta) - H_{res}(0^0)}{H_{res}(90^0) - H_{res}(0^0)} = Sin^2(\theta) \tag{8.9}$$

According to the magnetic resonance theory, the resonance condition for ferrimagnetic and ferromagnetic particles with uniaxial anisotropy is given by

$$H_{res} = \frac{\omega}{\gamma} - H_a P_2 \cos(\theta) \tag{8.10}$$

where Υ is the gyromagnetic ratio, ω is the angular frequency of microwave, H_a anisotropy field, P_2 (cos θ) is the second-order Legendre polynomial, and θ is the angle between the axis of the particle and field. According to Equation (8.10), for a completely aligned system, H_{res} for parallel (θ = 0°) and perpendicular (θ = 90°) configurations is given by

$$H_{res}(0°) = \frac{\omega}{\gamma} - H_a \tag{8.11}$$

$$H_{res}(90°) = \frac{\omega}{\gamma} + \frac{1}{2}H_a \tag{8.12}$$

For ferrofluids, no preferred orientation direction exists at room temperature, resulting in zero anisotropy in the magnetic fluid samples. But FMR results of ferrofiber confirm the preferential orientation giving rise to the uniaxial anisotropy field. The fitted observed data under the Morais model provides insight into ferrogel fibers intrinsic magnetic anisotropy behavior [14, 16]. The theoretical fit can be performed using the value of I obtained from the Langevin function fit using Equation (8.13) to get the value of uniaxial magnetic anisotropy energy constants.

$$H_{Res} = H_{EF} - H_x - H_{EK} \tag{8.13}$$

where $H_{EK} = \frac{K_{EK}}{I}(3\cos^2\theta - 1)$ is an effective uniaxial anisotropy field, H_{EF}: effective magnetic field, H_x: exchange anisotropy field, I: magnetization of nanoparticle and K_{EK}: effective anisotropy constant.

Table 8.3. Fitted FMR parameters.

Sample / Fitted Parameters	H_a (G)	$\left[\dfrac{K_{EK}}{I}\right]$ (G)	$H_{EF} - H_X$ (G)
B	266.4	132.9	3400
C	329.6	164.7	3407

Table 8.3 shows the quantitative effect of the increasing magnetic filler content in values. The increase in H_a and K_{EK}/I values is considerable. However, the difference in $[H_{EF} - H_X]$ values is not substantial.

8.4 Rheological behavior of ferrogel

8.4.1 Effect of shear rate on viscosity

The incorporation of micron-nano-sized magnetic particles significantly affects ferrogel materials' flow properties and durability. The viscoelastic properties of these gels show non-Newtonian behavior due to the alignment of magnetic nanoparticles along the field direction [17]. Figure 8.9 shows the viscosity (η) response as the function of shear rate at different magnetic fields for samples with different magnetic fluid concentrations. The response curve is composed of three regions (1) low shear rates ($\dot{\gamma} = 0.01$–500 s^{-1}) coalescence of polymeric-nanoparticles occurs, resulting in higher viscosity as it acts like a solid. The ferrogel resists flow strongly as polymeric bonding and magnetic forces dominate over shearing hydrodynamics. With the application of a magnetic field, the strength of the structure increases further, leading to an even higher viscosity. (2) Intermediate shear rates ($\dot{\gamma} = 500$–700 s^{-1}), here ferrogel starts the flow and an abrupt drop in viscosity are observed. The polymeric chins start to lose their structure under the shearing forces and the polymeric structure shows behavior similar to an elastic solid. However, beyond a critical strain, it acts as inelastic material. The increase in magnetic field values causes an increase in the strength of the polymeric structure leading to higher viscosity values [18]. (3) At high shear rates ($\dot{\gamma} = 700$–1000 s^{-1}), the polymeric structure breaks down completely into smaller units, thereby a sudden decrease in viscosity. Further, shear thinning occurs as structure breakdown continues with an increasing shearing rate. Now the hydrodynamic forces completely taking over magnetic forces between particles and complete inelastic behavior is observed. The magnetic content concentration plays a vital role in determining the flow and magnetic properties of ferrogels. As visible from Fig. 8.9, with the increase in ferrofluid concentration, the viscosity of ferrogel also escalates. When the magnetic field is applied, the magnetic nanoparticles embedded in the polymeric matrix start to align in the direction of the field, pulling along the polymer chains and thereby causing a hindrance to the flow of the gel.

8.4.2 Effect of shear rate on shear stress

Figure 8.10 shows the shear stress response as a function of the shear rate ($\dot{\gamma} = 0.01$–1000 s^{-1}) at different magnetic fields (H = 0 to 0.5T) for different ferrofluid

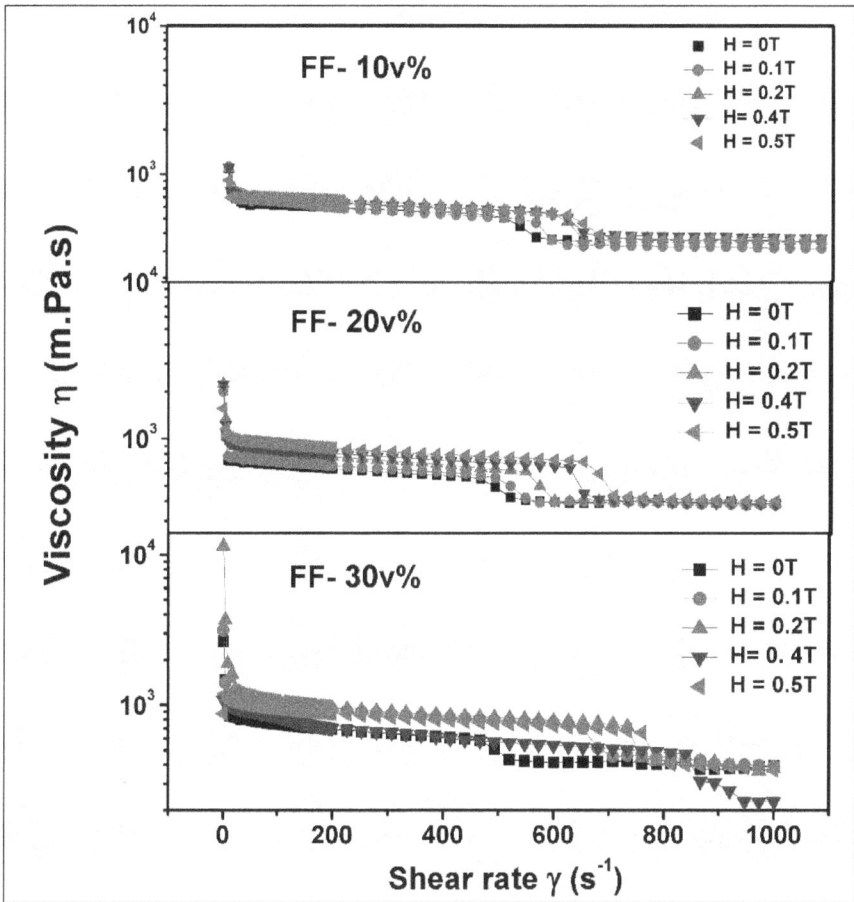

Figure 8.9. Shows the viscosity plot of the sample as a function of shear rate at the different magnetic fields [17].

concentrations of the ferrogel sample. It can be seen that stress-strain shows a no-linear behavior. Until a certain point known as the elastic point, the ferrogel's response curve (0-400 s-1) exhibits a linear behavior. After that, it shows a non-linear nature. A further increase in stain value causes non-linearity in the samples and after a critical strain value, the magnetic gel does not regain its original position. This point is defined as the yield point of the magnetic gel samples. From the flow curve, three different yield stress points can be defined: elastic–limit yield stress, static yield stress and dynamic yield stress. The samples show a linear response at low shear rate values, but beyond a certain point their is a drop in shear stress due to polymer chain breakage. This critical point is known as the elastic limit yield stress; up to this point, the material shows complete recovery when applied shear is removed.

Further, increasing the shear rate beyond this critical point, materials exhibiting non-linear inelastic characteristics as cross-linked bond magnetic fluid particles in a polymer matrix start to break. The minimum stress required to cause fluid flow

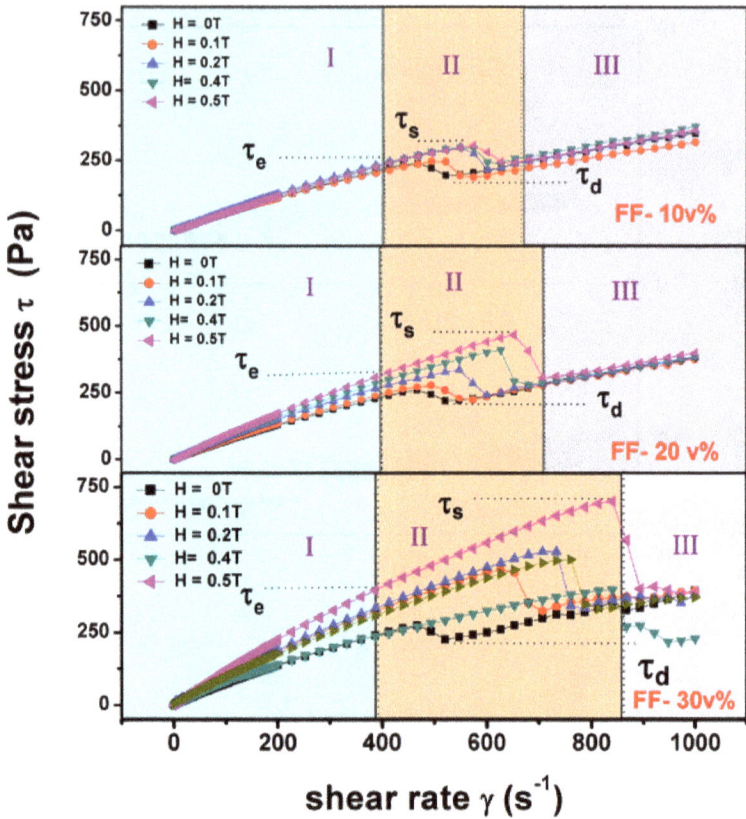

Figure 8.10. Shear stress response as a function of shear rate at the different magnetic fields divided into three regions [17].

is static yield stress. After this point, an increase in shear rate causes a decrease in shear rate due to the slip of particle aggregate plates in the direction of the applied force. As seen in Fig. 8.10; an increase in shear rate from ($\dot{\gamma} = 463-600 \text{ s}^{-1}$) causes a decrease in stress. This nonlinear characteristic is due to the shear-induced breakup of plate-shaped aggregates after a critical stress value. The inelastic nature of these deformations leads to a decrease in shear stress with the relaxation of particle aggregation, lowering the stress hindrance in the flow direction. After that, the dynamic yield point is defined as the saturation point of the macroscopic stress-strain curve. This value curve shows a saturation behavior, particle chain aggregates disappear instantaneously. It is clear that the material's yield stress is improved significantly (approximately twice) with an increase in particle concentration. It is due to the entangled magnetic nanoparticles forming chain aggregates in the field direction, which resist the smooth stress flow. This hindrance increases the yield stress of the material. The increase in a magnetic field, with the strength of these structures also expands; thus, shear stress also grows.

8.4.3 *Magnetoviscous effect*

The samples mentioned above responded to the magnetic sweep test of all the magnetic gel samples with a varying magnetic field (0 to 1 T) at a constant shear rate ($\dot{\gamma} = 100\ s^{-1}$) is shown in Fig. 8.11a. All the samples show an increase in viscosity with an escalation in a magnetic field, as magnetic nanoparticles align themselves due to magnetic interaction. These particles are embedded in the polymer matrix and move together with the matrix, thus causing a restriction to the magnetic gel motion. After a critical magnetic field, structures formed due to magnetic interactions are dominant over the applied shear forces and show a saturation behavior. Further, increasing particle volume concentration of magnetic fluid particles leads to higher viscosity due to a growth in magnetic interactions and causes a greater influence on the fluid flow. Saturation that is critical to the magnetic field shifts towards a higher field value with increasing particle concentrations. From these observations, it can be concluded that the increase in viscosity with field application gives a direct indication of the strong bonding between polymer matrix and magnetic fluid particles. The optimum value of magnetic particle concentration also allows maximum stability and enhanced magnetic properties. Embedded magnetic gels show lower stability for higher magnetic particle concentration due to loosely bonded magnetic particles. The addition of an optimum constriction allows maximum effectiveness of the magnetic gel.

The stability and relaxation behavior can be understood in magnetic gel samples with transient viscosity measurement at a fixed shear rate ($\dot{\gamma} = 100\ s^{-1}$). The transient behavior is studied in three regions: (i) H = 0 T (viscosity response in a steady-state constant shear without field application). In this region an almost linear viscosity response is observed in the sample, which signifies the Newtonian behavior of the sample. (ii) H = 0.18 T. Here a magnetic field of 0.18T is switched on while the shear rate remains the same. A significant increase in viscosity is observed due to

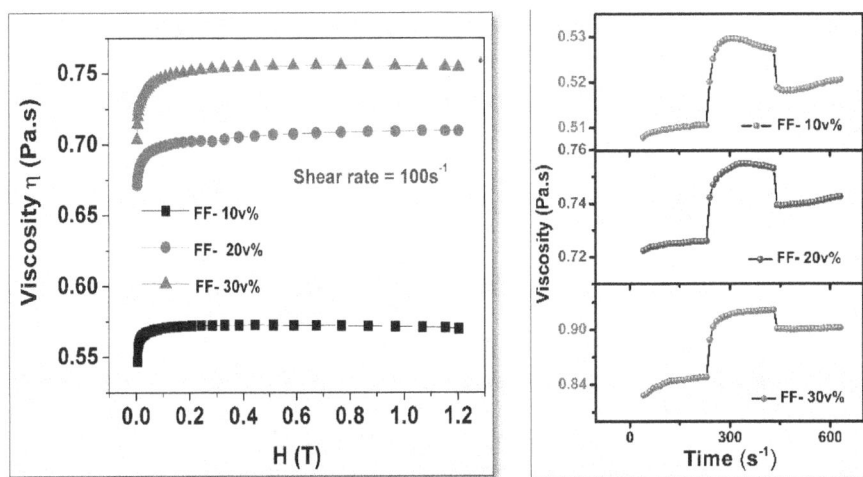

Figure 8.11. (a) Magneto sweep test of different volume fraction, (b) Transient viscosity response of PVA magnetic gel samples: 100 s^{-1}, region I: H = 0, region II: H = 0.18T and region III: H = 0 [17].

the field-induced structure formed in the field direction due to magnetic interactions. A straight line between regions (i) and (ii) shows a quick transient and efficient response of magnetic gel. (iii) H = 0 T, in this region the magnetic field is switched off; thus, viscosity decreases. However, the viscosity is slightly higher than initial values in region (i), due to the formation of smaller chain aggregates still present in the sample [17]. From Fig. 8.11b, it can be seen that all the magnetic gels show an efficient and quick magnetic response. Though the samples do not return to their actual pre-condition state due to changes in their polymeric structure, it is still in the expected range.

8.5 Summary

The chapter gave a brief insight on the magnetic composites, their synthesis and characteristics. The synthesis methods and characterization techniques discussed do not give a complete account of all methods and techniques used. The methods and techniques used depends on various factors such as raw material, cost of production, quality standards and final points of application. Here, the most generally used methods with a few examples to give an idea about ferrofluid based composites and their characteristics have been discussed. The selection of raw materials and desired properties depends entirely on the final application. Thus, the final list properties to be tested varies from industry to industry. But all these methods and techniques such as coprecipitation of nanoparticles, viscosity measurements, magnetization measurements, etc., have to follow ISO, ASTM or other standard procedures to form a quality product.

We have described various synthesis methods which included processing of both magnetic nanoparticles, polymer or MWCNTs. These methods can be generalized for many cases, but do not cover all. For example, functionalization of CNTs with –COOH using HNO_3 is not the only method available. Besides, the functionalization method changes depending on the functional group to be attached. Thereafter, we described some properties and their characterization methods such as morphology using SEM, TEM, magnetic properties, applying VSM, EPR and magneto-viscoelasticity using magneto-rheometer. These are a few properties from larger properties that need to be studied, tested and quantified for the final product. Hopefully, this chapter will be able to stipulate the basic foundation in the area of magnetic composites.

Acknowledgement

Authors thank Dr. Tejendra Kumar Gupta, Amity Institute of Applied Sciences, Amity University, Noida, India and Dr. Rajeev Kumar, CSIR- Advanced Materials and Processes Research Institute, Bhopal, India for providing MWCNT samples. AS acknowledges the financial support by IUAC UFR grant no. 73301, Inter-University Accelerator Center, New Delhi, India. KJ and RPP are thankful to CSIR for financial support (sanction no. 21(1124)/20/EMR-II).

References

[1] S. Pathak, K. Jain, P. Kumar, X. Wang and R. P. Pant. Improved thermal performance of annular fin-shell tube storage system using magnetic fluid. Applied Energy 239: 1524–35.

[2] S. Pathak, K. Jain, V. Kumar and R. P. Pant. Magnetic fluid based high precision temperature sensor. IEEE Sensors Journal 17(9): 2670–5 (2017).

[3] S. Pathak, R. Zhang, B. Gayen, V. Kumar, H. Zhang, R. Pant et al. Ultra-low friction self-levitating nanomagnetic fluid bearing for highly efficient wind energy harvesting. Sustainable Energy Technologies and Assessments 52: 102024 (2022).

[4] D. A. Allwood, G. Xiong, C. Faulkner, D. Atkinson, D. Petit and R. Cowburn. Magnetic domain-wall logic. Science 309(5741): 1688–92 (2005).

[5] K. Jain, S. Pathak and R. Pant. Enhanced magnetic properties in ordered oriented ferrofibres. RSC Advances 6(75): 70943–6 (2016).

[6] R. P. Pant, A. Shankar, J. Komal, M. Chand. Ferrofluid-MWCNT hybrid nanocomposite in liquid state. Google Patents (2017).

[7] V. Datsyuk, M. Kalyva, K. Papagelis, J. Parthenios, D. Tasis, A. Siokou et al. Chemical oxidation of multiwalled carbon nanotubes. Carbon 46(6): 833–40 (2008).

[8] P.-X. Hou, C. Liu and H.-M. Cheng. Purification of carbon nanotubes. Carbon 46(15): 2003–25 (2008).

[9] B. D. Cullity. Elements of X Ray Diffraction: Creative Media Partners, LLC (2018).

[10] I. Stamatin, A. Morozan, A. Dumitru, V. Ciupina, G. Prodan, J. Niewolski et al. The synthesis of multi-walled carbon nanotubes (MWNTs) by catalytic pyrolysis of the phenol-formaldehyde resins. Physica E: Low-dimensional Systems and Nanostructures 37(1-2): 44–8 (2007).

[11] D. Jiménez, X. Cartoixa, E. Miranda, J. Sune, F. A. Chaves and S. Roche. A simple drain current model for Schottky-barrier carbon nanotube field effect transistors. Nanotechnology 18(2): 025201 (2006).

[12] M. A. Correa-Duarte, M. Grzelczak, V. Salgueiriño-Maceira, M. Giersig, L. M. Liz-Marzán, M. Farle et al. (2005). Alignment of carbon nanotubes under low magnetic fields through attachment of magnetic nanoparticles. The Journal of Physical Chemistry B. 109(41): 19060–3.

[13] I. T. Kim, G. A. Nunnery, K. Jacob, J. Schwartz, X. Liu and R. Tannenbaum. Synthesis, characterization, and alignment of magnetic carbon nanotubes tethered with maghemite nanoparticles. The Journal of Physical Chemistry C. 114(15): 6944–51 (2010).

[14] A. Shankar, M. Chand, G. A. Basheed, S. Thakur and R. P. Pant. Low temperature FMR investigations on double surfactant water based ferrofluid. Journal of Magnetism and Magnetic Materials 374: 696–702 (2015).

[15] F. Gazeau, J. Bacri, F. Gendron, R. Perzynski, Y. L. Raikher, V. Stepanov et al. Magnetic resonance of ferrite nanoparticles: evidence of surface effects. Journal of Magnetism and Magnetic Materials 186(1-2): 175–87 (1998).

[16] P. Morais, M. Laraand K. S. Neto. Electron spin resonance in superparamagnetic particles dispersed in a non-magnetic matrix. Philosophical Magazine Letters 55(4): 181–3 (1987).

[17] S. Pathak, K. Jain and R. Pant. Improved magneto-viscoelasticity of cross-linked PVA hydrogels using magnetic nanoparticles. Colloids and Surfaces A: Physicochemical and Engineering Aspects 539: 273–9 (2018).

[18] L. J. Felicia and J. Philip. Effect of hydrophilic silica nanoparticles on the magnetorheological properties of ferrofluids: a study using opto-magnetorheometer. Langmuir 31(11): 3343–53 (2015).

9

Insight on Efficient Biocompatible Magnetic Nanoparticles for Cancer Treatment using Magnetic Hyperthermia Technique

Arjun Singh,[1,2,3] *Prashant Kumar,*[1,2,4] *Komal Jain,*[1]
Vidya Nand Singh,[1] *K.K. Maurya*[1,2,*] *and R.P. Pant*[1,2,*]

9.1 Introduction

Cancer is a deadly disease that claims millions of lives every year. In 2020, the GLOBOCAN database estimated 19.3 million new cancer cases and almost 10 million deaths worldwide. This number is estimated to increase to 28.3 million by 2040 [1]. Cancer is an abnormality inside the body in which the cells split in an unfettered way leading to the formation of tumors. Cancer cells have a category of harmless cells that grow close by without affecting healthy tissues. Other cancer cells include malignant cells, which attack the neighboring cells rapidly, leading to a dangerous and life-threatening situation. Cancer in various forms, such as breast cancer, lung cancer, throat cancer, prostate cancer, etc., has been diagnosed all across the globe. In the past few years, there has been a significantly improved survival rate of cancer patients. The advancement in survival shows the progress in the early diagnosis of cancers followed by enhanced treatment methods [2, 3].

[1] Indian Reference Materials Division, CSIR-National Physical Laboratory, New Delhi-110012, India.
[2] Academy of Scientific and Innovative Research (AcSIR), Ghaziabad, Uttar Pradesh 201002, India.
[3] Department of Physics, Indian Institute of Technology, Jammu-181221, India.
[4] School of Science, RMIT University, Melbourne, VIC 3000, Australia.
* Corresponding authors: kkmaurya@nplindia.org; rppant@nplindia.org

Medical procedures for cancer treatment are radiotherapy, chemotherapy, surgery, immunotherapy and gene therapy. Radiation therapy knocks high-energy photons or charged particles on targeted areas and breaks their DNA strands, killing the cancer cells or slowing their reproduction rate [4]. However, this treatment option has the drawback of affecting the surrounding healthy tissues, leading to several side effects such as fatigue, colon perforation, infertility and a significant cause of secondary cancers [5, 6]. Another widely employed treatment for cancer is chemotherapy which uses drugs to destroy cancer cells. Chemotherapeutic drugs attack the dramatically dividing cancer cells at their destination, but dismally, numerous normal healthy cells also fall prey to their destructive effects. These drugs are cytotoxic and affect other normal healthy cells [7]. One frequently used treatment for this deadly disease is surgery which also comprises several adverse effects such as inflammation, blood clots and several other problems during or after the removal of the cancer tissue. Sometimes, it may lead to tumor reappearance or metastasis if the tumor is not eradicated correctly [8]. The most challenging point encountered by the above-reported cancer treatments is the lack of potency in destroying all lethal cancerous cells. Due to the risk and inefficiency observed in these traditional methods, there is an increased demand for considering advanced cancer treatment strategies [9].

Another advanced treatment option is hyperthermia which involves heating the selected or complete body at an elevated temperature than the average body temperature of 37°C for a continued phase of time [10, 11]. The advantages of high body temperature in fighting infections and treating sickness has been perceived for ages. Heat energy aims at the affected region despite the whole body in the present hyperthermia treatment. It is under the supervision of an external agent to improve treatment outcomes and minimize side effects on the body. Hyperthermia can be used singularly or in combination with other therapy treatments such as radiotherapy or chemotherapy. It is claimed that a temperature greater than 43°C increases the cytotoxic response of cells or drugs. During local hyperthermia, neighboring tissues can be saved from getting being damaged because of controlled heat generation. The treatement of local hyperthermia's most widespread side effects include burns, blisters, pain and discomfort [12].

Local hyperthermia techniques adopted for treatments use radiofrequency, microwaves and ultrasound waves. These techniques are recognized as effective, but they cease to be singular therapy despite all the merits. Another factor observed during this treatment is the damage caused by the heat to the surrounding tissue. Before reaching the aimed tumor site, the wave currently involved in these methods traverses through the healthy region, which absorbs its energy, leading to unwanted damage due to heat or under dosage to the aimed area [3]. It shows the necessity for new techniques that overcome these problems and limit the heating only to the affected site.

Magnetic nanoparticle hyperthermia uses small-sized ferrite nanoparticles, typically 5–50 nm in diameter [9, 13, 14]. After synthesizing MNPs, a ferrofluid is prepared by mixing the nanoparticles with a dispersing medium. To ensure the colloidal stability of the fluid, the MNPs are coated with suitable biocompatible materials. The surfactant-coated materials harm the non-toxicity effect on viable cells and target the particle at the desired tumor site. On applying an alternating magnetic field, heat generates in the magnetic nanoparticles. The quantity of heat produced depends largely on features such as the spatial allocation of the nanoparticles, dimensions, coating, magnetic field strength, frequency, etc. [15]. To ensure the human safety limit, the product of field and frequency in a clinical trial should be less 5×10^9 Am^{-1} S^{-1}. Magnetite and maghemite are well renowned and the two most widely used materials for their biocompatibility with human tissues. Compared to traditional hyperthermia methods, magnetic nanoparticles prove more promising because they deliver sufficient thermal energy to deep-rooted tumors with disturbed geometries and do not give up power to the neighboring tissues.

To estimate the potential of MNPs for magnetic hyperthermia, the fundamental of magnetism needs to understand (as discussed next). The heating mechanism for calculating SAR values and experimental calorimetric SAR results are elaborated later. The cytotoxicity and biocompatibility of MNPs using different calorimetric, fluorometric and dye-based assays are discussed after that. At the end a summary of the work is presented.

9.2 Basics of magnetism in hyperthermia

To improve the performance of Magnetic Hyperthermia (MH), the fundamental concepts of magnetism need to be precise. The factor influencing magnetism at the nanoscale is of prime concern, including intrinsic and extrinsic parameters. When the material is placed in an external field, different materials act differently based on the intrinsic magnetic dipole and the net magnetization. The materials are categorized into diamagnetic, ferromagnetic, paramagnetic, antiferromagnetic and ferrimagnetic [16], as shown in Fig. 9.1 (a). In the case of diamagnetic materials, the magnetic dipoles are absent without a magnetic field. On application of the magnetic field, the magnetization is weakly aligned in the opposite direction. In paramagnetic materials, the dipoles weakly align in the same direction of the area and possess small moments.

In the case of ferromagnetic materials, magnetic dipoles are always present even if the external field is not applied. Such materials display a permanent magnetic moment. However, ferromagnetic materials differ from antiferromagnetic and ferrimagnetic materials [17]. Due to the applications in magnetic hyperthermia, magnetic drug delivery, MRI, etc., an important class of magnetic materials is ferromagnetic and ferrimagnetic [18–20]. When ferromagnetic or ferrimagnetic is reduced below a specific critical size (generally less than 50 nm, depending on the materials), they show a response similar to paramagnetic when the magnetic field is absent [19, 21–23]. However, with the rise of the magnetic field, a rapid increase in the magnetic moment is observed [24, 25]. These types of materials are called superparamagnetic materials. The characteristics measuring time in superparamagnetism are greater than the Neel relaxation time. Figure 9.1b represents the superparamagnetism below a certain size r_0 lies in the single domain region with a minimal coercive value. As the size of the magnetic nanoparticle increases in the single domain region, the value of coercivity increases reaching a maximum. It then starts declining with the formation of domain walls in the multidomain regions [26, 27].

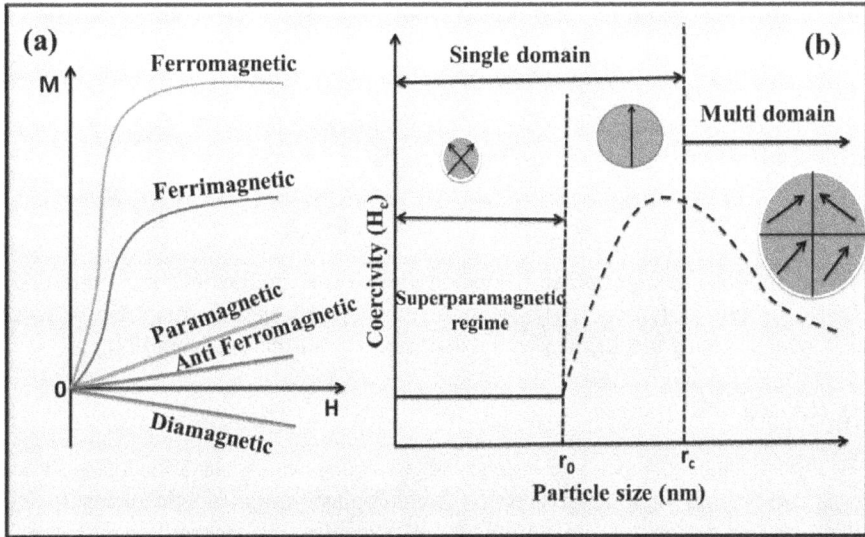

Figure 9.1. Schematic representation of magnetization (M) versus applied field strength (H) for different types of magnetic materials. (a) Schematic represents MH behavior of the material when exposed to an external field H. (b) Schematic shows that below a certain size r_0 magnetic nanoparticle shows superparamagnetism having a coercivity value that is very small. However, as the size increases, coercivity starts increasing and is maximum at r_c and starts decreasing as the energy utilized in forming domain or bloch walls is observed in a magnetic field's nonexistence.

9.3 Mechanism of heat transmission

When an alternating magnetic field applies to NPs, the moment rotates along the field, leading to the liberation of heat from the MNPs. Depending on the size of MNPs, three types of loss mechanisms are known [26, 28]. The first loss mechanism is the hysteresis loss in NPs with particle size greater than 100 nm, generally in multi-domain magnetic materials. The second loss mechanism is due to eddy, mainly in bulk particles. The third mechanism is relaxation loss, found in particles less than 50 nm, usually in a single domain or superparamagnetic [8, 27, 29, 30].

A. Hysteresis loss

Hysteresis is a loss mechanism observed in bulk and multi-domain ferro and ferrimagnetic materials. The hysteresis formation is due to two effects: one is the orientation of the magnetic moment, and the other is the change in the size of domains or movement of domain walls. When an Alternating Magnetic Field (AMF) applies to NPs, more than two times the order of coercive strength of the NPs, the magnetic moments tend to orient with an AMF leading to the liberation of a large amount of heat [31]. The hysteresis loss is calculated by measuring the area under the hysteresis loop and is written as.

$$A = \int_{-H_{max}}^{+H_{max}} \mu_0 \, M(H) dH \tag{9.1}$$

The specific absorption rate or power dissipation of MNPs can be measured as

$$SAR = A.f \qquad (9.2)$$

where f is the frequency of an AMF. M is the magnetization of NPs, H is the external ac field and A is the area under the hysteresis curve.

B. Eddy current

Eddy current is one other possible heating mechanism. In magnetic materials, eddy current heating is noticed only for bulk magnetic materials. Magnetic particles display this heating mechanism with a diameter greater than one mm [32].

C. Relaxation losses

The third mechanism having the potency of generating heat is relaxation losses shown by single-domain superparamagnetic, single-domain ferromagnetic nanoparticles. It is observed that under clinically bearable magnetic field strength and frequency combinations, the superparamagnetic and single-domain nanoparticles possess higher SAR than hysteresis or eddy loss in multi-domain particles [33]. Relaxation losses contain two characteristic modes commonly known as Néelian and Brownian relaxations.

D. Néel relaxation

Néel's relaxation mechanism is generally observed in single domain superparamagnetic or ferromagnetic nanoparticles. In Néel's relaxation, when an external field is applied to MNPs, the individual magnetic moment rotates inside the particle core against the crystal anisotropy energy. This rotation of moment against the anisotropy barriers produces heat in particles. The Néel relaxation mechanism is shown in Fig. 9.2 (a). The magnetic moment tends to align in the direction of the field and as the field switches in the other direction, the moment starts following the field direction. The Néel relaxation time is denoted by τN and is given by the following Equation [29].

$$\tau_N = \frac{\sqrt{\pi}\, \tau_0 e^{\Gamma}}{2\sqrt{\Gamma}} \qquad (9.3)$$

$$\Gamma = \frac{KV_m}{k_B T} \qquad (9.4)$$

where τ_0 is the attempt time, Γ is the anisotropy energy to thermal energy ratio, K is the crystalline magnetic anisotropy constant, V_m is the core magnetic volume and is written as

$$V_m = \frac{4\pi r^3}{3} \qquad (9.5)$$

where r is the radius of the magnetic core. Thus Néelian relaxation depends on the magnetic anisotropy of the materials and the size of the magnetic core.

E. Brown relaxation

When the MNPs mix in a dispersion medium, they tend to have a random or zigzag motion, known as Brownian motion. The external AMF on the MNPs leads to the rotation of the whole particle, resulting in its alignment with the magnetic field [30, 34, 35]. The rotational movement of the particles is opposed by the suspending medium leading to heat production. The time consumed by the magnetic nanoparticle to be in line with the external magnetic field is known as Brownian relaxation time and can be represented mathematically as:

$$\tau_B = \frac{3\eta V_H}{k_B T} \qquad (9.6)$$

$$V_H = V_M \left(1 + \frac{\delta}{r}\right) \qquad (9.7)$$

where η is the coefficient of viscosity of the matrix fluid, V_H is the hydrodynamics volume greater than magnetic volume, and δ is the thickness of the absorbed surfactant or coating layer. k_B is the Boltzmann constant, and T is the absolute temperature.

Thus Brownian relaxation depends on the size of the particles and the thickness of the surfactant or coating layer. Moreover, it also depends on the viscosity of the suspended medium. Both Néel and Brown's relaxation depend on the temperature. In a solution, Néel and Brown's relaxation occurs parallell when applying an external AMF [5, 12]. The combined effect of both can write as

$$\tau = \frac{\tau_N \cdot \tau_B}{\tau_N + \tau_B} \qquad (9.8)$$

Thus, the Néelian and Brownian relaxation are essential parameters that must be tuned to utilize MNPs for hyperthermia technique effectively.

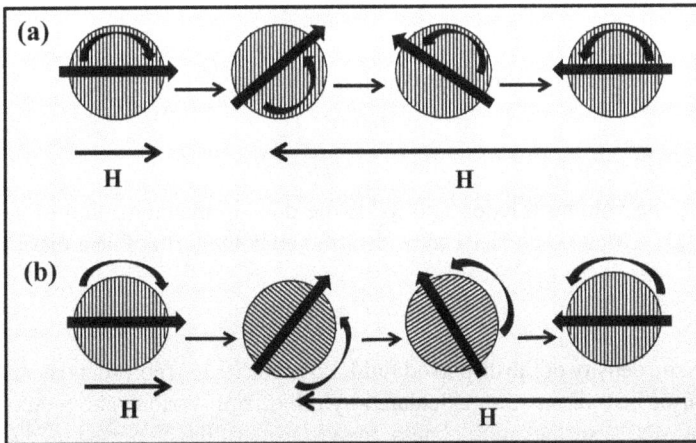

Figure 9.2. (a) Represents Néel relaxation mechanism, when AMF is switched on, the magnetic moment rotates with the direction of AMF, without the rotation of the particle. (b) shows the Brownian relaxation, where the magnetic moment rotates with the movement of particle direction of AMF.

9.4 Analytical model for measuring Specific Loss Power (SLP)

To access the heating efficiency of MNPs, the estimation of SLP is necessary. Rosensweig proposed the first theoretical model of heat treatment in 2002 [33]. It is based on a linear response theory and only applies to the small superparamagnetic region. However, with time, various analytical methods have been proposed for measuring SLP, which have broad applicability. Carrey et al., 2011 presented a function (ξ) [36] as

$$\xi = \frac{\mu_0 M_s H_0 V_m}{k_B T} \qquad (9.9)$$

where μ_0 represents the magnetic permeability in free space, M_s represents saturation magnetization and H_0 is the maximum field strength. If the function $\xi \ll 1$, the Linear Response Theory (LRT) is applicable for measuring the loss power of MNPs. The LRT is valid for a single domain, particularly for the superparamagnetic region. In LRT, the SLP is directly proportional to the applied field 'H' square. If $\xi > 1$, the calculation of the loss power of MNPs is based on Rayleigh models and is applicable for ferromagnetic materials. The SLP is directly proportional to the applied field 'H' cube in the Rayleigh model [36, 37]. The Stoner-Wohlfarth Model (SWM) is also applicable to those ferromagnetic materials where the condition $\mu_0 H_{max} > 20 \, H_c$ is satisfied and H_c is the coercivity of MNPs. By combining all these models, the heat dissipation power (P) can write as:

$$P = \pi \mu_0 H^n \chi_0 f \left[\frac{2\pi f t}{1 + (2\pi f t)^2} \right] \qquad (9.10)$$

where f is the frequency of ac magnetic field, χ_0 is the static equilibrium susceptibility and can be written as

$$\chi_0 = \chi_i \frac{3}{\xi} \left(\coth \xi - \frac{1}{\xi} \right) \qquad (9.11)$$

where χ_i is the initial susceptibility, which is given as:

$$\chi_i = \frac{\mu_0 \, \varphi M_d^2 \, V_m}{3 k_B T} \qquad (9.12)$$

where φ is the volume fraction and M_d is the domain magnetization of suspended particles ($M_d = M_s/\varphi$). The SLP can be determined in the form of heat dissipation as :

$$SLP = \frac{P}{\rho} \qquad (9.13)$$

where ρ is the density of the dispersed fluid. Equation (9.13) represents the theoretical estimation of heat dissipation calculated by optimizing various parameters such as the size of MNPs, surfactant thickness, the viscosity of the dispersed medium, AMF amplitude and frequency [2, 38–40].

9.5 Experimental SAR characteristics

SAR can be determined experimentally by placing a small ferrofluid in a resonance coil setup. The picturization of the design is shown in Fig. 9.3. The sample is covered in a styrofoam container and placed inside the coil. An external AMF is applied to the coil arrangement. The thermometer or temperature probe is used for measuring the heat. The quantity of heat measured per unit mass of the MNPs is expressed in a specific absorption rate (SAR) [15, 28, 41, 42]. The equation governing SAR is written as

$$SAR = (m_{np} c_{np} + m_w c_w) \frac{dT}{dt} \tag{9.14}$$

where m_{np} is the mass and c_{np} is the specific heat capacity of MNPs. m_w and c_w are the mass and specific heat of water, respectively. The term dT/dt is the initial slope of temperature versus time cover. The SAR value linearly depends on the slope. Further, the SAR value is estimated from fitting parameters obtained under different thermodynamic approaches. In the literature reported , researchers employed models such as the Initial Slope Method (ISM), Box Lucas Method (BLM), Newton's cooling approach, etc. [43].

A. Initial slope method (ISM)

In the initial slope, the method is based on the assumption that during an initial period (< 60 s), the heat loss due to the surroundings is almost negligible as the

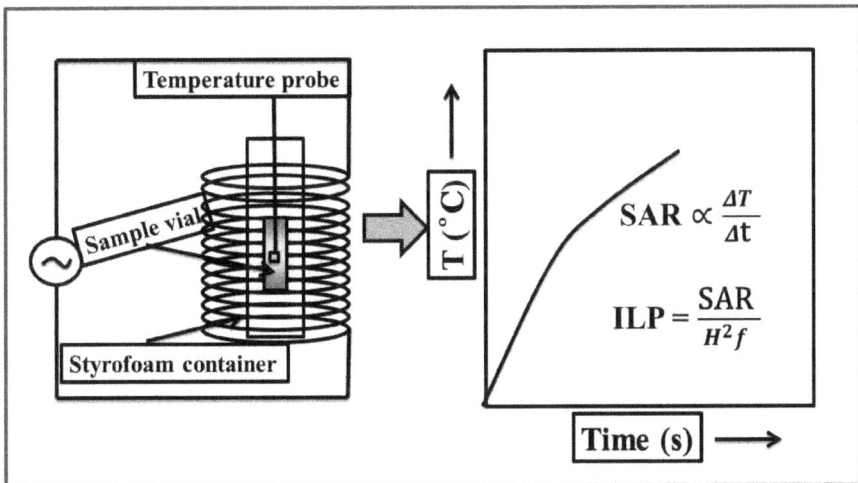

Figure 9.3. Left panel represent the schematic illustration of magnetic hyperthermia setup. Right panel shows the time dependance temperature curve.

system acts like an adiabatic. The temperature versus time curve is almost linear in the initial 1 min. The equation for SAR according to ISM is written as

$$SAR_{ISM} = \frac{\Delta T}{\Delta t} \frac{C\rho}{m_{Fe/MNPs}}$$

(9.15)

where $\dfrac{\Delta T}{\Delta t}$ is the slope of the first 1 min on exposure to AMF, C represents heat capacity of dispersant, ρ denotes density. $m_{Fe/MNPs}$ is the concentration of iron or MNPs in the solution. Most researchers reported SAR assuming the ISM method.

B. Box Lucas Method (BLM)

The heating profile curves of MNPs shows nonlinearity concerning time under applied AMF as the system follows the non-adiabatic condition. In that case, one can use BLM. The heating curve is fitted as:

$$\Delta T = a(1 - e^{(-b(t - t_0))})$$

(9.16)

The extracted parameters a and b after appropriate fitting are used to estimate SAR value as

$$SAR_{BLM} = \frac{abc}{m_{Fe/MNPs}}$$

(9.17)

C. Newton's cooling approach (NCA)

Newton's cooling approach can also be used by assuming that the change in temperature of the magnetic sample over time t follows the non-adiabatic experimental condition under an alternating magnetic field can be expressed by the following equation:

$$T = T_0 + \Delta T_{max}(1 - \exp(\tfrac{-t}{\tau})$$

(9.18)

The temperature-dependent plot has been fitted using Equation (9.19). The obtained fitted parameters ΔT_{max}, τ are required to calculate the SAR value using the equation:

$$SAR = \frac{C}{m_{sample}} \frac{\Delta T_{max}}{\tau}$$

(9.19)

The SAR values depend linearly on the frequency (f) and quadratically on the field (H). Different research groups measure SAR by choosing different sets of field amplitude (H) and frequency (f), so it is difficult to compare the heat efficiency based on SAR values. The Intrinsic Power Loss (ILP), a standard alternative to SLP, is introduced by normalizing SLP with the concerning field amplitude (H) and frequency (f) [4, 7, 44–46].

$$ILP = \frac{SLP}{H^2 f}$$

(9.20)

Table 9.1. Summarizes the synthesis technique, average particle size, coating material used, Magnetic characteristics at Room Temperature (RT), Alternating Magnetic Field (AMF) amplitude, frequency and SAR value.

Material	Size (nm)	RT magnetization	AMF (kA/m)	Frequency (kHz)	SAR (W/g)	References
Fe_3O_4	19	$M_s = 80$ emu/g	15	320	502	[35]
	5	$M_s = 34.39$ emu/g	26	265	39	[41]
	12	Ms = 12 emu/g	15.9	62	14	[47]
	10	Ms = 25.6 emu/g	-	150	64.6	[38]
$MgFe_2O_4$	20	Ms = 35 emu/g	26.7	265	85.57	[31]
						[48]
$Zn_{0.9}Fe_{2.1}O_4$	11	Ms = 12 emu/g	3	700	6.86	[49]
$CoFe_2O_4@MnFe_2O_4$	15	110 emu/g	37.3	500	2280	[3]
$CoFe_2O_4@Fe_3O_4$	15	105 emu/g	37.3	500	1120	
$MnFe_2O_4@CoFe_2O_4$	15	Ms = 108 emu/g	37.3	500	3034	
Zn0.4Fe2.6O4@ Zn0.4Mn0.6Fe2O4	15	Ms = 150 emu/g	37.3	500	3886	

It is reported that the SAR value can be increased by optimizing the magnetic properties of MNPs. Lee et al. reported that magnetic nanoparticles by varying core and shell structures such as $CoFe_2O_4@MnFe_2O_4$, Zn0.4Fe2.6O4@Zn0.4Mn0.6Fe2O4 and measuring the SAR parameters [3]. The core and shell structure exchange forms the magnetic interaction between hard and soft magnetic materials. The SAR values increased up to a large extent, as shown in Table 9.1.

9.6 Biocompatibility and cytotoxicity of MNPs

Superparamagnetic nanoparticles (SPNPs) have a broad scope in future healthcare applications such as magnetic hyperthermia, targeted drug delivery, magnetic bioseparation, cancer detection or as a contrast agent in Magnetic Resonance Imaging (MRI), etc. When these SPNPs interact with the culture media, the nutrient or protein present in the media is absorbed by these particles and forms a complex layer on its surface, which induces cytotoxicity. Therefore, it is essential to check the interaction between SPNPs and the culture media, as it depends on both the composition of SPNPs and the type of culture media [50–53].

Various tests are performed to check the biocompatibility (means not being toxic or harmful to living tissues) of these SPNPs. To fulfill the biomedical requirements of SPNPs in targeted drug delivery, targeted chemotherapy, magnetic hyperthermia, etc., SPNPs should have hemocompatibility and cytocompatibility. The biomaterials (MNPs) are hemocompatible if there occurs the minimum formation of blood clotting (thrombus) and blood coagulation and have less hemolysis (destruction of RBCs). On the other hand, biomaterials are cytocompatible. If it occurs, there no induction of the cell death or cell lysis on particular tissue interacting with MNPs [52, 54].

The first step is to measure cytotoxicity characteristics such as RBC lysis assay, cell viability, cell proliferation or cell lysis using *in vitro* cytotoxicity assays. There are various types of cytotoxic assays reported in literature. They are categorized into cell lysis assay and cell viability assay. In the cell lysis assay, the lactate dehydrogenase compound reduces in the cellular medium on cell death. The cytotoxicity percentage of cell death is calculated as

$$\% \text{ cytotoxicity} = \frac{\text{The absorbance of control cells-Absorbance of treated cells}}{\text{The absorbance of control cells}} \times 100$$

(9.21)

The number of live cells escalates with an increase in absorbance. Besides LDH, WST-8 or Hoechst 33342 methods are also used to investigate SPNPs using different cell lines. The assay for cytotoxicity or cell viability is broadly based on (i) Calorimetric assays, (ii) Fluorometric assays, and (iii) Dye exclusion assays [55–57].

1. Colorimetric assays
In colorimetric cell viability assays, various tetrazolium dye-based assays are utilized, as they are simple, fast and less expensive.

(a) MTT assay:
MTT assay is one of the most widely used techniques for determining cell viability and proliferation. In the MTT assay, tetrazolium salt is transformed into violet formazan crystals in the presence of viable cells. These crystals are then dissolved in isopropanol or dimethyl sulfoxide (DMS) and exposed under a spectrophotometer or plate reader (ELISA) at 490 nm. The percentage of the viable cell using dye-based colorimetric assay is calculated as

% cell viability =

$$\frac{\text{Absorbance of test (T)-Absorbance of control positive (cell killed)}}{\text{The absorbance of control negative (without any treatment)}} \times 100 \quad (9.22)$$

(b) XTT assay:
In the XTT assay, the tetrazolium salt reduces to a water-soluble formazan compound. It saves an additional step of dissolving as in the case of MTT. The optical density is then measured using a spectrophotometer via a plate reader. The more the number of living cells, the more the mitochondrial dehydrogenase activity will increase and thus more absorbance.

(c) WST-1 assay:
In this assay, the tetrazolium salt reduces to water-soluble formazan using mitochondrial dehydrogenase activity through an intermediate electron acceptor. This assay is similar to the XTT assay, where formazan does not require to dissolve separately and is hence valuable when screening occurs in bulk amounts.

(d) WST-8 assay:
In this assay, the yellow color tetrazolium salt is reduced into an orange color formazan under the presence of a viable cell. The product formation largely depends

on the number of live cells. It can be helpful in the range of 200 and up to 25000 cells/wall. Further, the product is water-soluble and hence no additional step is required.

(e) LDH assay:

In this case, the nicotinamide adenine dinucleotide (NADH) is used to convert yellow tetrazolium salt into red formazan, easily dissolved in water. The quantity of formazan obtained is measured at an absorbance of 490 nm, which increases with a growth in the number of dead cells

2. Fluorometric assays

The fluorescent assays are slightly more sensitive than colorimetric assays. In this assay, the reaction occurs either with the production of the fluorescent compound from nonfluorescent or vice-versa. For example, the use of non-fluorescent compounds such as fluorescein diacetate undergoes hydrolase and changes to fluorescent.

(a) Resazurin (Alamar blue) assay

In this method, the Alamar blue or resazurin undergoes reduction to fluorescent resorufin. The excitation wavelength between 530–570 nm and emission between 580–620 nm are used.

(b) 5-CFDA-AM assay

This technique is similar to Alamar blue, where 5-CFDA-AM convert into a fluorescent substance such as carboxyfluorescein, which is nonpermeable to the living cell membrane.

3. Dye exclusion assay

This assay is based on the exclusion principle, where several dyes are nonpermeable to living cells. A few examples are described below.

(a) Trypan blue (TB) stain assay:

The principle underlying in TB assay is that living cells have intact cell membranes and thus exclude TB dyes and appear white. On the other hand, the non-viable cell absorbs dyes and appears blue. It is one of the most specific assays to analyze the cell viability in the culture media. The % cell viability is calculated as

$$\% \text{ cell viability} = \frac{\text{total no. of living cells per ml sample}}{\text{total no. of cells per ml sample}} \times 100 \quad (9.23)$$

(b) Erythrosine B stain assays, Eosin and Congo red:

The erythrosine B, Eosin and Congo red stain assays have a similar mechanism as TB, where the stain depends on the integrity of the cell membrane. Further, erythrosine B stain assays are more advantageous over TB as being nontoxic, non-binding serum protein and require no incubation period.

The interaction of MNPs and response to particular cell lines is critical for therapeutic development. The interaction may harm the cells (cytotoxicity), promote cell division or be ineffective. The cytotoxicity of MNPs depends on the dose employed, as a higher concentration causes a decrease in the mitochondrial activity of human fibroblasts. The results obtained from cytotoxicity are difficult to compare because the toxicity of MNPs is influenced by various parameters such as the morphology of the particle, surfactant coating and other magnetic properties. Furthermore, different cell lines interact in different ways with the same particle. Therefore, working on the cell line for a particular MNPs to access toxicity is essential. Kia et al. reported the cytotoxic effect of Bare Fe_3O_4, $OA-Fe_3O_4$ and C-Fe using the human hepatoma BEL-7402 cell line [51]. The cell toxicity was analyzed using a tetrazolium salt-based (WST-1) assay. After 24 hr of exposure at varying doses of bare, oleic acid and carbon-coated Fe_3O_4 MNPs, the viable cell's percentage was calculated at 490 nm. It is found that the cell viabilities of BEL-7402 are above 60% for all three types up to 0.1 mg/ml concentration. When the concentration is increased further, cell viability drops below 60%. Moreover, the cell death rate for all types of MNP follows the order of carbon-coated $Fe_3O_4 > Fe_3O_4 > OA-Fe_3O_4$ in the same concentration [51].

Cristian Lacovita et al. studied the cytotoxicity of iron oxide nanoparticles and compared them with Mn and Zn ferrite nanoparticles [50]. The synthesized different MNPs were exposed to four different cell lines, namely: human epithelial cells (D407), lung cells (A549), melanoma cells (MW35) and mouse melanoma cells (B16F10) for 24 hr. After incubation, the cells patterns were observed under transmission electron microscopy. They analyzed that no cytotoxicity is observed at the lowest concentration of 50 μg/ml. As the concentration is greater than 200 μg/ml, the cell viability starts decreasing in both ZnFe2O4 and MNFe$_2$O4. Further, they compared the cytotoxicity of $ZnFe_2O_4$ and $MNFe_2O_4$ with uncoated and coated Fe_3O_4 and found that coated materials show less toxicity than bare particles [50].

9.7 Summary

Cancer is one of the deadly diseases that claim many lives worldwide. New research and technology open the challenge of introducing a minimal side effect technique that can cure cancer easily. The present work reviews the efficiency of MNPs for their effective use in cancer treatment using the Magnetic Hyperthermia (MHT) Technique. The MHT is a new initiative approach that uses biocompatible MNPs to destroy cancer cells. In MHT, the MNPs are placed in a resonance coil-based arrangement where an external AMF (with radiofrequency in the range of 100KHz to 500KHz and providing an ac field H = 10KA/m to 50KA/m) is applied, the MNPs release from 42–46°C, which are used to kill the cancer cell. The heating of MNPs is mainly due to hysteresis loss and relaxation losses (Néelian and Brownian relaxation). The hysteresis loss is observed in multidomain ferromagnetic nanoparticles whereas relaxation loss is dominant in single-domain ferromagnetic or superparamagnetic nanoparticles (SPNPs). The amount of heat released by MNPs per unit mass is expressed in terms of a Specific Absorption Rate (SAR) or Specific Loss Power (SLP). The SAR value of MNPs increases with a growth in frequency and applied field strength. The maximum tolerable limit is obtained from the product of frequency and field strength as per the Brezovich criterion should not exceed 5×10^8 $Am^{-1}s^{-1}$. The SAR value of MNPs is estimated using various theoretical models such as linear response theory, stoner Wohlfarth model, Rayleigh model, etc. The linear response theory is valid in the case of single domain superparamagnetic nanoparticles, where SAR is proportional to the square of the applied field H., the Stoner Wohlfarth and Rayleigh models are applicable in the case of ferromagnetic nanoparticles, where SAR is proportional to the cube of the applied field. The experimental SAR is determined based on a resonance coil-based setup, where the sample is placed inside the coil and heat is measured using a thermometer. The temperature versus time curve is plotted and fitted using various models such as the Box-Lucas method, Newton's cooling approach, Initial slope method, etc. It is observed that SAR values of MNPs are optimized by tailoring various magnetic characteristics such as saturation magnetization, magnetic anisotropy, spin relaxation time, etc. To have effective utilization of MNPs for hyperthermia treatment, it is essential to check their biocompatibility and cytotoxicity. The MNPs should have hemocompatibility and cytocompatibility. The MNPs are hemocompatible if there is the minimum formation of blood clotting (thrombus) and blood coagulation and have less hemolysis (destruction of RBCs). On the other hand, the magnetic materials are cytocompatible, if there occurs no induction of cell death or cell lysis on particular tissue interacting with MNPs.The assay for cytotoxicity or cell viability is broadly based on the calorimetric assay, dye exclusion assays and fluorescent assay. The interaction of MNPs and response to particular cell lines is critical for therapeutic development. The interaction may harm the cells (cytotoxicity), promote cell division or ineffective cytotoxicity. The MNPs are dose-dependent and higher concentrations induce cell cytotoxicity. Further, the MNPs tend to add coating agents, which may reduce cytotoxicity and be effectively used for targeted drug delivery, chemotherapy and magnetic hyperthermia applications.

References

[1] H. Sung et al. Global Cancer Statistics 2020: GLOBOCAN estimates of incidence and mortality worldwide for 36 cancers in 185 countries. CA: A Cancer Journal for Clinicians 71(3): 209–249 (2021).

[2] J. Jose et al. Magnetic nanoparticles for hyperthermia in cancer treatment: an emerging tool. Environmental Science and Pollution Research 27(16): 19214–19225 (2020).

[3] J.-H. Lee et al. Exchange-coupled magnetic nanoparticles for efficient heat induction. Nature Nanotechnology 6(7): 418–422 (2011).

[4] P. Hatthakarnkul et al. Systematic review of tumour budding and association with common mutations in patients with colorectal cancer. Critical Reviews in Oncology/Hematology 167: 103490 (2021).

[5] N. P. Praetorius and T. K. Mandal. Engineered nanoparticles in cancer therapy. Recent Pat. Drug. Deliv. Formul. 1(1): 37–51 (2007).

[6] Maier-Hauff, K. et al. Intracranial thermotherapy using magnetic nanoparticles combined with external beam radiotherapy: results of a feasibility study on patients with glioblastoma multiforme. Journal of Neuro-Oncology 81(1): 53–60 (2007).

[7] P. G. Corrie Cytotoxic chemotherapy: clinical aspects. Medicine 36(1): 24–28 (2008).

[8] I. Hilger, R. Hergt and W. A. Kaiser. Towards breast cancer treatment by magnetic heating. Journal of Magnetism and Magnetic Materials 293(1): 314–319 (2005).

[9] E. A. et al. Périgo. Fundamentals and advances in magnetic hyperthermia. Applied Physics Reviews 2(4): 041302 (2015).

[10] Kandasamy, G. et al. Functionalized hydrophilic superparamagnetic iron oxide nanoparticles for magnetic fluid hyperthermia application in liver cancer treatment. ACS Omega 3(4): 3991–4005 (2018).

[11] S. Jha, P. K. Sharma and R. Malviya. Hyperthermia: Role and risk factor for cancer treatment. Achievements in the Life Sciences 10(2): 161–167 (2016).

[12] S. Toraya-Brown and S. Fiering. Local tumour hyperthermia as immunotherapy for metastatic cancer. International Journal of Hyperthermia 30(8): 531–539 (2014).

[13] C. Martinez-Boubeta et al. Learning from nature to improve the heat generation of iron-oxide nanoparticles for magnetic hyperthermia applications. Scientific Reports 3(1): 1652 (2013).

[14] H. M. Williams. The application of magnetic nanoparticles in the treatment and monitoring of cancer and infectious diseases. Bioscience Horizons: The International Journal of Student Research 10: p. hzx009 (2017).

[15] L. Kafrouni and O. Savadogo. Recent progress on magnetic nanoparticles for magnetic hyperthermia. Progress in Biomaterials 5(3): 147–160 (2016).

[16] C. Kittel. Introduction to Solid State Physics. Wiley Eastern (2018).

[17] K. Jain et al. Dynamic magneto-optical inversion in magnetic fluid using NanoMOKE. Journal of Magnetism and Magnetic Materials 475: 782–786 (2019).

[18] P. Kumar et al. Low-temperature large-scale hydrothermal synthesis of optically active PEG-200 capped single domain MnFe2O4 nanoparticles. Journal of Alloys and Compounds 904: 163992 (2022).

[19] S. Pathak et al. Facile synthesis, static, and dynamic magnetic characteristics of varying size double-surfactant-coated mesoscopic magnetic nanoparticles dispersed stable aqueous magnetic fluids. Nanomaterials 11(11) (2021).

[20] A. Singh et al. Tuning the magnetocrystalline anisotropy and spin dynamics in CoxZn1-xFe2O4 (0 ≤ x ≤ 1) nanoferrites. Journal of Magnetism and Magnetic Materials 493: 165737 (2020).

[21] P. Kumar et al. Enhanced static and dynamic magnetic properties of PEG-400 coated CoFe2–xErxO4 (0.7 ≤ x ≤0) nanoferrites. Journal of Alloys and Compounds 887: 161418 (2021).

[22] Mishra, A. et al. Measurement of static and dynamic magneto-viscoelasticity in facile varying ph synthesized CoFe$_2$O$_4$-Based magnetic fluid. IEEE Transactions on Magnetics 55(12): 1–7 (2019).

[23] S. Pathak et al. Improved thermal performance of annular fin-shell tube storage system using magnetic fluid. Applied Energy 239: 1524–1535 (2019).

[24] S. Mornet et al. Magnetic nanoparticle design for medical diagnosis and therapy. Journal of Materials Chemistry 14(14): 2161–2175 (2004).

[25] I. M. Obaidat, B. Issa and Y. Haik. Magnetic properties of magnetic nanoparticles for efficient hyperthermia. Nanomaterials 5(1): 63–89 (2015).

[26] R. Hergt, S. Dutz and M. Röder. Effects of size distribution on hysteresis losses of magnetic nanoparticles for hyperthermia. Journal of Physics: Condensed Matter 20(38): 385214 (2008).

[27] R. Hergt, S. Dutz and M. Zeisberger. Validity limits of the Néel relaxation model of magnetic nanoparticles for hyperthermia. Nanotechnology 21(1): 015706 (2009).

[28] R. Hergt et al. Magnetic particle hyperthermia: nanoparticle magnetism and materials development for cancer therapy. Journal of Physics: Condensed Matter 18(38): S2919–S2934 (2006).

[29] R. Hergt et al. Maghemite nanoparticles with very high AC-losses for application in RF-magnetic hyperthermia. Journal of Magnetism and Magnetic Materials 270(3): 345–357 (2004).

[30] R. Hergt et al. Enhancement of AC-losses of magnetic nanoparticles for heating applications. Journal of Magnetism and Magnetic Materials 280(2-3): 358–368 (2004).

[31] A. B. Salunkhe, V. M. Khot and S. H. Pawar. Magnetic hyperthermia with magnetic nanoparticles: a status review. Curr. Top. Med. Chem. 14(5): 572–94 (2014).

[32] P. Fannin and S. Charles. On the calculation of the Neel relaxation time in uniaxial single-domain ferromagnetic particles. Journal of Physics D: Applied Physics 27(2): 185 (1994).

[33] R. E. Rosensweig. Heating magnetic fluid with alternating magnetic field. Journal of Magnetism and Magnetic Materials 252: 370–374 (2002).

[34] J. Giri et al. Investigation on T c tuned nano particles of magnetic oxides for hyperthermia applications. Bio-Medical Materials and Engineering 13: 387–399 (2003).

[35] P. Guardia et al. Water-soluble iron oxide nanocubes with high values of specific absorption rate for cancer cell hyperthermia treatment. ACS nano 6(4): 3080–3091 (2012).

[36] J. Carrey, B. Mehdaoui and M. Respaud. Simple models for dynamic hysteresis loop calculations of magnetic single-domain nanoparticles: Application to magnetic hyperthermia optimization. Journal of Applied Physics 109(8): 083921 (2011).

[37] B. Mehdaoui et al. Optimal size of nanoparticles for magnetic hyperthermia: a combined theoretical and experimental study. Advanced Functional Materials 21(23): 4573–4581 (2011).

[38] D. H. Kim et al. Targeting to carcinoma cells with chitosan-and starch-coated magnetic nanoparticles for magnetic hyperthermia. Journal of Biomedical Materials Research Part A: An Official Journal of The Society for Biomaterials, The Japanese Society for Biomaterials, and The Australian Society for Biomaterials and the Korean Society for Biomaterials 88(1): 1–11 (2009).

[39] A. Rajan and N. K. Sahu. Review on magnetic nanoparticle-mediated hyperthermia for cancer therapy. Journal of Nanoparticle Research 22(11): 1–25 (2020).

[40] R. Sappey et al. Nonmonotonic field dependence of the zero-field cooled magnetization peak in some systems of magnetic nanoparticles. Physical Review B 56(22): 14551 (1997).

[41] R. Ghosh et al. Induction heating studies of Fe3O4 magnetic nanoparticles capped with oleic acid and polyethylene glycol for hyperthermia. Journal of Materials Chemistry 21(35): 13388–13398 (2011).

[42] C. S. Kumar and F. Mohammad. Magnetic nanomaterials for hyperthermia-based therapy and controlled drug delivery. Adv. Drug Deliv. Rev. 63(9): 789–808 (2011).

[43] S. K. Paswan et al. Optimization of structure-property relationships in nickel ferrite nanoparticles annealed at different temperature. Journal of Physics and Chemistry of Solids 151: 109928 (2021).

[44] M. Albino et al. Role of Zn2+ Substitution on the magnetic, hyperthermic, and relaxometric properties of cobalt ferrite nanoparticles. The Journal of Physical Chemistry C 123(10): 6148–6157 (2019).

[45] I. J. Bruvera et al. Determination of the blocking temperature of magnetic nanoparticles: The good, the bad, and the ugly. Journal of Applied Physics 118(18): 184304 (2015).

[46] M. Gonzales-Weimuller, M. Zeisberger and K. M. Krishnan. Size-dependant heating rates of iron oxide nanoparticles for magnetic fluid hyperthermia. Journal of Magnetism and Magnetic Materials 321(13): 1947–1950 (2009).

[47] X. Zhou et al. Adsorption of sodium oleate on nano-sized fe3o4 particles prepared by coprecipitation. Current Nanoscience 3(3): 259–263 (2007).

[48] D.-L. Zhao et al. Inductive heat property of Fe3O4/polymer composite nanoparticles in an ac magnetic field for localized hyperthermia. Biomedical Materials 1(4): 198 (2006).

[49] A. Hanini et al. Zinc substituted ferrite nanoparticles with Zn0. 9Fe2. 1O4 formula used as heating agents for in vitro hyperthermia assay on glioma cells. Journal of Magnetism and Magnetic Materials 416: 315–320 (2016).

[50] C. Iacovita et al. Hyperthermia, cytotoxicity, and cellular uptake properties of manganese and zinc ferrite magnetic nanoparticles synthesized by a polyol-mediated process. Nanomaterials 9(10): 1489 (2019).

[51] W. Kai et al. Cytotoxic effects and the mechanism of three types of magnetic nanoparticles on human hepatoma BEL-7402 cells. Nanoscale Research Letters 6(1): 480 (2011).

[52] S. Kamiloglu et al. Guidelines for cell viability assays. Food Frontiers 1(3): 332–349 (2020).

[53] J. S. Kim et al. Toxicity and tissue distribution of magnetic nanoparticles in mice. Toxicological Sciences 89(1): 338–347 (2005).

[54] R. Vakili-Ghartavol et al. Toxicity assessment of superparamagnetic iron oxide nanoparticles in different tissues. Artificial Cells, Nanomedicine, and Biotechnology 48(1): 443–451 (2020).

[55] N. Malhotra et al. Potential Toxicity of Iron Oxide Magnetic Nanoparticles: A Review. Molecules 25(14): 3159 (2020).

[56] V. Vinodhini and C. Krishnamoorthi. Effect of dispersants on cytotoxic properties of magnetic nanoparticles: a review. Polymer Bulletin (2021).

[57] G. S. Zamay et al. Aptamers increase biocompatibility and reduce the toxicity of magnetic nanoparticles used in biomedicine. Biomedicines 8(3): 59 (2020).

10

Novel Applications of Magnetically Controllable Fluids in Energy Harvesting, Sensing, and Thermal Applications

Saurabh Pathak,[1,2,]* *Bishakhadatta Gayen,*[6] *Vinod Kumar,*[4]
Xu Wang,[5] *Sang-Koog Kim*[2,]* and *R.P. Pant*[3]

◇◇

10.1 Introduction

Magnetic manipulation of materials has shown immense potential due to their unique physical, chemical and optical properties that can be altered by an externally applied magnetic field [1, 2]. These properties, make them a good selection for a wide variety of applications [3–5]. One of the essential parts of both basic research and applications of these advanced materials depends on understanding the structural, flow, magnetic, thermal, mechanical and electrical properties of these materials to attain the anticipated assets [6]. The tunable properties of these magnetic materials make them suitable for many scientific, industrial and commercial applications. Nanotechnology has opened a whole new horizon of several new applications for these materials as Nano-sized particles act entirely differently from the bulk phase of these materials [7–9]. Magnetic materials appear primarily in solid forms and

---

[1] Department of Mechanical Engineering, University of Melbourne, Parkville, VIC, 3010 Australia.
[2] National Creative Research Initiative Center for Spin Dynamics and SW Devices, Department of Materials Science and Engineering, Seoul National University, Seoul 151-744, South Korea.
[3] CSIR-National Physical Laboratory, New Delhi-110012, India.
[4] Netaji Subhash Technology Institute, New Delhi-110078, India.
[5] School of Engineering, RMIT University, Melbourne 3001, Australia.
[6] CAOS, Indian Institute of Science, Bangalore, India 560012.
* Corresponding authors: pathak@snu.ac.kr, sangkoog@snu.ac.kr

none of the fluids in nature shows magnetic properties at room temperature [10]. It is because magnetism is an asset associated with the crystalline nature and ordering at the atomic level in materials lacking in fluids. Solid materials which also show that magnetic properties lose their magnetism well before melting due to significant disturbances at the atomic scale due to temperature rise, which restricts magnetic ordering [11]. To overcome this challenge and prepare fluid that the magnetic field can control, Steve Papell, (1963) prepared a colloidal suspension of magnetically polarizable particles in a carrier medium [12]. These were the first synthetic liquid magnets and escalated the interest of researchers from different fields as they present fascinating properties [13].

Although the concept of the Magnetic Fluid (MF) came from Papell, the fluid he prepared was not stable and far from helpful for practical applications. R. E. Rosensweig and colleagues added a significant development in the field. They gave the name ferrofluid (FF) and created a new domain or research named ferro-hydrodynamic (FHD) [14]. Their work escalated to achieve highly magnetic liquid with improved stability pioneering the stage for the development of various applications of FF in the next decade [15]. Multiple applications of FFs are well established, such as magnetic seals, heat transfer in miniature devices, energy harvesting and semi-active dampers [11]. However, consistent research effort is still ongoing for these applications to improve their efficiency [16]. In the last two decades, numerous new application domains have also escalated, such as biomedical applications (magnetic resonance imaging contrast agent, hyperthermia, targeted drug delivery, etc.), adaptive optics shape-shifting magnetic mirror, thermomagnetic convection, analytical instrumentation and precision measurements which are very promising for wide interdisciplinary applications [17]. The focus of the research in the field has been directed on maximizing the magnetic moments of these fluids with enhanced stability to improve the performance of the existing application [12, 18].

MFs are colloidal suspensions of magnetic nanoparticles (MNPs) in the size range of 2–30 nm in the desired carrier liquid. It displays unique properties such as low frictional and sealing with a glass substrate and capillary, magnetic levitation by forming magnetic bearing at the poles, magneto-viscous effects, improved thermal conductivity, variable density and magnetic field-controlled rheology [19, 20]. It exhibits Newtonian behavior when no external magnetic field is applied, whereas, under the influence of an external magnetic field, it shows a non-Newtonian behavior [21]. In the absence of an externally applied field, the viscosity of MFs does not depend on operating conditions such as applied shear, oscillation, etc. It remains constant for a fixed temperature [19]. However, as soon as it comes under the influence of an external magnetic field, the viscosity of MFs starts to vary nonlinearly with different operative conditions and displays a shear-thinning behavior for a low magnetic field and high shear rates [22]. In static loading conditions (low applied shear rates), it acts like a semi-solid due to the chain structure's formation, which prevents the fluid's smooth, streamlined motion. When the applied shear rate is increased, the chains in the structure start to cease causing a decrease in viscosity and MF acts like a liquid. This behavior of a reduction in viscosity at a high shear rate is labeled as thinning behavior [18, 22, 23].

MF sticks to permanent magnets and forms a bearing around the pole. The MNPs form a layer around the pole and the thickness of these layers depends on the magnetic strength of the permanent magnet and MNPs characteristics. MF flows towards high-to-low magnetic flux density with a gradient magnetic field application. When interacting with the permanent magnet, it starts to move in the direction of the poles due to the higher magnetic field at the pole [17]. MFs accumulate near the pole and attach to it, creating an upward pressure gradient causing the lifting of the magnets from the base surface [24]. This lifting of the magnet from the base surface due to the attachment of the MNPs is labeled as the formation of magnetic bearing. The formation of magnetic bearing on the poles of the permanent magnets provides a passive levitation, which does not require any power as in the case of other magnetic levitation. This magnetic bearing provides an ultra-smooth surface for the movement of the magnets with minimum friction making it advantageous for energy harvesting applications where frictional losses are one of the major drawbacks [17,25]. Thus, MF bearings are very effective as they suffer from less frictional fluctuations and are more controllable systems that can be developed. The high value of the magnetic saturation of the MFs keeps them within the cavity under centrifugal shocks and vibrational forces to ensure long service life and low environmental contamination [10].

One of the most critical features of nanomagnetic fluid is the relative change in their viscosity with a gradient magnetic field (∇H), termed Magneto-Viscous Effect (MVE). MFs exhibit Newtonian behavior without an external applied magnetic field [22]. However, in the presence of a magnetic field, a dipole moment is induced in each of the particles, resulting in the chain-like structure formation in the direction of the applied field and behaving as non-Newtonian fluid [23]. Such structures are critical factors in controlling the rheological properties of these fluids. These structures allow the material to act like a solid material until a critical stress value depends on the induced dipolar interaction. Usually, the Bingham model describes the flow properties of such fluids below a critical yield stress value, above which chain-structure brakes and material starts to flow and acts like Newtonian Fluid (NF) [26].

Effective thermal management of the systems is significant for technical challenges faced by numerous industries such as microelectronics, automobiles, sensors and actuators, solid-state lighting, waste heat recovery and manufacturing as it governs the system's efficiency. The thermal conductivity of these fluids increases significantly with the application of a magnetic field and can reach up to four-times higher values than without a magnetic field [24]. Philip et al. reported that $k_f/k = 4$, where k_f is the thermal conductivity of fluid at magnetic field intensity 200 mT and k is thermal conductivity without a magnetic field [27]. The main reason for the enhancement in thermal conductivity is the formation of nanochannels which propagates heat at higher rates. It contributes to higher heat transfer properties, eventually increasing thermal efficiency. MFs can be controlled by applying an external magnetic field, which gives flexibility to take variable loads and load-dependent localized cooling for effective thermal management. MFs treasure trove many applications in wide variety of field as shown in Fig. 10.1 [1, 10, 19, 21, 27].

MFs have had a wide variety of application areas in the past 50 yr since the first MF synthesis was reported. It has engaged the attention of multidisciplinary

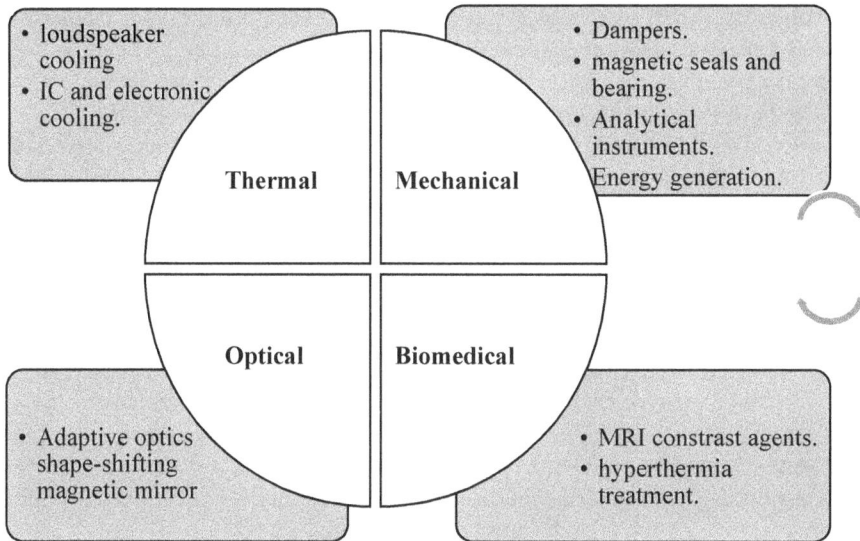

Figure 10.1. Sectoral branching of various applications of MF.

researchers as these advanced materials' properties can be altered by the commonly used magnetic field setups without any major underpinning requirements [3]. Although the primary challenge still lies in effectively controlling their properties which has been undertaken from the latest development in nanotechnology. These applications have gained importance in everyday life and represent a significant commercial value. One of the earliest applications of MF is a magnetic seal, which has been the critical application in hard disk drives to provide free spinning to the shaft as a small amount of MF is placed between the shaft and housing, which are contained inside magnets [10, 25]. These magnetic seals have received a lot of attention as they can withstand large loads at high speed and with high efficiency. The other primary application of MF is as a coolant in loudspeakers and small electronic devices, as they provide a highly efficient cooling performance [28]. In the early 80s, MF was used in loudspeakers to remove heat from the voice coil. Even with miniaturizing trends of the current system, MF is still used as a coolant in speakers due to its ability to withstand high pressure. It can remove large heat loads over a small space at higher thermal efficiency [24]. Another vital application of MF is in semi-active-controlled dampers, as they show viscoelastic behavior and can be used in automobile and space applications. These dampers utilize the change in density of the MF with a change in the magnetic field as a shock absorber and reliever gradually require very low power consumption [29].

The sensation and actuation property of MF has also been successfully utilized in optical applications such as adaptive optics shape-shifting magnetic mirrors and analytical instruments [26, 30]. Low frictional property and magnetic levitation with permanent magnets have been efficiently used for electricity generation [31]. The energy application of MF is quite an intriguing prospect as they present higher efficient energy generation with slight modification in the current systems [18, 22, 23, 32]. A more recent application of MF is in the biomedical field, where Magnetic Fluid

Hyperthermia (MFH) treatment is used to treat Lyme disease and cancer [9]. MFH is a method that uses the high-temperature treatment of the affected cells by generating heat from the highly anisotropic materials' energy losses. The flipping of spins in the external magnetic field generates heat for the treatment [9, 23].

Even after five decades of consistent efforts in the area by numerous research groups from around the world, MF still remains within the realm of possibility to come to its flourishing. However, there is a large gap between the advancement in materials and their translation to the device, which restricts it from reaching its full potential [3, 33–35]. Although, the recent development in the field of nanotechnology has opened new thriving interdisciplinary applications of MFs. With the progress of sophisticated characterization methods and precise measurement tools, MF-based technologies have opened a new horizon for applications with improved performance. To bridge the gap between the materials to the device, a fundamental understanding of the properties is vital [36]. In this chapter, the unique properties of the MFs will be described first, followed by energy, sensing and thermal applications later, respectively. Different applications have been discussed with operative principles and mechanisms, along with the latest developments.

10.2 Unique properties

MF demonstrates unique sensations and actuation properties in the magnetic field, which can be tuned externally. The dispersion of the superparamagnetic particles in the carrier liquid allows MF to display non-Newtonian characteristics in the presence of the field and Newtonian in the absence of the field. This unique characteristic makes them exciting prospects for active control devices. Here the magneto-rheology, enhanced thermal characteristics, hydrodynamics and self-levitation properties of the MFs will be described.

10.2.1 MF bearing formation and passive levitation

MFs stick to the permanent magnet and form a bearing around the poles, providing a frictional surface due to the improvement of the permanent magnet. Due to this pressure-induced upliftment, the solid-to-solid contact between the permanent magnet and substrate is restricted and the interface between them works as a liquid interface [37]. This liquid interface provides a smooth, frictionless motion to the permanent magnets, which can be effectively used to develop novel devices. The improvement of the permanent magnet due to pressure-induced body forces is termed self-levitation or passive levitation (Fig. 10.2 (a)). These frictionless bearings provide a leak-proof sealing and are widely used in the markets for sealing purposes [17]. A magnetic coil is generally placed under the magnet with an outer diameter equivalent to the diameter of the magnet within the bearing (Fig. 10.2 (b)) for energy harvesting applications.

Further, Figs. 10.3 (c) and 10.3 (d) depict the energy harvester developed by capitalizing on the self-levitation properties of the MFs. A combination of six MF levitation-based permanent magnets moves towards the fixed coil, which produces

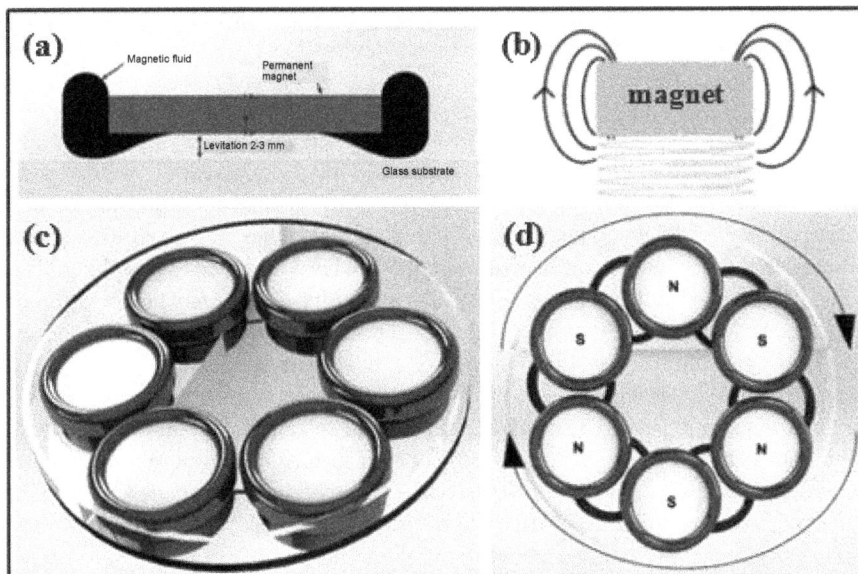

Figure 10.2. (a) Levitation of the permanent magnet on the glass substrate (b) Magnetic field representation of the permanent magnet over the inductor coil (c-d) Permanent magnet configuration coated in magnetic fluid with inductor coils shown on the underside at the static and moving position (Reproduced with permission from Ref. [38]).

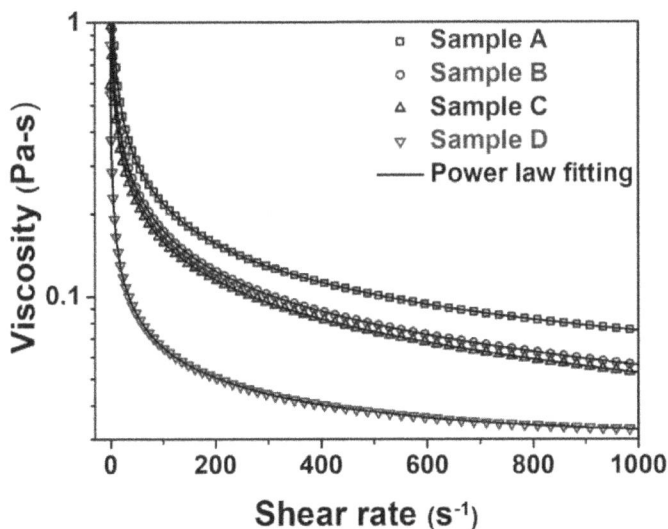

Figure 10.3. Viscosity as a function of shear rate for MFs at 0.18T applied field. MFs display a shear thinning behavior at higher applied shear rates (Reproduced with permission from Ref. [37]).

electricity. The self-levitation and magnetic bearing properties are extensively used to develop many novel and high-performance devices [38].

10.2.2 *Rheological properties*

Rheology or viscoelastic properties of the MF play an essential role in many technological applications. In the case of MFs, it is crucial to determine how an external magnetic field modifies the rheological properties of fluids. Rheological properties become more complicated when the fluid is a mixture of several liquids or contains dispersed solid particles. MFs usually exhibit Newtonian behavior without a magnetic field where shear stress is linearly dependent on the shear rate. In contrast, its viscosity is modified in the presence of a field. The application of an external magnetic field causes the increase in viscosity of the MF and the nonlinearity of the stress-strain relationship. The change in density due to nanochannel formation causes a hindrance in the fluid streamline flow, which in turn cause the non-Newtonian behavior [18, 22, 23, 32]. In the absence of the field, the viscosity of the MFs remains constant; however, as the magnetic field is turned on, the viscosity of the MFs shows a non-linear behavior that changes with the shear rate Fig. 10.2.

10.2.3 *Hydrodynamic properties of MFs*

Hydrodynamics deals with the laws of motion of liquids and gases and their interaction with solids. Hydrodynamics may be termed magneto-hydrodynamics, electro-hydrodynamics or hydrodynamics, depending on the body's forces. In hydrodynamics, the body force is gravity, while in magneto-hydrodynamics, the body force is j x B, i.e., the Lorentz force. Assuming that magnetization induced in the fluid is always parallel to the direction of the applied field, the body force in MFs becomes $F_m = \dfrac{V}{4\pi MH}$ [39]. MFs motion is mainly governed by the ferrohydrodynamic equations and can be controlled by the external magnetic field. It allows the development of applications such as cooling systems with localized and directional cooling, damping systems and soft robotics [23].

10.2.4 *Thermal properties*

The thermal properties of the MFs can be altered and tuned using the external magnetic field. MF shows enhanced thermal properties in the presence of the magnetic field and is thus widely studied and employed for active and semi-active cooling systems. With the trend of militarization in electronics, the growing need for a more efficient cooling system evolved rationally. The thermal conductivity of the MF increases four-fold in the presence of the magnetic field, resulting in enhanced heat transfer (Fig. 10.4). The system's thermal efficiency with MF as a working media is very high due to the exceptional heat transfer characteristics. Thermal properties of the MF are dependent firstly on the MNPs size, shape and concentration and secondly on the carrier liquid [24]. A wide range of applications, such as localized and self-propelled cooling, cooling, energy storage, transportation and conversion devices, have been

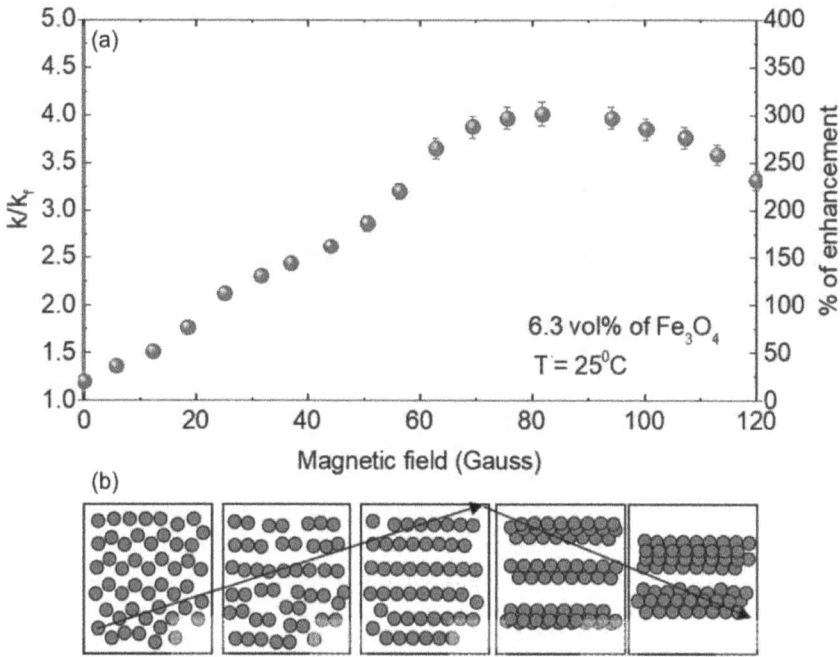

Figure 10.4. (a) Variation of thermal conductivity ratio (k/k_f) and its percentage enhancement as a function of the applied magnetic field (Replotted from Ref. [27]), (b) Schematic representation of the chain-like structures formed within the fluid at different magnetic fields (Adapted from [Ref. 41]). The arrow depicts the increasing and decreasing trend of thermal conductivity with an escalating magnetic field (Reproduced with permission from Ref. [36]).

developed and recently employed the enhanced heat transfer characteristics of MFs [40].

10.3 Energy applications

The energy harvesting applications of MFs have been gaining significant attention in the past few decades [42]. The unique and exciting properties of MFs have been extensively used for developing highly efficient energy production by scavenging various energy sources such as wind, thermal, vibration and ocean currents [38, 43]. Energy harvesters employing MFs are developed explicitly in compact sizes targeting the mW to μW power generation. With the miniaturization trends, many wearable sensors, wireless sensor networks for environmental and weather monitoring, etc., require self-sustained power in restricted spaces [44, 45]. However, recently, an effort has been made towards developing large-scale energy harvesting using the self-levitation properties of the MFs propelled by wind energy [37, 38]. Here the recent developments in the field of MF-based energy harvesting will be described broadly on the basis of; (1) Energy harvester based on the motion of the MFs (2) based on the self-levitation properties of the permanent magnet-MF geometry and (3) based on the magneto-thermal effects of the MFs. All these energy harvesters were developed targeting a specific type of application and thus the basic principal, different mechanisms and target applications of each of the three categories will be described.

10.3.1 Energy harvesting based on the motion of the MFs

MFs display superparamagnetic characteristics and thus have a large magnetic moment compared to the paramagnetic material [46]. It allows the development of an energy harvester by fixing the conductor coil around the MFs motion (results in a change in magnetic flux), which is generally generated by vibration or oscillation [43]. The energy harvester under this classification works on the basic principle of electromagnetism, i.e., Faraday's law which states that the change in the magnetic flux across a conductor results in the generation of the electromotive force (e.m.f.) and its magnitude is proportional to the rate of change of the magnetic flux [37]. These energy harvesters mainly target small-scale energy production from mW to μW [47, 48]. Most of the past research on the energy harvester was focused on the motion of the MFs, capitalizing on the mechanical vibrations [49].

An electromagnetic vibration energy harvesting based on the sloshing motion of the MFs was done by Bibo et al. for micro-power generation [50]. The energy harvester capitalizes on the sloshing motion of the MFs inside the column of the seismically-excited tank, as shown in Fig. 10.5. The MF's magnetic times are randomly oriented, as seen in Fig. 10.5 (a), which is aligned using a permanent

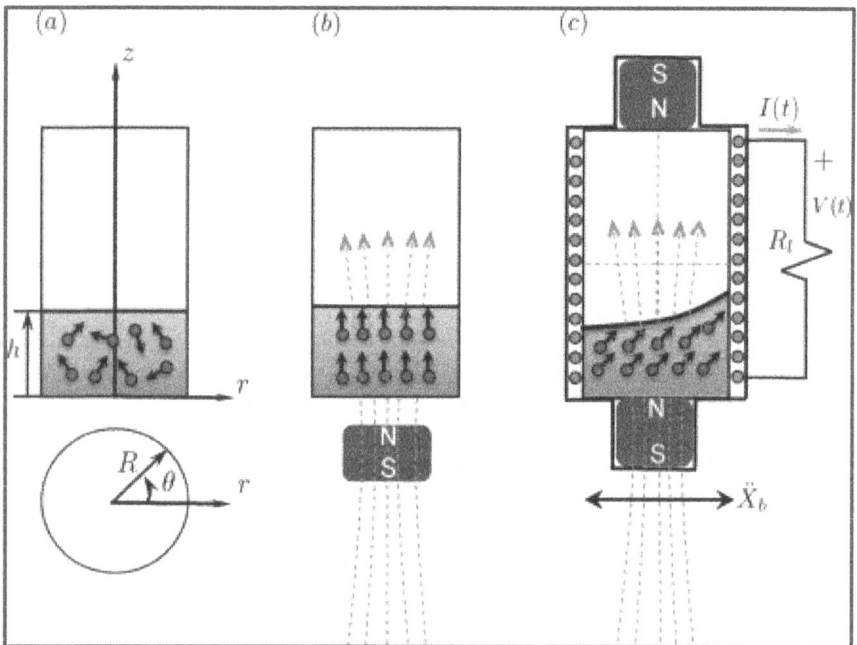

Figure 10.5. Schematic diagram of the basics of the MF-based energy harvester. (a) Randomly oriented MNPs with no net magnetic moment. (b) A permanent magnet is placed at the bottom, which aligns the MNPs resulting in a net magnetic moment. (c) Sloshing motion of the MFs, resulting in a change in magnetic flux (Reproduced with permission from Ref. [50]).

magnet as depicted in Fig. 10.5 (b). MFs display superparamagnetic characteristics; thus, their net magnetic moment is non-zero only when an external magnetic field is applied. The major advantage of using MFs for energy harvesting is their ability to capture even the smallest variation that permanent magnets can not capture due to their inertia. A conductor coil set is placed wrapping the seismically-excited tank, as shown in Fig. 10.5 (c). When the seismic excitations match the frequency of the infinite modal frequencies of the MFs column (resonance conditions), surface waves with large amplitude in the horizontal and rotational direction are generated. The motion of the sloshing liquid changes the orientation order of the magnetic dipoles resulting in a time-varying magnetic flux. The change in the magnetic flux due to the motion of the MFs inside the seismically-excited tank results in a revision in the magnetic flux relative to the coil, which results in electricity generation [50].

Further, a hybrid concept of the triboelectric-electromagnetic generator using MFs was proposed by Seol et al., which took advantage of both the constituent of the MFs, i.e., carrier liquid and magnetic nanoparticles [47]. The electrostatic induction along the side walls made of polymer materials and carrier liquid generates a triboelectric component. The electromagnetic induction among the suspended MNPs and an outer conductor coil causes the electromagnetic component. The schematic of both the energy harvesting components is depicted in Fig. 10.6. It is the first report on the triboelectric-electromagnetic generator's hybrid concept, which indicates its highly responsive nature to the smallest of vibrations, its ability to be

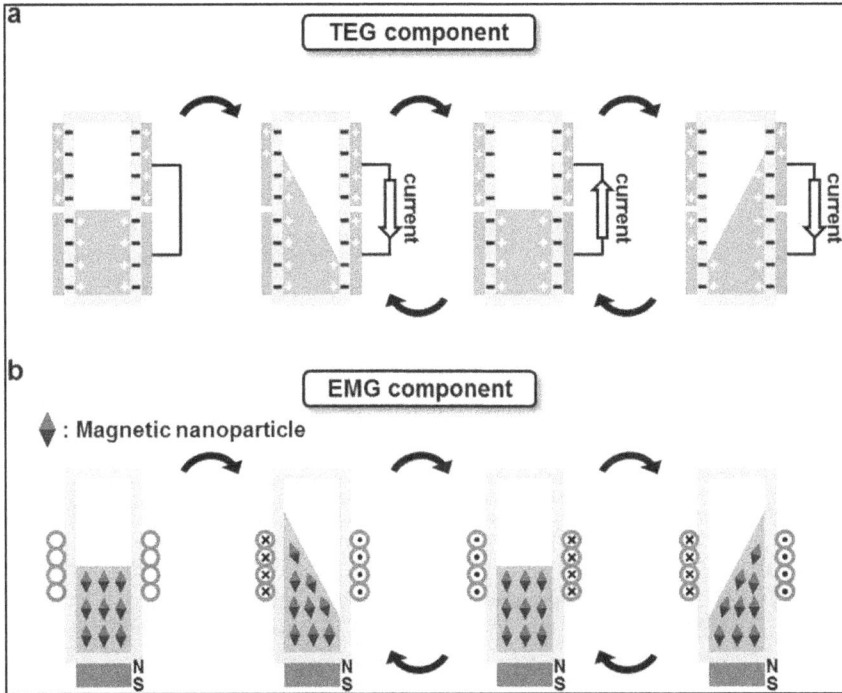

Figure 10.6. Operating principle of the MF-based triboelectric-electromagnetic generator. (a) Operating principle of the triboelectric nano generator component. (b) Operating principle of the electromagnetic generator component (Reproduced with permission from Ref. [47]).

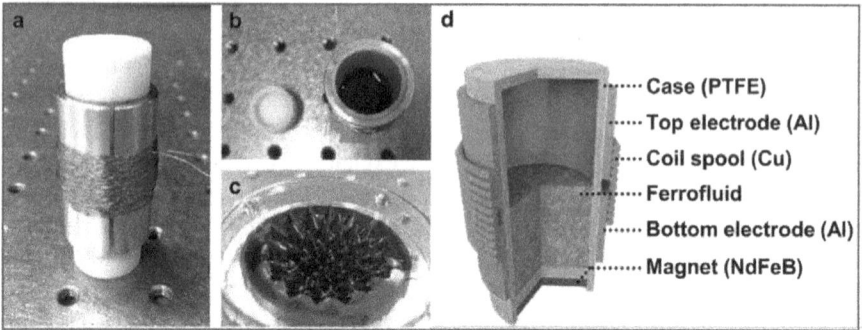

Figure 10.7. Structure of the MF-based triboelectric-electromagnetic generator: (a) Image of the outer body of the device. (b) Top view image inside of the device. (c) MFs spike when a magnet is placed just below the MF placed in the glass container (d) Section view of the energy harvester showing different layers (Reproduced with permission from Ref. [47]).

tuned for different types of input energy and its highly durable characteristics during operation. The significant advantage of the hybrid system is a tremendously low threshold amplitude and a wide range of operating frequencies [47].

Figure 10.7 shows the MF-triboelectric-electromagnetic generator made of a hollow cylinder with inner walls made of polytetrafluoroethylene. The main reason for choosing polytetrafluoroethylene is its high value of negative order of electrification. The cylinder is half-filled with MFs, as shown in Fig. 10.7 (b). Aluminium electrodes surround the outside walls of the polytetrafluoroethylene cylinder and a wire coil of 200 turns is wrapped around it. The NdFeB permanent magnets are placed only at the bottom of the cylinder to polarize the MNPs and concentrate all the MFs at the bottom. The aluminium electrode is formed in a c shape to minimize the distance between the magnet and coil, as depicted in Fig. 10.7 (a), which restricts the eddy currents and subsequently in a damping effect. The thickness of the cylinder is 1mm with outer and inner diameters of 30 and 28 mm, respectively, with a total height of 80 mm. Figure 10.7 (d) shows the section view of the energy harvester depicting different layers with material descriptions. However, the generated power is in the order of micro-watts which can be scaled up by adopting the design modification and parameter optimization. Some other investigations have also reported adopting a similar device structure through simulation and experimental investigations [42, 47, 51]. The main focus of the research on these devices is inclined toward scalability for large-scale energy production [47].

10.3.2 *Based on the self-levitation of the permanent magnet*

Passive levitation or self-levitation is the phenomenon of the upliftment of permanent magnets from the base surface due to the pressure induced by the MNPs on the base surface [17]. In the case of cylindrical magnets, MFs tend to move towards the edges of the magnets, forming a layer of MFs of around 3 mm, which separates the permanent magnets from the base surface and supports the load [37]. The formation of the magnetic bearing around the loop of six permanent magnets of 30 mm diameter and 15 mm thickness is shown in Fig. 10.8. Other shapes of magnets were

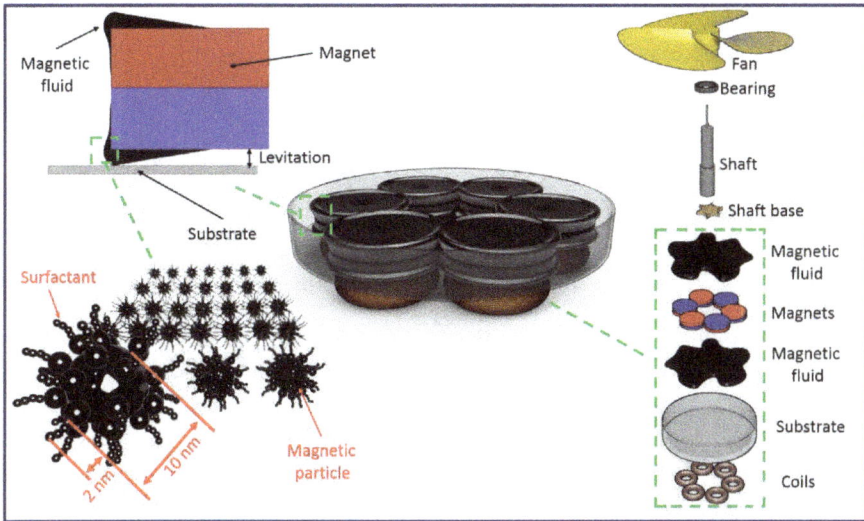

Figure 10.8. Schematic representation of the composition of the magnetic fluid-bearing-based wind energy harvester (Reproduced with permission from Ref. [37]).

also studied, such as rectangular with the self-levitation phenomenon; however, they were limited due to the specific nature of applications in which they can be used. The cylindrical magnets are more widely studied as they can be replaced in the existing systems without major modifications.

An MF-based energy harvester was developed by Pathak et al. capitalizing on the self-levitation properties of the cylindrical magnet-MF geometry [37, 38]. The ultra-low friction surface is formed between the magnet and bottom glass substrate to reduce the friction significantly, which greatly improves the energy conversion efficiency. The device reportedly works on the basic principle of Faraday's law of electromagnetic induction, where a combination of the self-levitated magnets moved by the input wind energy for a fixed coil produces e.m.f. in the coils that are connected to the external electrical circuit. Wind energy was chosen as the input by Pathak et al. as it is one of the cheapest, cleanest and most economical energy resources [38]. They presented a detailed numerical and experimental investigation of the MF-based wind energy harvester and displayed a significant improvement in the energy conversion efficiency. MF-based wind energy harvester provides a remedial to the most encountered problems in wind energy generation, i.e., the requirement of high wind speed for efficient performance [37]. Mechanical friction is a notable loss in most systems that limits their performance.

Figure 10.8 shows the schematic of the prototype wind energy harvester, where the center part shows the magnetic bearing formed by the six magnets and coils placed at the bottom of the magnets. The levitation of the magnets using MFs by the pressure-induced upliftment is depicted on the left side, and the structure of the wind energy harvester is shown on the right side of Fig. 10.8. In addition, the experimental setup of the prototype wind energy harvester is shown in Fig. 10.9, which consists of magnetic bearing rings of six magnets, six coils each with one-thousand turns, a Hylam shaft support structure for the shaft base, glass substrate, fans and energy

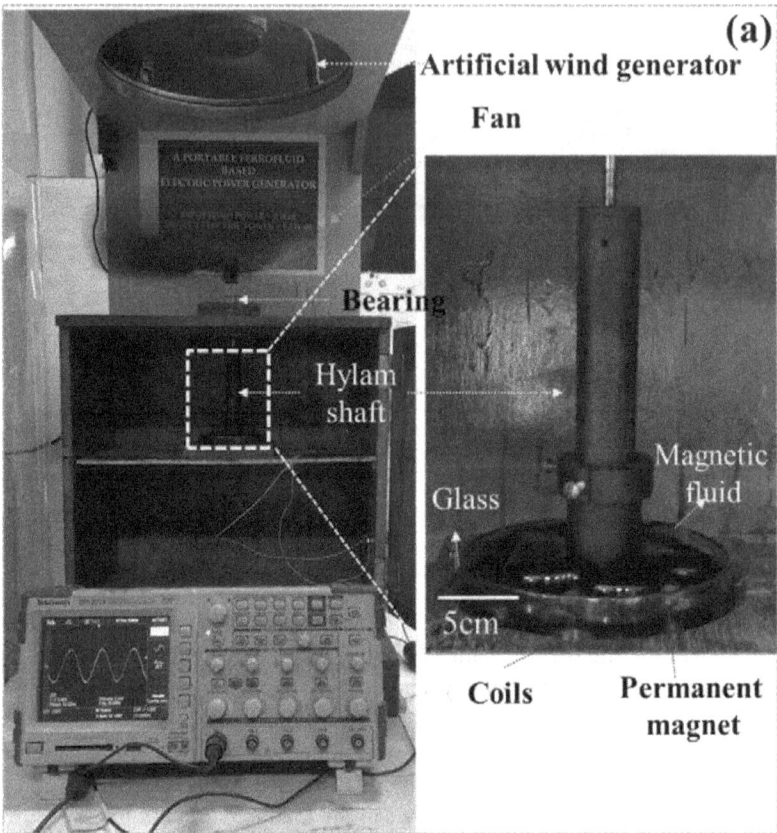

Figure 10.9. (a) Experimental setup of the prototyped magnetic fluid wind energy harvester. (b–d) Load bearing capacity measurement of the permanent magnet with a magnetic fluid bearing where (b) Permanent magnet and circular ceramic alumina boat. (c) Permanent magnet with the nanomagnetic fluid bearing on top of the ceramic alumina boat for measuring the bearing load capacity of individual magnets. (d) The nanomagnetic fluid bearing of six magnets forms a nanomagnetic fluid ring on the permanent magnet to measure and calculate the bearing load capacity by increasing the calibrated loads (Reproduced with permission from Ref. [37]).

harvesting power electronics. An enlarged view of the magnetic bearing and Hylam shaft is also shown on the right side of Fig. 10.9. This wind energy harvester can be efficiently operated at a low wind speed of 1.8 m/s, and its efficiency grows tremendously with the increase in wind speed. The prototype system reported a 26% efficiency at a wind speed of 4m/s and can efficiently generate around 3 W power [37]. The friction factor of MF also shows the lowering of shear-dependence and frictional power reduces significantly at the higher shear rate. It shows that frictional power reduces greatly at higher operating torque, which leads to higher efficiency [37, 38].

In addition, they used Maxwell simulation to verify the prototype results numerically and the dependency of the MF's rheological properties of the system. MF-based energy conversion systems are very efficient and can provide an effective solution to the current needs of wind energy systems. Magneto-viscoelasticity of

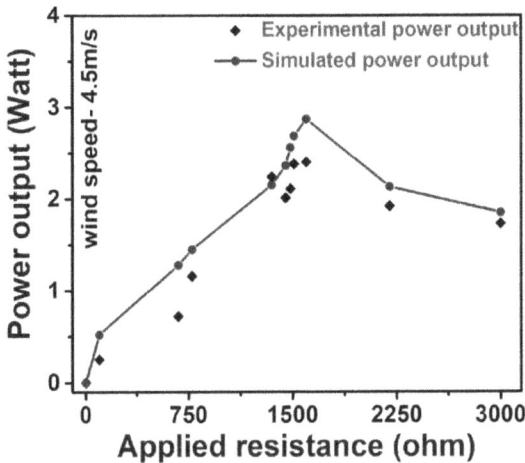

Figure 10.10. Experimental and simulated power output values for the nanomagnetic fluid-based wind energy harvester at different input wind speeds (Reproduced with permission from Ref. [37]).

MFs is the major parameter that directly affects the performance, as shear-induced viscoelasticity directly modifies the frictional power loss in the magnetic bearings. Further, the experiments were performed for the MF-based energy harvester with varying wind speeds and magnetic bearing configurations to probe the effect of each quantity on the power output. Figure 10.10 shows the power output curve of the MF-based wind energy harvester with varying wind speeds at different applied loads. The maximum power in the energy harvester is obtained when the externally applied load matches the internal resistance of the coils. Machine learning approaches such as the Response Surface Method (RSM) and Genetic Algorithm (GA) have also been employed to obtain the optimal geometrical parameters [38].

Additionally in another approach, Wu and colleagues developed MF self-levitation-based electromagnetic wearable three degrees of freedom (DoF) resonance human body motion energy harvester [52]. Wearable sensors are gaining significant attention in the recent past decade and these devices require self-powered energy for convenience [53]. These devices are vital for our daily life as they are part of health monitoring and tracking the system and thus, a reliable energy source is necessary [54]. Many of these devices work on lithium-ion batteries, which require frequent charging; however, recently, many other self-powered sources such as thermoelectric generators, triboelectricity, solar power, etc., have been extensively studied to power these devices without the need for frequent charging [55]. Working towards this direction, Wu et al. developed an MFs-based electromagnetic energy harvester capitalizing on the motion of the human body. The schematic of the developed energy harvester is shown in Fig. 10.11, which consists of a rectangular permanent magnet with a magnetic bearing on the edges. It provides the magnets with a smooth motion inside the rectangular box, wrapped by a copper coil from outside to scavenge electricity. The permanent magnet connects the two elastic springs within a rectangular box, forming a 3-DoF vibrator [52].

The energy harvester developed has been probed by simulation and experiments with different operating and loading conditions. A simulation model is established,

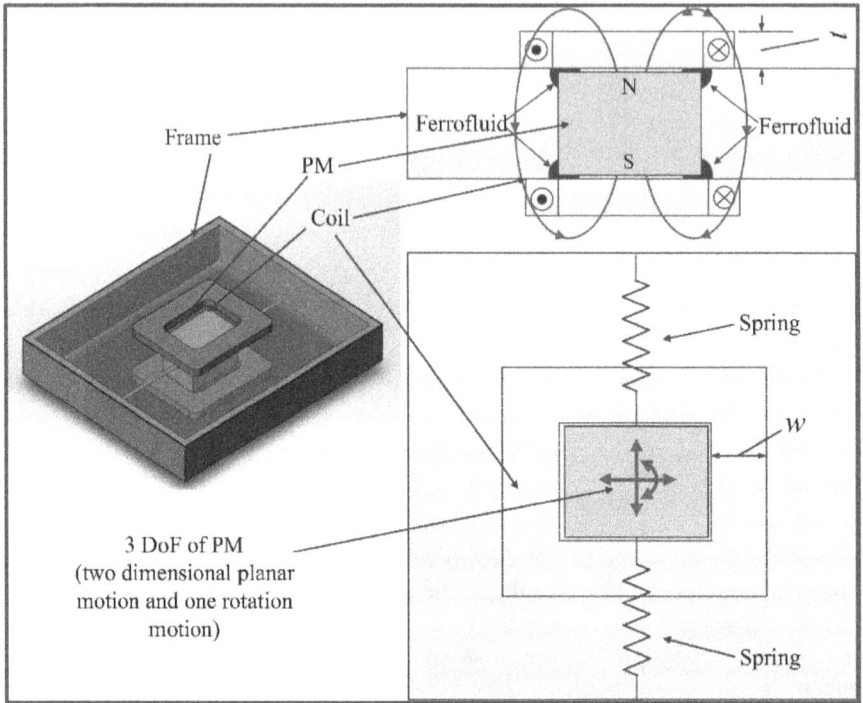

Figure 10.11. Schematic of the MF self-levitation-based electromagnetic wearable three Degrees of Freedom (DoF) resonance human body motion energy harvester (Reproduced with permission from Ref. [52]).

which is the first reported for the interface of the magnet-MF-solid substrate. Experimental results at different resistance loading also resulted in peak power of 1.1 mW and 2.28 mW in walking and running conditions, respectively. Further, the energy generated is stored by making an energy storage circuit that transfers the generated low AC voltage to 5 V DC, resulting in an average storage power of 0.014 mW and 0.149 mW while walking or running , respectively. Further, Fig. 10.12 shows the developed energy harvester attached to the shoe to offer a continuous power supply for wearable sensors and devices. The resonance frequency has been adjusted by optimizing the energy harvester's structure and the springs' stiffness. The proposed energy harvester can also be attached to human hands by geometry optimization and scaling down the device's size. The low energy efficiency of the developed device is also due to several loss factors, which can be improved by efficient trapping of the magnetic flux changes [52].

10.3.3 Electrochemical, thermal energy storage

The conversion of waste heat to electricity has long been an emerging prospect for tackling energy scarcity worldwide [57]. Thermoelectric (TE) materials are one of the

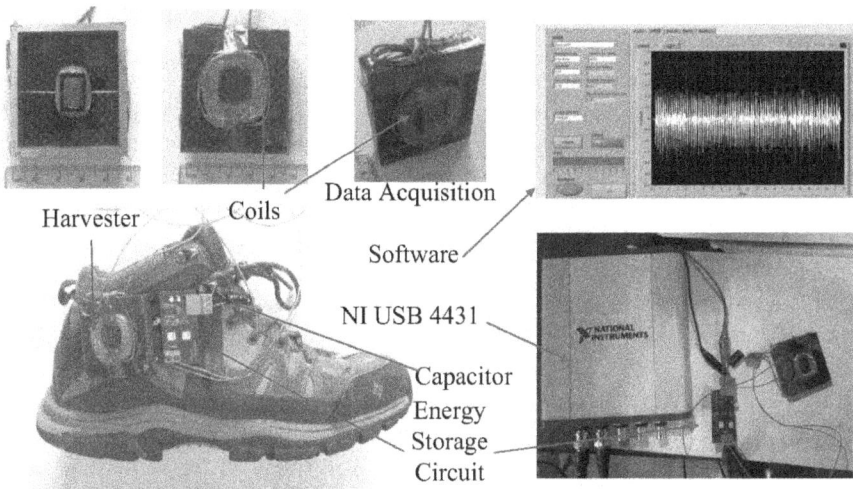

Figure 10.12. Prototype of the MF self-levitation-based electromagnetic wearable three degrees of freedom resonance human body motion energy harvester (Reproduced with permission from Ref. [52]).

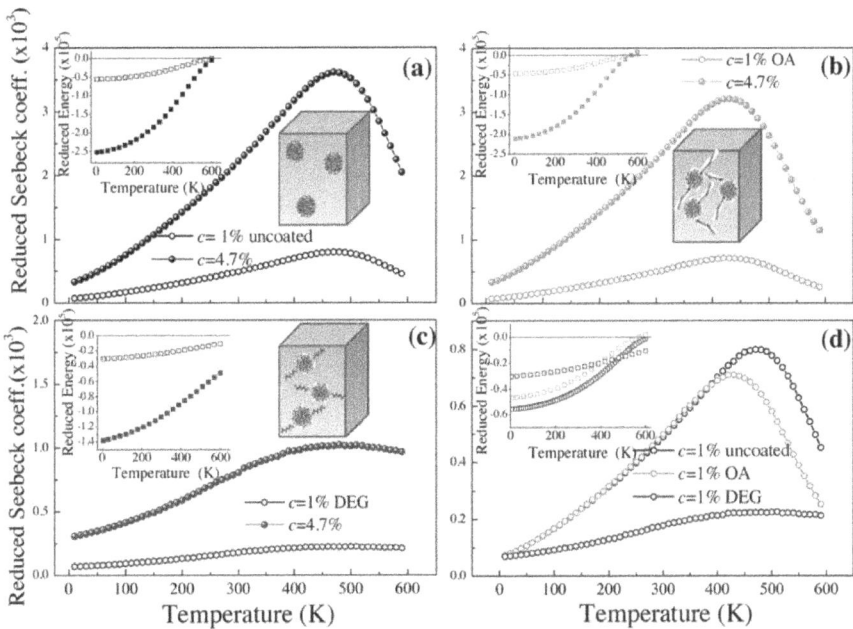

Figure 10.13. Reduced Seebeck coefficient of $CoFe_2O_4$ nanoparticles as a function of the average temperature T for two concentrations $c = 1\%$ (open symbols) and $c = 4.7\%$ (closed symbols) for (a) uncoated nanoparticles (black), (b) oleic acid (OA) coated, (c) diethylene glycol (DEG) coated nanoparticles and (d) comparison of the different coatings for $c = 1\%$. Insets: The average thermal energy $<E>$ includes the dipolar energy and the anisotropy energy terms with temperature-dependent g and k for the two-particle concentrations (Reproduced with permission from Ref. [56]).

most popular materials which convert low-to-medium grade waste heat into useful electricity [58]. The ionic liquid-based thermo-electrochemical cells were recently reported by Vasilakaki et al. as an alternative to the solid-state TE generators for waste heat to electricity scavenging applications [56]. They have shown that using the MFs as liquid-based TE material augments the Seebeck effect (Fig. 10.13). It opens new outlooks for developing a novel TE device with improved efficiency and cost-effectiveness. It is the first report on the TE application of magnetically polarizable particles dispersed in liquid media. They have adopted a thermodynamic approach to obtain the Seebeck coefficient's analytical expression. The Seebeck coefficient is mainly a function of magnetic interactions (dipolar, magnetostatic, super-exchange, etc.) and magnetic characteristics of the MNPs (saturation magnetization, coercivity, anisotropy, etc.). They calculated the Seebeck coefficient of the cobalt ferrite and γ- iron oxide by implementing a mesoscopic scale modeling with the Monte Carlo Metropolis algorithm. The algorithm is implemented by adopting dilute MFs. The results indicate that the Seebeck coefficient escalates with an increase in the anisotropy of the MF system. The magnetic interaction among the particles also plays a crucial role in controlling the TE characteristics of the MFs. They presented an exciting discovery that for a wide range of temperature values, thermally induced switching of the magnetization vector of each MNPs is blocked by the increase in magnetic anisotropy. It results in an overall improvement in the response of the chemical potential variation as a function of thermal gradient and, accordingly, an increase in the Seebeck coefficient. This study demonstrates the opening of a new perspective in electrochemical and thermal energy storage using the magneto-thermal properties of the MFs [56].

10.4 Sensing applications

The use of the MFs as a working or sensing element in various types of sensors and transducers is gaining significant attention due to the miniaturization trend and development of novel sensors to meet the current demand for improved performance [59]. MFs displays systematic variations with the change in different physical properties and characteristics, which serves as a basis for developing the sensors employing MFs as sensing/working elements. In general, the MFs sensors and transducers work on various principles such as inductive, capacitive or inductive–capacitive detecting, exploiting the magnetic levitation, optical, thermal, acoustic, viscous and magneto-electric properties [60]. Here the focus is mainly on the recent developments in temperature, magnetic field and the tilt sensor with their key performance parameters.

10.4.1 Temperature sensor

Temperature is a widely used physical quantity that plays a key role in our daily lives in large-scale industries. Precision temperature sensing is important in many applications as it works as a trigger to start the reaction [61]. To achieve precision temperature sensing, calibration plays a key role, as the lower the calibration system's uncertainty, the higher the reliability and precision. However, the current calibration setups are expensive and their operation-maintenance cost is significantly higher, which restricts their use as the reference standard [62]. They are very complicated, bulky and require highly skilled manpower, making them inefficient for day-to-day operations. The growing need for standardization in the systems requires more efficient and cost-effective methods to lower the calibration cost, ultimately leading to overall cost reduction of the final product [17].

Pathak et al. developed a novel and highly sensitive MF-based temperature sensor for precision temperature measurement. The experimental setup of the temperature sensor is shown in Fig. 10.14 [17]. The developed device is a potential candidate for calibration as it is inexpensive, small in size and easy to operate. Using MF bearing-based sealing as a piston for temperature sensing is a novel method and a very sensitive temperature sensor has been developed. The coefficient of friction between the magnet and the support structure is largely reduced to 0.002 using MF bearing [14]. The basic principle of the MF-based temperature sensor is the same as any expansion-based thermometer where the increase in the temperature of the materials causes expansion and this is generally marked as a function of temperature. In the expansion-based thermometer, mercury, ethanol, etc., are used for expansion, while temperature is used for the measurement [17]. However, these materials have a low coefficient of volume expansion which lowers their sensitivity. The MF-based temperature reported by Pathak et al. uses air in place of mercury or ethanol, which has a 20 times higher coefficient of volume expansion. Thus, the air is trapped inside the glass bulb by sealing it from the top using a magnetic bearing-based piston which moves by air on expansion or contraction with temperature change [17].

The sensitivity of the developed temperature sensor based on air trapping is very high compared to any expansion-based thermometer as the motion of the magnets

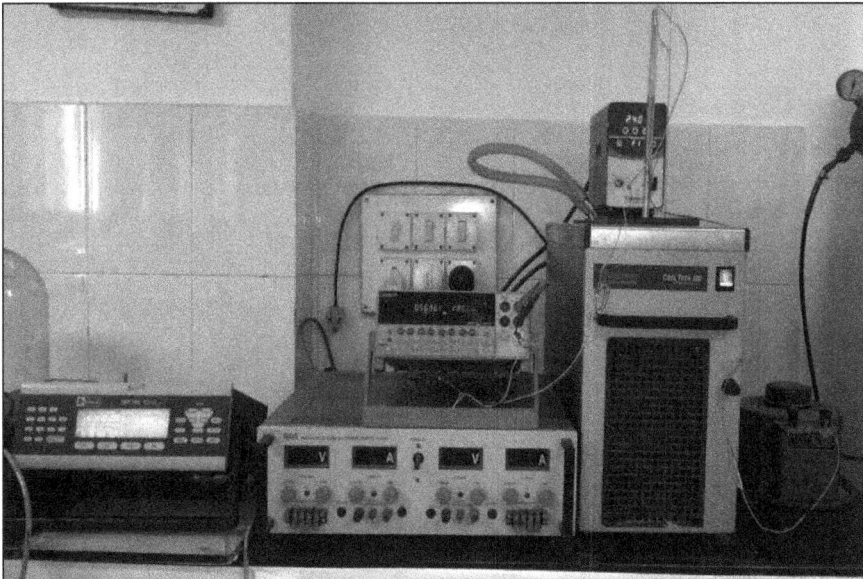

Figure 10.14. Experimental setup of the magnetic fluid-based temperature sensor (Reproduced with permission from Ref. [17]).

is converted into electrical signals by placing primary and secondary coils wrapping the capillary. Additionally, in the temperature sensor, at a constant atmospheric pressure, a temperature variation of 1°C causes a change of 20 mm in the position of the liquid column, which corresponds to a variation of 267 (±12) mV in the output voltage, as depicted in Fig. 10.15. Thus, the device shows high sensitivity of 3.7 (± 0.2) mK and can be used where high accuracy in temperature measurements is required. Further, the detailed uncertainty calculation has also been presented for the MF-based temperature sensor, which suggests that the variability of the device falls well within the range of 3-sigma limits and gives a highly repeatable set of results.

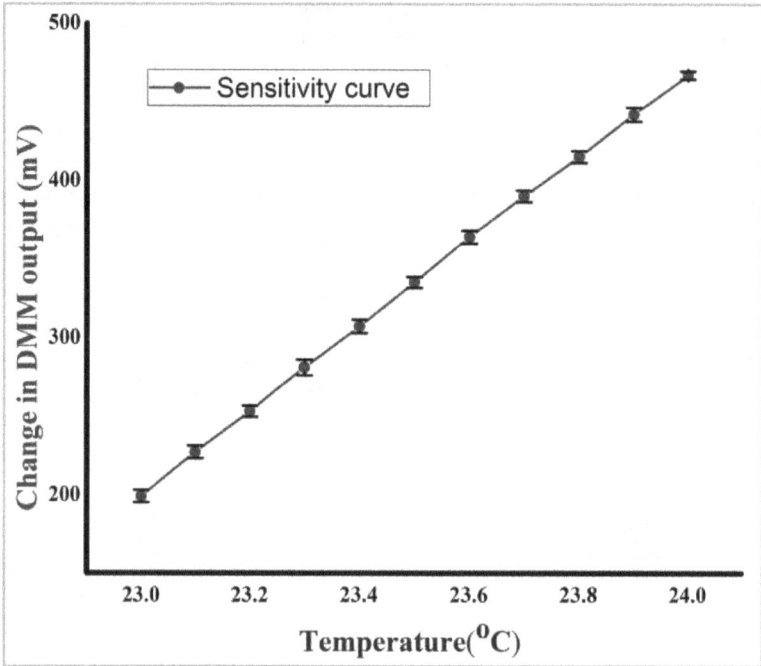

Figure 10.15. Experimental results of the magnetic fluid-based temperature sensor showing the change in 1°C of the temperature sensor results in around 267 mV change in digital multimeter (Reproduced with permission from Ref. [17]).

The uncertainty estimation has been carried out by the universal method and both repeatability and reproducibility have been probed [17].

10.4.2 *Magnetic field sensor*

Magnetic field sensing plays an important role in the measurement of the electrical current, which is vital for many industries (metallurgy, power, oil, aviation, etc.) and biomedical fields (imaging, detection and therapeutics) [33]. Various mechanisms such as magneto-transistor, magneto-resistive and the Hall Effect were used to measure the magnetic field. Although, with the recent development in nanotechnology, magnetic field sensors with low power consumption, compact, remote monitoring capabilities and free from electromagnetic interferences have gained significant attention. Sensors based on the magnetic field present all these characteristics and are very suitable for magnetic field sensing.

Optical fiber-based magnetic field sensors are highly reliable, sensitive and developed on two kinds of mechanisms. The first relies on the state polarization of light with the magnetic field and the second is based on the magnetostrictive properties of materials in combination with Mach–Zehnder interferometry. However, many optical fibers in conjunction with MF configurations have recently been developed based on their change in refractive index with magnetic field change. Azad et al. recently reported the design and development of a rapid and highly sensitive photonic crystal fiber-based magnetic field sensor in which MFs penetrated through the nanoholes in the crystals [63]. The sensor's design is based on the polyethylene glycol-based MFs incorporation inside the photonic crystal fiber-containing air holes of nano-sizes. The uniform insertion of the MFs inside the photonic crystal was done by using a setup as shown in Fig. 10.16 (a) following the Poiseuille law which suggests that the constant pressure gradient among the two ends of the fiber results in a laminar fluid flow. Further, an SEM micrograph of the cross-section of the photonic crystal is depicted in Fig. 10.16 (b), showing air holes of an average diameter of 480 nm oriented in a hexagonal lattice pattern with a 1.4 mm pitch. The SEM images of the bare and infiltrated photonic crystal fiber are also shown in Figs. 10.16 (c) and 10.16 (d), which confirm the uniform infiltration of the MNPs inside the air holes.

The experimental setup of the MFs infiltrated photonic crystal-based magnetic field sensing consists of a NIR laser, flat magnets near the sensing, a Hall probe for calibration, mirrors, a linear polarizer and a CCD camera, as shown in Fig. 10.17. The developed sensor demonstrated high reliability for magnetic field detection range 0–350 G and could detect as low as 20 G in the transmission mode measurements. Further, an analytical validation (by waveguide light transmission model adopting finite-element method simulations) has been presented, which indicates a good fit ($R \geq 0.996$) with the Langevin function. The developed hybrid optical photonic crystal fiber infiltrated magnetic field sensors can be advantageous for sensing in biochemical and environmental applications.

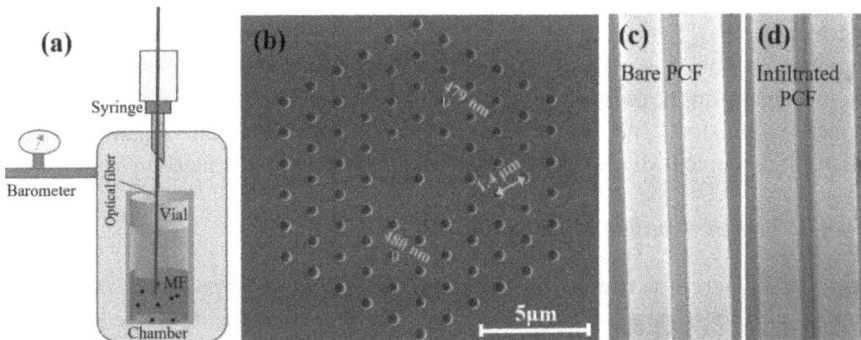

Figure 10.16. (a) Schematic of the experimental setup for filling the photonic crystal fiber. (b) Cross-section SEM image. (c) Optical microscope image of Bare PCF, and (d) infiltrated side views of the photonic crystal fiber (Reproduced with permission from Ref. [63]).

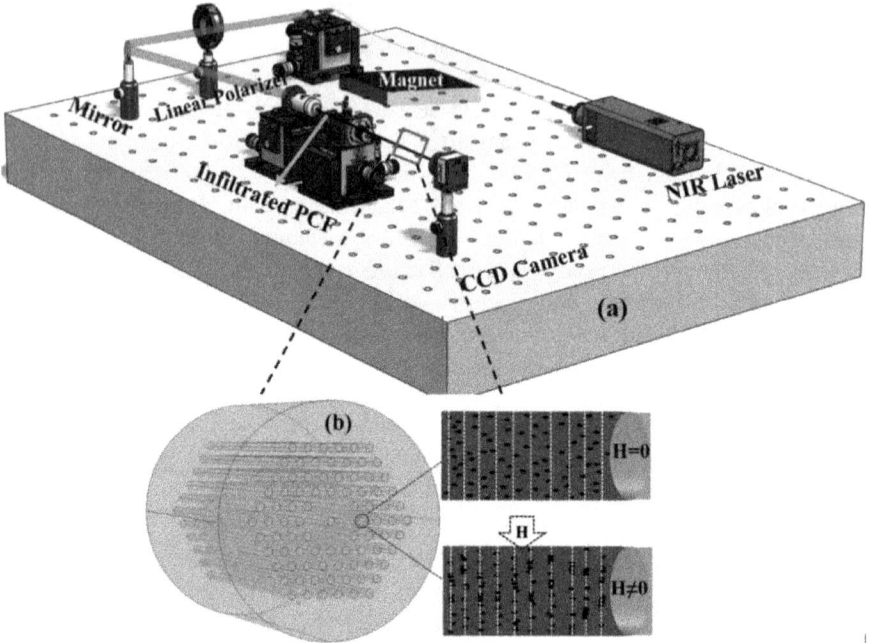

Figure 10.17. Schematic illustration of (a) experimental setup, (b) magnetic nanoparticles arrangement within the holes of the PCF with (bottom image) and without (top image) the application of an external magnetic field (Reproduced with permission from Ref. [63]).

Luo et al. also developed a magnetic field sensor showing high sensitivity based on the microfiber coupled with MFs [64]. They capitalized on the sensitivity of the microfiber couplers to the surrounding refractive index and the refractive index of the MFs is sensitive to the magnetic field [65]. Thus, the coupling of the two has been realized for sensing the magnetic field. The magnetic field strength is measured by calculating the dip wavelength shift and transmission loss change of the transmission spectrum. They investigated the magnetic field strengths at different wavelengths and established that the microfiber coupled-MF sensor displays wavelength-dependent characteristics and the maximum sensitivity of 191.8 pm/Oe is achieved at 1537 nm wavelength [64]. Sensitivity of -0.037 dB/Oe was also reported by quantifying the variation of the fringe visibility. The microfiber coupled-MFs magnetic field sensor has a potential application in tuneable all-in-fiber photonic devices such as magneto-optical modulators, filters and sensing [65]. A schematic representation of the microfiber coupled-MF sensors and microscopy images are depicted in Fig. 10.18.

10.4.3 Tilt and inclination sensor

Apart from energy harvesting, the low frictional and magnetic properties of the MFs can be used for inclination and tilt sensing. The sensor can be developed either by the motion of MFs or permanent magnet MF conjunction, which creates an ultra-smooth surface or by the direct motion of the MFs. The smallest perturbation due to the tilt or inclination results in the movement of the permanent magnet MFs conjunction, which is converted into electrical signals by placing a coil wrapped

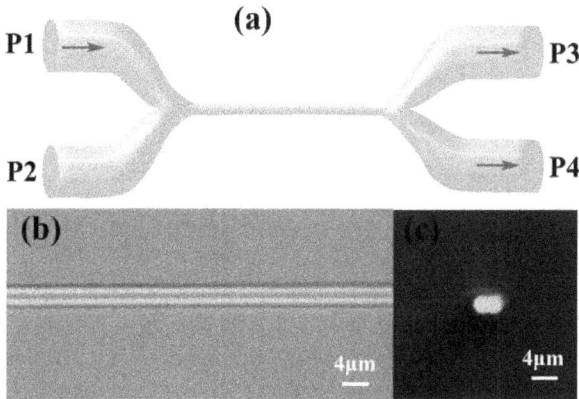

Figure 10.18. (a) Schematic of the proposed microfiber coupler and (b) and (c) microscope images of coupling region and cross-section of the microfiber coupler cutting at the uniform waist region (Reproduced with permission from Ref. [64]).

around the geometry. DeGraff et al. developed a tilt sensor based on the movement of the MFs inside the transformer setup, which induces a change in voltage [66]. This change in voltage has been calibrated to develop a highly sensitive tilt sensor. The developed tilt sensor has been designed and tested within the tilt angles range of 0°–70°. A high voltage output of 140–200 mV is observed, making the developed sensor highly sensitive. The motion of the mechanical mass leads to a change in the tilt angle, which causes the MFs to move across the coil generating a voltage output. The usual problem with the MFs system is leakage which has been restricted by the efficient design of the cylinder. The developed tilt sensors are capable of responding to the smallest variation.

In addition, permanent magnet magnetic fluid conjunction can be used for inclination sensing. The ultra-low friction surface formed at the bottom of the permanent magnet by MFs remains stationary only when the surface is leveled. A permanent magnet responds to even the smallest perturbation in the level with a movement that can be tapped off by placing the coils at the bottom of the magnets. Many researchers have depicted the concept of the inclination sensor with this phenomenon, but no experimental system has been built yet.

10.5 Thermal applications

Cooling is an essential part of most systems and efficient thermal management results in improvement in the performance of the system [67]. The ever-growing technological inventions require a high-performance, active and localized control cooling system [31, 68]. MFs demonstrate significantly improved thermal properties in the presence of the magnetic field. Thus, it is a suitable candidate to meet the current demands of new technological developments [24]. The thermal conductivity of MFs can be increased four-fold at the optimized applied field strength [27]. At low magnetic fields, the particles start to form chain structures, which increment the thermal conductivity as these chains propagate heat at a faster rate [27, 69]. However, at very high field strengths, the chains start to aggregate, causing a hindrance in thermal energy transport, resulting in the lowering of thermal conductivity. Thus, optimal values of the applied field strengths result in the highest increase in the thermal conductivity of the MFs [70]. Here two main areas of the cooling applications of MFs will be focused; self-propelled cooling and thermal energy storage devices.

10.5.1 *Magnetic fluid-based cooling application*

A cooling system that the external magnetic field can control provides a significant advantage as it allows the tenability of the cooling load. Cooling system-based on MFs are mainly designed on the principle of thermomagnetic convection [70]. The magnetization of the MFs modifies with a change in the temperature of the MFs. The lower temperature of the MFs results in a higher magnetization, whereas, at a higher temperature, the magnetization of the MFs is low. Thus, a driving force is generally created by applying the external magnetic field, which can be used for heat transfer [71]. As the driving force is applied passively using a permanent magnet, a self-pumping cooling system was built by Chaudhary et al. acting as a semi-active controlled cooling system [72]. This system takes advantage of the active cooling systems, however, as the driving force is passive and does not require any additional power, making the active-controlled cooling systems expensive [73].

A schematic representation of the working of the cooling system is depicted in Fig. 10.19, where the cold MFs particles with non-zero magnetization values move towards the hot region from where the heat has to be removed by applying a magnetic field by a permanent magnet [73]. In the hot region, MFs take up the heat load and moves towards the heat sink. At this point, MFs either lose their magnetization (heated above Curie temperature) or magnetization reduces significantly. It allows the movement of the MFs to the heat sink as they can no longer be attracted to the permanent magnet [73]. At the heat sink, the MFs again gain the finite value of magnetization by dumping the heat into the heat sink resulting in a reduction of temperature. Furthermore, the MFs move toward the permanent magnet in the hot region, and the self-propelled cooling cycle goes on without requiring additional power [73].

Chaudhary et al. studied the water-based Mn–Zn ferrite MNPs dispersed MF employing the self-propelled cooling system by experimental investigation and validated the results with simulation performed by COMSOL Multiphysics [73]. They found that the applied magnetic field strength and MNPs content mainly influence the heat removal capacity of the cooling system. They reported that the temperature difference of 20°C and ~ 28°C was observed for the applied field strength of 0.3 T when the initial temperature of the hot region was maintained at 64°C and 87°C, respectively. This also agreed well with the experimental and simulation results [73]. Further, Pattanaik et al. designed the cooling system employing the MFs thermomagnetic convection. They studied the effect of choosing the optimal MFs on the heat loads of the cooling system [70]. They analyzed the cooling performance of various ferrites and metallic nanoparticles and suggested that in ferrites γ-Fe_2O_3, Fe_3O_4 and $CoFe_2O_4$ and metallic particles, FeCo MNPs dispersed MFs displayed the best cooling performance. They found that when the temperature of the hot region

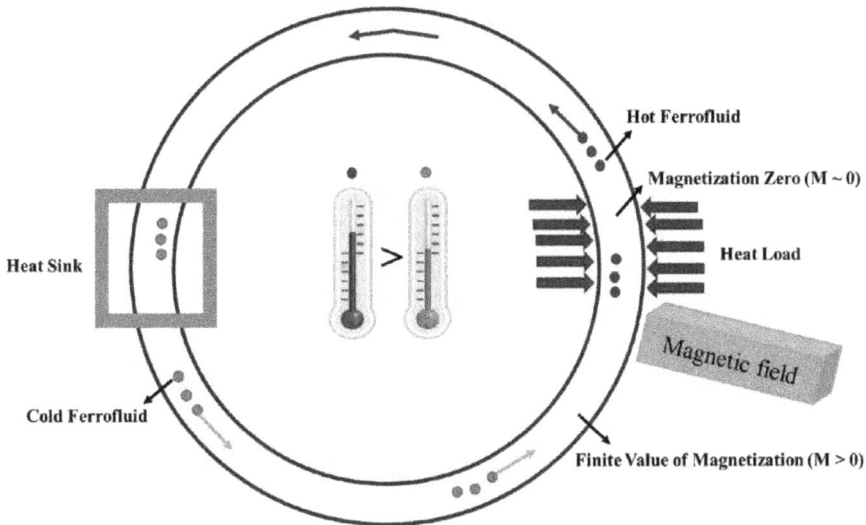

Figure 10.19. The schematic of the self-pumping cooling system employing MFs thermomagnetic convection (Reproduced with permission from Ref. [73]).

is closer to the Curie temperature, the best performance from the cooling system is observed [73].

Further, the improvement in the thermomagnetic convection for ferrites and metallic MNPs has been confirmed by different non-dimensional numbers such as the average Nusselt number, non-dimensional magnetic Rayleigh number and the Peclet number [70]. The investigation significantly improves the performance of the self-pumping cooling system devices. Some other reports were also published that probed the self-propelled cooling system both by experiment and simulation [71, 72].

10.5.2 Thermal energy storage using magnetic fluid

Thermal energy storage devices are vital to many energy conversion and transport systems. Most thermal energy storage devices are based on the cylindrical design where the phase change materials are stored inside the cylinder. However, the heat transport characteristics of the cylinder cannot be controlled, resulting in a drawback and reducing the system's efficiency. To overcome this, Pathak et al. have designed and developed a novel energy storage device based on the annular fin-MF geometry, which improved heat transfer characteristics, and the external magnetic field can control its cooling performance. The developed device is a hybrid concept that takes advantage of the improved and tuneable thermal properties of the MFs in conjunction with the cost-effectiveness of the natural convection system. However, most of the systems use natural convection systems as they are cost-effective and do not have any moving parts, which is one of the major drawbacks of the forced convection system. The developed energy storage has all the features of the forced convection system; but, it does not have any moving parts and is cost-effective, making it an ideal choice for energy storage and transport systems.

The MNPs dispersed in MFs form a chain-like structure in the field direction, which propagates heat faster, resulting in improved thermal performance. The developed device is based on the annular fin geometry and the MFs are attached to the top surface of the annular fins with the help of small magnets. These small magnets attached to the top surface of the annular fins make a series of magnets in opposing polarization to form a self-sustained magnetic ring, as shown in Fig. 10.20. Then the MFs are added to the magnetic ring, forming a uniform layer of magnetic fluid around the magnets. This layer of magnetic fluids enhances the heat transfer due to the formation of the nanochannels of the MNPs. It propagates heat rapidly, working as a chill block used for directional solidification. To investigate the thermal performance of this energy storage device, they designed an in-house setup, as shown in Fig. 10.20. The experimental setup used for the measurement is shown in Fig. 10.21, which depicts the temperature bath, annular fin-MF geometry and temperature probes. Pathak et al. probed two types of fluid; water and kerosene-based fluid and reported that a 60% increase in the overall heat transfer is observed using water-based MFs, whereas a 55% increase is observed for the kerosene-based fluid.

In addition, the fin efficiency increased by 20% and 18% for water and kerosene-based fluid, respectively. They also plotted the temperature distribution along the fin length and observed a significant temperature drop at the annular fins' top surface. It indicates that the overall heat transfer increase is due to the presence of the MFs. In addition, annular fins are part of many heat transfer systems and an increment in their heat transfer efficiency can be vital for improving the overall performance of these systems.

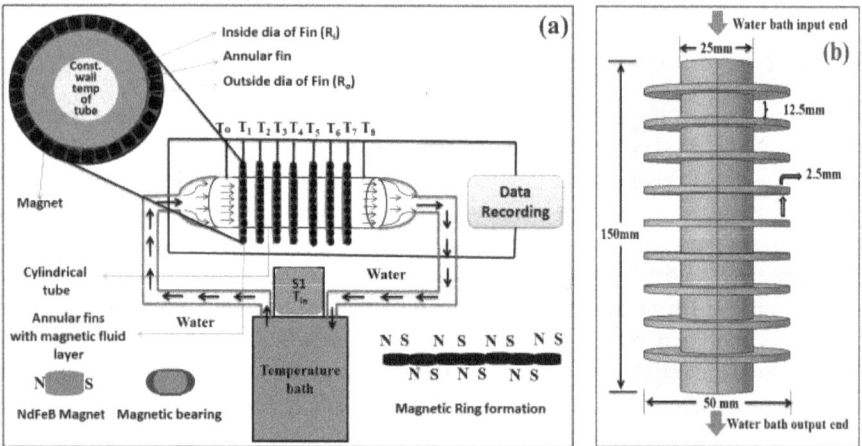

Figure 10.20. (a) Schematic of the experimental setup showing different measurement system components (b) schematic of the test section (Reproduced with permission from Ref. [24]).

Figure 10.21. In-house experimental setup with different components used to measure the heat transfer characteristics (Reproduced with permission from Ref. [24]).

10.6 Summary

MFs display excellent and unique properties and are thus used in a wide variety of fields. However, this chapter only focused on their energy harvesting, sensing and thermal engineering applications. These devices are designed based on the enhanced thermal, magneto-viscous and self-levitation properties of the MFs and rely on the tenability of the external magnetic fields. The energy applications of MFs discussed here were based on three broad classifications (based on self-levitation, MFs motion and magneto-thermal effect). Their operating principles, mechanisms and recent developments have been described.

Further, we focused on the sensing applications of the MFs and discussed in detail the high precision temperature, magnetic field and tilt/inclination sensor. The sensing application of MFs takes advantage of the systematic variations of the MF's properties with changes in the surrounding. At the end, the thermal application of the MFs, which have been developed based on the enhanced thermal characteristics of the MFs in the presence of an external magnetic field and their tenability to meet the variable loads have been described. We mainly focused on the self-pumping semi-active MFs-based cooling system, thermal energy storage and transport system using annular fins-MF geometry. All these applications have made significant advancements in the last decade, contributing to the effort to meet the ever-growing demand of designing a system with improved performance to meet the challenges that appear with technological advancements.

References

[1] G. Giakisikli and A. N. Anthemidis. Magnetic materials as sorbents for metal/metalloid preconcentration and/or separation. A review. Analytica Chimica Acta. 2013/07/30/;789: 1–16 (2013).

[2] S. Pathak, R. Verma, S. Singhal et al. Spin dynamics investigations of multifunctional ambient scalable Fe3O4 surface decorated ZnO magnetic nanocomposite using FMR. Scientific Reports. 2021/02/15;11(1): 3799 (2021).

[3] A. Singh, S. Pathak, P. Kumar et al. Tuning the magnetocrystalline anisotropy and spin dynamics in CoxZn1-xFe2O4 ($0 \leq x \leq 1$) nanoferrites. Journal of Magnetism and Magnetic Materials. 2020/01/01/;493: 165737 (2020).

[4] M. Wierucka and M. Biziuk. Application of magnetic nanoparticles for magnetic solid-phase extraction in preparing biological, environmental and food samples. TrAC Trends in Analytical Chemistry. 2014/07/01/;59: 50–58 (2014).

[5] A. Speltini, M. Sturini, F. Maraschi et al. Recent trends in the application of the newest carbonaceous materials for magnetic solid-phase extraction of environmental pollutants. Trends in Environmental Analytical Chemistry. 2016/04/01/;10: 11–23 (2016).

[6] Á. Ríos and M. Zougagh. Recent advances in magnetic nanomaterials for improving analytical processes. TrAC Trends in Analytical Chemistry. 2016/11/01/;84: 72–83 (2016).

[7] M. Trojanowicz. Impact of nanotechnology on design of advanced screen-printed electrodes for different analytical applications. TrAC Trends in Analytical Chemistry. 2016/11/01/;84: 22–47 (2016).

[8] F. Sanchez and K. Sobolev. Nanotechnology in concrete – A review. Construction and Building Materials. 2010/11/01/;24(11): 2060–2071 (2010).

[9] O. M. Koo, I. Rubinstein and H. Onyuksel. Role of nanotechnology in targeted drug delivery and imaging: a concise review. Nanomedicine: Nanotechnology, Biology and Medicine. 2005/09/01/;1(3): 193–212 (2005).

[10] X. P. Do and S.-B. Choi. High loaded mounts for vibration control using magnetorheological fluids: Review of design configuration. Shock and Vibration. 2015: 18 (2015).

[11] S. R. Hong, N. M. Wereley, Y. T. Choi et al. Analytical and experimental validation of a nondimensional Bingham model for mixed-mode magnetorheological dampers. Journal of Sound and Vibration. 2008/05/06/;312(3): 399–417 (2008).

[12] S. Mørup and M. F. Hansen. Frandsen C. 1.04 Magnetic nanoparticles. pp. 89–140. *In*: D. L. Andrews, R. H. Lipson and T. Nann (eds.). Comprehensive Nanoscience and Nanotechnology (Second Edition). Oxford: Academic Press (2019).

[13] R. E. Rosensweig. Ferrofluids: Introduction. pp. 3093–3102. *In*: K. H. J. Buschow, R. W. Cahn, M. C. Flemings et al. (eds.). Encyclopedia of Materials: Science and Technology. Oxford: Elsevier (2001).

[14] R. E. Rosensweig. Ferrofluids: Introduction. Reference Module in Materials Science and Materials Engineering: Elsevier (2016).

[15] N.-T. Nguyen. Chapter 3 - Computational transport processes for micromixers. pp. 73–112. *In*: N.-T. Nguyen (ed.). Micromixers (Second Edition). Oxford: William Andrew Publishing (2012).

[16] A. Dominguez, R. Sedaghati and I. Stiharu. Modeling and application of MR dampers in semi-adaptive structures. Computers & Structures. 2008/02/01/;86(3): 407–415 (2008).

[17] S. Pathak, K. Jain, N. Jahan et al. Magnetic fluid based high precision temperature sensor. IEEE Sensors Journal 17(9): 2670–2675 (2017).

[18] A. Mishra, S. Pathak, P. Kumar et al. Measurement of static and dynamic magneto-viscoelasticity in facile varying pH synthesized CoFe2O4-based magnetic fluid. IEEE Transactions on Magnetics. 55(12): 1–7 (2019).

[19] Genc, S. and B. Derin. Synthesis and rheology of ferrofluids: a review. Current Opinion in Chemical Engineering. 2014/02/01/;3: 118–124 (2014).

[20] W. Luo, S. R. Nagel, T. F. Rosenbaum et al. Dipole interactions with random anisotropy in a frozen ferrofluid. Physical Review Letters. 1991 11/04/; 67(19): 2721–2724.

[21] A. Chandna, D. Batra, S. Kakar et al. A review on target drug delivery: magnetic microspheres. Journal of Acute Disease. 2013/01/01/;2(3): 189–195 (2013).

[22] N. Jahan, G. A. Basheed, K. Jain et al. Dipolar interaction and magneto-viscoelasticity in nanomagnetic fluid. Journal of Nanoscience and Nanotechnology //;18(4): 2746–2751 (2018).

[23] N. Jahan, S. Pathak, K. Jain et al. Improved magneto-viscoelasticity of cross-linked PVA hydrogels using magnetic nanoparticles. Colloids and Surfaces A: Physicochemical and Engineering Aspects. 2018/02/20/; 539: 273–279 (2018).

[24] S. Pathak, K. Jain, P. Kumar et al. Improved thermal performance of annular fin-shell tube storage system using magnetic fluid. Applied Energy. 2019/04/01/;239: 1524–1535 (2019).

[25] K. Raj and R. Moskowitz. A review of damping applications of ferrofluids. IEEE Transactions on Magnetics. 16(2): 358–363 (1980).

[26] K. Jain, S. Pathak, P. Kumar et al. Dynamic magneto-optical inversion in magnetic fluid using NanoMOKE. Journal of Magnetism and Magnetic Materials. 2019/04/01/;475: 782–786 (2019).

[27] J. Philip, P. D. Shima and B. Raj. Enhancement of thermal conductivity in magnetite based nanofluid due to chainlike structures. Applied Physics Letters. 2007/11/12; 91(20): 203108 (2007).

[28] P. Kumar, S. Pathak, A. Singh et al. Optimization of cobalt concentration for improved magnetic characteristics and stability of CoxFe3-xO4 mixed ferrite nanomagnetic fluids. Materials Chemistry and Physics. 2021/03/06/:124476 (2021).

[29] H. W. Müller and M. Liu. Structure of ferrofluid dynamics. Physical Review E. 11/27/;64(6): 061405 (2001).

[30] P. Sharma, V. V. Alekhya, S. Pathak et al. A novel experimental approach for direct observation of magnetic field induced structuration in ferrofluid. Journal of Magnetism and Magnetic Materials. 2021/04/19/:168024 (2021).

[31] J. Philip and J. M. Laskar. Optical Properties and Applications of Ferrofluids—A Review. Journal of Nanofluids. 1(1): 3–20 (2012).

[32] N. Jahan, S. Pathak, K. Jain et al. Enchancment in viscoelastic properties of flake-shaped iron based magnetorheological fluid using ferrofluid. Colloids and Surfaces A: Physicochemical and Engineering Aspects. 2017/09/20/;529: 88–94 (2017).

[33] S. Pathak, R. Verma, P. Kumar et al. facile synthesis, static, and dynamic magnetic characteristics of varying size double-surfactant-coated mesoscopic magnetic nanoparticles dispersed stable aqueous magnetic fluids. Nanomaterials 11(11): 3009 (2021).

[34] R. Verma, S. Pathak, A. K. Srivastava et al. ZnO nanomaterials: Green synthesis, toxicity evaluation and new insights in biomedical applications. Journal of Alloys and Compounds. 876: 160175 (2021).

[35] P. Kumar, S. Pathak, A. Singh et al. Microwave spin resonance investigation on the effect of post processing annealing of CoFe2O4 nanoparticles [10.1039/D0NA00156B]. Nanoscale Advances. (2020).

[36] M. Kole and S. Khandekar. Engineering applications of ferrofluids: A review. Journal of Magnetism and Magnetic Materials. 2021/11/01/;537:168222 (2021).

[37] S. Pathak, R. Zhang, B. Gayen et al. Ultra-low friction self-levitating nanomagnetic fluid bearing for highly efficient wind energy harvesting. Sustainable Energy Technologies and Assessments. 2022/08/01/;52: 102024 (2022).

[38] S. Pathak, R. Zhang, K. Bun et al. Development of a novel wind to electrical energy converter of passive ferrofluid levitation through its parameter modelling and optimization. Sustainable Energy Technologies and Assessments. 2021/12/01/;48: 101641 (2021).

[39] M. Song, Y. Zhang, S. Hu et al. Influence of morphology and surface exchange reaction on magnetic properties of monodisperse magnetite nanoparticles. Colloids and Surfaces A: Physicochemical and Engineering Aspects. 2012/08/20/;408: 114–121 (2012).

[40] X. Zhang, L. Sun, Y. Yu et al. Flexible ferrofluids: design and applications [https://doi.org/10.1002/adma.201903497]. Advanced Materials. 2019 2019/12/01;31(51): 1903497.

[41] J. Philip, P. Shima and B. Raj. Evidence for enhanced thermal conduction through percolating structures in nanofluids. Nanotechnology 19(30): 305706 (2008).

[42] Q. Liu, S. F. Alazemi, M. F. Daqaq et al. A ferrofluid based energy harvester: Computational modeling, analysis, and experimental validation. Journal of Magnetism and Magnetic Materials 2018/03/01/;449: 105–118 (2018).

[43] S. Alazmi, Y. Xu and M. F. Daqaq. Harvesting energy from the sloshing motion of ferrofluids in an externally excited container: Analytical modeling and experimental validation. Physics of Fluids. 2016/07/01;28(7): 077101 (2016).

[44] V. R. Challa, M. Prasad, Y. Shi et al. A vibration energy harvesting device with bidirectional resonance frequency tunability. Smart Materials and Structures 17(1): 015035 (2008).

[45] B. Mann and N. Sims. Energy harvesting from the nonlinear oscillations of magnetic levitation. Journal of Sound and Vibration 319(1-2): 515–530 (2009).

[46] S. Alazemi, A. Bibo and M. F. Daqaq. A ferrofluid-based energy harvester: An experimental investigation involving internally-resonant sloshing modes. The European Physical Journal Special Topics. 224(14): 2993–3004 (2015).

[47] M.-L. Seol, S.-B. Jeon, J.-W. Han et al. Ferrofluid-based triboelectric-electromagnetic hybrid generator for sensitive and sustainable vibration energy harvesting. Nano Energy 2017/01/01/;31: 233–238 (2017).

[48] Y. Wang, Q. Zhang, L. Zhao et al. Vibration energy harvester with low resonant frequency based on flexible coil and liquid spring. Applied Physics Letters 2016/11/14; 109(20): 203901 (2016).

[49] M. P. Soares dos Santos, J. A. F. Ferreira, J. A. O. Simões et al. Magnetic levitation-based electromagnetic energy harvesting: a semi-analytical non-linear model for energy transduction. Scientific Reports. 2016/01/04; 6(1): 18579 (2016).

[50] A. Bibo, R. Masana, A. King et al. Electromagnetic ferrofluid-based energy harvester. Physics Letters A. 2012/06/25/; 376(32): 2163–2166 (2012).

[51] Y. Kim. Induced voltage characteristics of back-iron effect for electromagnetic energy harvester using magnetic fluid. IEEE Transactions on Applied Superconductivity. 28(3): 1–4 (2018).

[52] S. Wu, P. C. K. Luk, C. Li et al. An electromagnetic wearable 3-DoF resonance human body motion energy harvester using ferrofluid as a lubricant. Applied Energy. 2017/07/01/;197: 364–374 (2017).

[53] S. Gbadamasi, M. Mohiuddin, V. Krishnamurthi et al. Interface chemistry of two-dimensional heterostructures – fundamentals to applications [10.1039/D0CS01070G]. Chemical Society Reviews (2021).

[54] H. Wang, Z. Zeng, P. Xu et al. Recent progress in covalent organic framework thin films: fabrications, applications and perspectives [10.1039/C8CS00376A]. Chemical Society Reviews. 48(2): 488–516 (2019).

[55] Y. Ding, Y. Qiu, K. Cai et al. High performance n-type Ag2Se film on nylon membrane for flexible thermoelectric power generator. Nature Communications. 2019/02/19;10(1): 841 (2019).

[56] M. Vasilakaki, I. Chikina, V. B. Shikin et al. Towards high-performance electrochemical thermal energy harvester based on ferrofluids. Applied Materials Today 2020/06/01/;19: 100587 (2020).

[57] T. Mao, P. Qiu, X. Du et al. Enhanced thermoelectric performance and service stability of Cu2Se Via tailoring chemical compositions at multiple atomic positions. Advanced Functional Materials 30(6): 1908315 (2020).

[58] Y. Lu, Y. Ding, Y. Qiu et al. Good Performance and Flexible PEDOT:PSS/Cu2Se Nanowire thermoelectric composite films. ACS Applied Materials & Interfaces 2019/04/03;11(13): 12819–12829 (2019).

[59] R. Verma, S. Pathak, K. K. Dey et al. Facile synthesized zinc oxide nanorod film humidity sensor based on variation in optical transmissivity [10.1039/D1NA00893E]. Nanoscale Advances (2022).

[60] M. A. Khairul, E. Doroodchi, R. Azizian et al. Advanced applications of tunable ferrofluids in energy systems and energy harvesters: A critical review. Energy Conversion and Management. 2017/10/01/;149: 660–674 (2017).

[61] P. Kumar, H. Khanduri, S. Pathak et al. Temperature selectivity for single phase hydrothermal synthesis of PEG-400 coated magnetite nanoparticles [10.1039/D0DT01318H]. Dalton Transactions 49(25): 8672–8683 (2020).

[62] M. A. P. Pertijs, A. L. Aita, K. A. A. Makinwa et al. Low-cost calibration techniques for smart temperature sensors. IEEE Sensors Journal 10(6): 1098–1105 (2010).

[63] S. Azad, S. K. Mishra, G. Rezaei et al. Rapid and sensitive magnetic field sensor based on photonic crystal fiber with magnetic fluid infiltrated nanoholes. Scientific Reports. 2022/06/11;12(1): 9672 (2022).

[64] L. Luo, S. Pu, J. Tang et al. Highly sensitive magnetic field sensor based on microfiber coupler with magnetic fluid. Applied Physics Letters 2015/05/11;106(19): 193507 (2015).

[65] L. Luo, S. Pu, J. Tang et al. Reflective all-fiber magnetic field sensor based on microfiber and magnetic fluid. Opt. Express. 2015/07/13;23(14): 18133–18142 (2015).

[66] A. DeGraff and R. Rashidi. Ferrofluid transformer-based tilt sensor. Microsystem Technologies. 2020/08/01; 26(8): 2499–2506 (2020).

[67] R. A. Kishore and S. Priya. A review on design and performance of thermomagnetic devices. Renewable and Sustainable Energy Reviews 81: 33–44 (2018).

[68] V. B. Varma, M. S. Pattanaik, S. K. Cheekati et al. Superior cooling performance of a single channel hybrid magnetofluidic cooling device. Energy Conversion and Management. 2020/11/01/;223: 113465 (2020).

[69] J. M. Laskar, J. Philip and B. Raj. Light scattering in a magnetically polarizable nanoparticle suspension. Physical Review E. 09/10/;78(3): 031404 (2008).

[70] M. S. Pattanaik, V. B. Varma, S. K. Cheekati et al. Optimal ferrofluids for magnetic cooling devices. Scientific Reports. 2021/12/17;11(1): 24167 (2021).

[71] H. Yamaguchi and Y. Iwamoto. Energy transport in cooling device by magnetic fluid. Journal of Magnetism and Magnetic Materials. 2017/06/01/;431: 229–236 (2017).

[72] L. Phor and V. Kumar. Self-cooling by ferrofluid in magnetic field. SN Applied Sciences. 2019/11/27;1(12): 1696 (2019).

[73] V. Chaudhary, Z. Wang, A. Ray et al. Self pumping magnetic cooling. Journal of Physics D: Applied Physics. 2016/12/19;50(3): 03LT03 (2016).

11

Magnetic Nanoparticles Change the Properties of Traditional Materials and Open up New Application Possibilities

Peter Kopcansky,[1,*] *Lucia Balejcikova,*[2] *Matus Molcan,*[1]
Oliver Strbak,[3] *Ivo Safarik,*[4] *Eva Baldikova,*[5]
Jitka Prochazkova,[5] *Ralitsa Angelova,*[6] *Kristyna Zelena*
Pospiskova,[7] *Michal Rajnak,*[8] *Katarina Paulovicova,*[1]
Maksym Karpets,[8] *Milan Timko,*[1] *Natalia Tomasovicova,*[1]
Katarina Zakutanska,[1] *Veronika Lackova*[1] and *Peter Bury*[9]

11.1 Introduction

Magnetic fluids as one of the pioneers of modern nanotechnologies are attractive for basic as well as applied research. They have prompted further research, especially in the field of nanoparticles, especially magnetic ones, giving them the ability to be manipulated in an external magnetic field, thus changing the properties required in various applications. Magnetic fluids and their composite systems are constantly being studied intensively.

This chapter presents the latest results obtained in the study of (i) bioinspired magnetic nanoparticles such as magnetosomes obtained using magnetotactic bacteria and ferritin derivatives such as magnetoferritin, (ii) magnetic fluid modification with high aspect ratio materials such as textiles, fibers, etc., (iii) transformer oil-based

---

Affiliations at the end of the chapter.

magnetic fluids with respect to their interaction with the electric field instead of the magnetic field, as the electric field in transformer technology is always present, (iv) composite systems of nematic liquid crystals and magnetic nanoparticles (so called ferronematics) to prepare composites with higher sensitivity to the external magnetic field.

The bioinspired magnetic particles can be used in all applications that provide magnetic fluids. Their uniqueness lies in the fact that they were created by nature or were inspired by nature. New results associated with these nanoparticles such as the possibility of hyperthermia and help with MRI imaging in medical diagnostics and the interactions of magnetoferritin with amyloid aggregates responsible for neurovegetative disorders and with natural pollutants for various environmental applications will be presented here.

Fibers, rods, wires or tubes modified by nanoparticles (especially by magnetic nanoparticles) are typical examples of high aspect ratio (nano)materials which are both found in nature or are prepared by human activity. Both native (nano)materials and their derivatives (e.g., textile and nanotextile) can be used in a wide variety of scientific and technological applications. Magnetic modification of the above mentioned (nano)materials enables to prepare magnetically responsive materials, having various interactions with a magnetic field. Magnetic modification with ferrofluids represent an easy-to-perform technique, applicable for the preparation of magnetically responsive (bio)materials.

Magnetic fluids based on transformer oils can help to improve the transfer of electricity in terms of efficiency and safety of transmission and thus contribute to the economy by reducing costs. They have been studied mainly for the response in the magnetic field, but here their interaction with the electric field, which is always present during transmission of electric energy is presented.

Today people cannot imagine life without magnetism or liquid crystals. Due to their anisotropic electrical properties, liquid crystals are well and easily controlled by an external electric field. Everyone is aware about liquid crystal displays (LCD), which are based on the optical response of liquid crystals to an electric field. Liquid crystals have a very low response to a magnetic field. The sensitivity to magnetic field can be enhanced by combining liquid crystals with magnetic nanoparticles and enable their new potential applications.

The influence of different sizes, shapes and origins of magnetic nanoparticles on magnetic sensitivity as well as a new modern method of investigating the structural properties of liquid crystals compositions with magnetic fluids by the Surface Acoustic Waves (SAW) method will be presented.

11.2 Bio-inspired magnetic nanoparticles for biomedical and environmental nanotechnology development

The Earth's current problems are associated with waste accumulation, especially from industrial production. Environmental pollution is related to diseases contributing to reducing the quality of life. Iron-based magnetic nanoparticles can be suitable, economically advantageous and effective candidates for solving many problems in the current global situation. This subchapter focuses on summarization of knowledge about bio-inspired iron-based magnetic nanoparticles (MNPs). The recent physicochemical properties and possibilities of their applications, especially for medical and environmental technological progress, are described here.

MNPs of iron oxides (often magnetite: Fe_3O_4) are solid objects whose size ranges from 1 to 100 nm. Such a dimension provides high surface reactivity with different thermodynamic features, compared to macroscopic systems of the same material type [1]. Unique physicochemical behavior offers applying MNPs in various fields of science, biomedicine or technology with increasing popularity [2]. At present, one can encounter magnetic nanoparticles daily (recording media, loudspeakers, seals in electronic devices, vibration dampers in automobiles, etc.) [3]. Laboratory-prepared magnetic nanoparticles coated with various (bio) polymers, surfactants or other stabilizing agents (e.g., proteins) dispersed in dispersant (e.g., water) form magnetic fluids with an exotic combination of fluid and solids properties [4, 5]. Iron oxide nanoparticles can occur naturally in the air, in many rocks and in soil and allows the Earth's magnetism development study and the determination of the age of rocks [6–8]. MNPs occur naturally in living systems [9, 10]. Simple magnetotactic bacteria (*Magnetospirillum magnetotacticum*) can biomineralize a chain of magnetic nanocrystals (magnetosomes) [11]. Magnetic nanoparticles are present in magnetotactic algae [12], honey bee, dolphin [9], the pigeon's beak [13], salmon [10]. Almost all living systems, including bacteria, plants, animals and humans, can create iron-based nanoparticles coated with apoferritin. Such stored iron forms ferritin, 10–12 nm metalloprotein nanoparticles. Magnetic nanoparticles could be prepared in the constrained apoferritin hollow sphere, forming unusual ferritin derivates (e.g., magnetoferritin) [5].

11.2.1 *Magnetotactic bacteria and magnetosomes*

Magnetotactic bacteria (MTB) are a class of microorganisms able to biomineralize iron-based mineral crystals inside the cell, known as magnetosomes. Magnetosomes are formed in chain-like structures and serve them like a compass needle to orient along geomagnetic field. Such behavior in seeking the organisms' optimal conditions for growth is called magnetotaxis [14]. The magnetosome chains comprise the ferrimagnetic crystals of magnetite (Fe_3O_4) or greigite (Fe_3S_4) and phospholipid bilayer membrane. The membrane ensures chain elasticity and encapsulates the crystals, thus preventing their coagulation. At the same time, it allows the possibility to bind bioactive substances on the chain surface [11]. The magnetosome crystal's size depends on the type of MTB as well as the surrounding conditions of their growth. It is possible to regulate their size in a targeted manner in laboratory conditions. In general, their size ranges between 35–120 nm. In this size, each magnetosome acts like a single-domain particle [15].

Despite the widespread occurrence and amounts of MTBs in their natural environment, culturing MTBs in laboratory conditions is challenging. Problems with cultivation result from their way of life adapted to chemically stratified (regional) sediments. The existence of MTB is accompanied by complex redox reactions that are difficult to simulate in the laboratory. Almost all MTB cultures are microaerophilic, but strictly anaerobic species or both, are also known [16]. To date, only a limited number of MTBs have been isolated in pure culture.

The possibility of magnetotactic bacteria growth under laboratory conditions and the ability to isolate magnetosomes from bacteria open the possibility for experiments at unique magnetic bioinspired material mainly in biotechnological application [11]. Their natural lipid-based capsule around the magnetic cores eliminates the problem of particle coating, as is the case with synthetically prepared particles, e.g., by coprecipitation. Another advantage is their chain-like formation, which increases their sensitivity to externally applied magnetic fields. In the field of magnetosome research, a number of inspiring work has been done, and a lot of papers have been published on medical imaging [17], drug delivery [18], cancer treatment (magnetic hyperthermia) [19], pathogen detection [20] and gene research [21]. However, when researching them, it is important to remember that magnetosomes are generally a complex system even after isolation. The bending stiffness determined by the magnetic interactions between the magnetosomes impacts a formation of complicated chains structures (linear as well as closed loops after isolation to minimize magnetic stray field energy) [22]. Therefore, a thorough analysis of the initial samples or their additional modification, such as shortening them, to obtain more uniform properties is very necessary.

11.2.1.1 Summary of recent findings in the field of modified magnetosomes

Next, the most important characteristics of isolated magnetosomes from magnetotactic bacteria: *Magnetospirillum magneticum* sp. *AMB-1* grown in our laboratory conditions are listed. The composition of the culture medium as well as the detailed isolation procedure, can be found in the work of Molcan et al. [23]. The magnetosome's physical properties largely depend on crystal purity which is determined by the biomineralization conditions. The next important factor determining their properties is chain formation, e.g., chains length, magnetic crystal core size and their configuration along the magnetic easy axis of magnetization. From this point of view, the parameters/properties of magnetosome particles can be modified in two ways in general. The 1st is by adjusting the composition of the culture medium in synthesis [24]. The 2nd is the modification of particles after synthesis by using the sonication method [23]. In the case of magnetosomes modification by sonication such prepared chains systems were analyzed in terms of morphology [25], magnetic hyperthermia [22, 25] and AC susceptibility properties [26]. If one wants to study the morphology of nanosystems, then electron microscopy is suitable in the first place. However, most experimental tests are performed on colloidal suspensions. For this reason, one often needs to know the real properties of nano-objects, in our case magnetosomes, in the form of a suspension. Such a solution is, therefore, for example, research using neutron techniques. The Small-Angle Neutron Scattering (SANS) technique can give information about size distribution and degree of polydispersity and data about the stability of the system (i.e., the presence of the non-disrupted organic shell on the particle surface). The sonication consequences on magnetosome chain parameters and their arrangement in the volume was studied in a paper [25]. Magnetosomes prepared in the usual way with a "standard" length (IM) and magnetosomes with a "short" length (SM) truncated by sonication treatment (at 20 kHz at a power of 120 W for 3 hr in a constant power mode) were compared. Heavy water-based samples were prepared to decrease non-coherent scattering from hydrogen and to increase the scattering contrast between the solvent and shell. In contrast the variation method (varying D_2O/H_2O content in the system) we wanted to provide the opportunity to analyze lipid shell and magnetic core of magnetosomes. The SANS experiment at magnetosome suspension was done for different D_2O content in the solvent (0, 15, 30 and 50% of D_2O). The neutron scattering measurements of magnetosome IM and SM samples (after subtraction of scattering from the buffer) indicate the presence of polydisperse particles in the sample volume. The initial scattering intensities confirm the differences in the average of IM and SM. This fact is also presented in Fig. 11.1 (a), where scattering intensities from IM and SM (dissolved in pure HEPES) are compared. Therefore, it can be summarized that the sonication provides a notable effect on the magnetosome chains length parameters. The lowest contrast of magnetosome dispersion was found from the scattering intensity at the smallest values of the scattering vector in the match point (Fig. 11.1 (b)). The effective match points (minimum of the curves) were estimated for IM = 0.318 and SM 0.348. From the values of match points, we estimated the volume fraction of magnetite.

It was found that a large amount of light components in the sample volume are trapped in the magnetosome chain structures. Such findings are important because what can be seen, for example with microscopy, does not give a realistic picture of a sample in which there may be components that can significantly affect its other properties. Measurements of heat development in an alternating magnetic field were performed on this type of sample (magnetosomes in the form of classical long chains as well as sonicated) [26]. However, it was taken into account to simulate real conditions, thus the so-called tissue-mimicking phantom was used to replicate properties of biological tissues. Figure 11.2 presents the results of heat development in AC magnetic field for 180 s at 356 kHz. The highest temperature increase was observed at 16 kA·m^{-1} (about 4°C). The temperature increases at 13 and 10 kA·m^{-1} was significantly lower. It can be assumed that these field intensities do not provide enough energy to activate the Brownian mechanism and rotate the magnetosomes. The measured increase in temperature originates only from Neel's relaxation. This conclusion can be proved by measuring magnetosomes in agar. For 16 kA·m^{-1}, there is a noticeable difference in the temperature rise for the suspension and the agar phantom. This temperature decrease is due to the partial elimination of Brownian motion in the agar gel structure.

By comparing heat evolution in the shortened sonicated chain system, a higher temperature increase was observed in the suspension compared to agar for each field strength. The higher values in the colloidal suspension can be again explained by the restriction of the Brownian motion in the agar structure. By comparing the magnetic hyperthermia of long and short chains, the shorter chains produced a higher heating effect and specific absorption rate values were also better. The explanation is that due to their smaller size, they are more likely not to be constrained by the gel structure and, as a result, will have a better chance of rotation due to the Brown mechanism.

Important knowledge about magnetosomes is that their response to externally applied magnetic fields also changes related with the length of the chains [22]. The

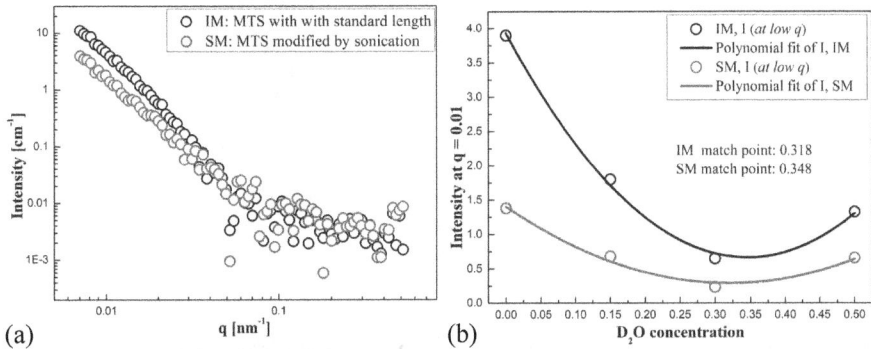

Figure 11.1. SANS curves of IM and SM magnetosomes (a), the scattering intensity at angle q = 0.01 as a function of the D_2O content (b).

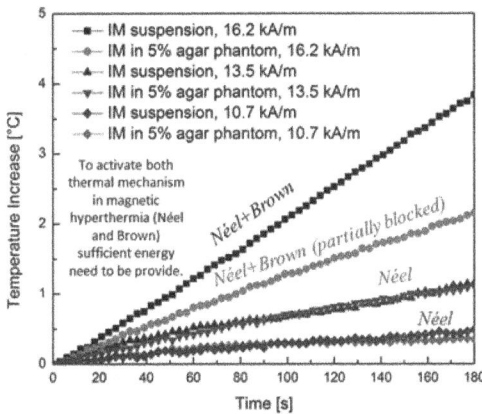

Figure 11.2. Heat evolution of magnetosomes suspension and magnetosomes in agar phantom under AMF.

magnetic susceptibility is investigated on samples of magnetosomes that differed as before—by the length of the chains. The different nature of the frequency dependence was shown by measuring the real and imaginary parts of magnetic susceptibility, In the case of long chains, we observed the monotonic nature of susceptibility. A hint of relaxation maximum appeared in the 1 Hz range. In the case of short chains, a fully developed Debay relaxation peak in the 1 kHz region was detected, indicating Brown's relaxation. The experimental data were fitted to an extended multi-core model, obtaining hydrodynamic dimension values at 640 nm for long chains, which is comparable to DLS measurements.

11.2.2 Ferritins function and modifications

Iron is toxic to higher organisms, including humans. Redox reactions of free iron followed by reactive radicals creation (Fenton and Haber-Weiss reactions) can lead to chemical bonds destroying within living cells or macromolecules [27]. Metalloprotein ferritin represents the safe storage of iron in non-toxic form. Disruption of the structure and function of ferritin or imbalance of metabolisms with

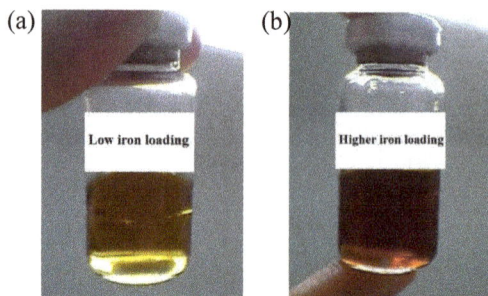

Figure 11.3. Photography of magnetoferritin colloidal solution with protein concentration ~ 6 g/l with (a) low iron loading ~ 168 Fe atoms per the one protein biomacromolecule; and (b) higher iron loading ~ 532 Fe atoms per the one protein biomacromolecule.

ferritin and iron role, causes many diseases [28]. Iron accumulation is closely related to neurodegenerative disorders [29]. Not fully defined pathological and physiological roles and *in vivo* transformations of the inorganic ferritin core to other phases (e.g., magnetite) are the main reason for extensive research studies of this material.

Specific set-up of physicochemical conditions for required reaction allows the synthesis and design of other inorganic phases inside the ferritin coat (apoferritin). Artificial ferritins (e.g., magnetoferritin) belong to the bio-inspired materials group, useful for biomedical, nanotechnological or environmental engineering. Magnetoferritin can serve as an ideal biomacromolecule for pathological magnetite simulation. It is prepared by an *in vitro* laboratory procedure using controlled thermo-oxidation conditions adapted to the formation of magnetite (Fig. 11.3a,b), i.e., temperature 65°C and alkaline pH 8.6 [30]. The theoretical mechanism involves the slow nucleation of dissolved Fe ions to solid nanocrystals inside the apoferritin shell. The process includes four main steps: entry of Fe^{2+} ions through protein hydrophilic channels into the apoferritin shell driven by an electrostatic gradient, oxidation of Fe^{2+} to Fe^{3+}, nucleation and inorganic core growth [5].

11.2.2.1 Ferritin derivates for biomedical applications

Biomedical application research requires a thorough structural characterization of materials. The structure of proteins is closely related to their function. Various modifications of the protein surface for drug binding are often necessary for targeted drug delivery with the help of carriers, e.g., magnetoferritin. The binding mechanism may involve pH changes of the protein solution. The effect of pH on the magnetoferritin structure was studied using small-angle X-ray and neutron scattering, dynamic light scattering and zeta potential measurement (as a colloidal stability parameter). The shell shape was affected greater by the iron content than pH changes. Protein dissociation to subunits at pH around 11 was the consequence of the alkaline hydrolysis of proteins. As the pH decreased to neutral values, the protein structure was packed. The perforated protein structure was related probably to the chemical conditions of the solution/synthesis. Strongly acidic pH caused protein denaturation and destruction (i.e., acid hydrolysis of proteins). The polymorphism: large fractal structures with the simultaneous presence of intact apoferritin macromolecules, unaffected by the medium conditions, acidity, synthesis conditions or iron, were observed. Visible sedimentation at neutral pH measured after magnetoferritin synthesis with higher applied amounts of iron was not the result of pH changes [31]. A recent work confirmed magnetoferritin protein shell destruction caused due to the markedly high amount of iron ions applied during the magnetoferritin synthesis [32].

Technical parameters set up for the materials synthesis are very important for specific application areas before the procedure is implemented. These parameters can be adjusted by the physical synthesis conditions change. Synthesis temperature for magnetoferritin preparation changed the state and thus the magnetic properties of the magnetoferritin inorganic cores. Kinetic energy regulation by the temperature increase resulted in the highest measured [33] magnetization. Consequently, the high temperature near the "melting point" (for apoferritin 80°C) initialized the protein denaturation process. The higher iron content of magnetoferritin synthesized at a lower synthesis temperature did not contribute to the formation of iron phases with high magnetization [34]. This material was amorphous with lower magnetization, similar to native ferritin [35].

The standard procedure for magnetoferritin synthesis (65°C, pH 8.6, including pH lowering and excessive oxidation as a result of synthesis intermediates) led to the lepidocrocite formation, confirmed by the crytallographic study. This major phase resulted in the weak hyperthermic effect [36]. For the magnetic hyperthermia treatment (local heating and subsequent destruction of tumor tissues), it is necessary to improve the parameters of this material: increase the size of magnetoferritin nanoparticles, expanding the particle density or transform magnetoferritin inorganic core into magnetite. The dehydration mechanism by increasing the pH during the reaction allows magnetite formation (addition of an alkaline, usually NaOH). The laboratory combination of iron accumulation, strongly alkaline conditions and the high temperature required for magnetite formation is not common in mammals. Magnetite, observed during pathological phenomena (such as neurodegenerative and cancer), could be crystallized *in vivo* with the help of another contributing factor (e.g., exposure of a living system to high-energy gamma radiation, the presence of a catalyst or electron donor) [35].

11.2.2.2 Medical diagnostic methods development

Ferritin and derived materials are suitable model systems for pathological ferritin comparisons. Magneto-optical and Magnetic Resonance Imaging (MRI) methods can distinguish standardized reference materials (e.g., ferritin derivatives) versus imaged tissues of living organisms for early non-invasive diagnostic purposes. MRI uses the quantum-mechanical properties of particles in the atomic nucleus in a magnetic field. Ferritin derivatives in low and high magnetic fields have already been tested, imaged and analyzed by MRI as *in vitro* comparison model materials for pathological tissues [35].

Magneto-optical properties of the studied materials allow the experimental study of the asymmetric processes after *in vitro* physicochemical reduction and reconstruction of ferritin cores [37]. The magneto-optical detailed studies and standardization of comparative samples of ferritin derivates can help understand the phases of formation mechanisms or release and entry of iron processes. Technical design of the magneto-optics based devices can sensitively and non-invasively help detect various unknown iron phases in ferritin as a (bio)sensor. Direct magneto-optical monitoring of metabolic changes over time associated with iron kinetics in cells is a promising tool for medical diagnostics. Observation of iron-induced response of organisms to the chosen treatment procedure can be a significant milestone in the medical environment with custom-designed drugs for the patient.

11.2.2.3 The role of ferritin and its derivates in neurodegeneration

The neurodegenerative disease's progress is closely related to iron management and ferritin function. Inspired by this idea, the interaction of native ferritin with native Aβ peptide was investigated. The structural changes of the Aβ peptide are the main hallmarks of recognizing Alzheimer's disease. The ferrozine assay showed the ability of Aβ peptide to reduce Fe^{3+} to Fe^{2+} from the ferritin core during the formation of disease-typical fibrils (deposits). During the formation of Aβ fibrils, almost 50% more Fe^{2+} ions, at a given time than was naturally released from ferritin in the direction of the electrostatic gradient in the diluted medium, were reduced. The fluorescence intensity and thus also the number of fibrils in the presence of ferritin decreased slightly compared to pure peptide. The thioflavin fluorescence test, together with the imaging of the shape and size of the fibrils by atomic force microscopy, showed minimal differences between the Aβ fibrils and their mixture with ferritin. During fibril formation in the presence of ferritin, the Aβ peptide probably served as an electron donor for the formation of Fe^{2+} ions as one of the essential source components for magnetite creation [38]. Ferritin can be a precursor of *in vivo* magnetite formation [38–40]. A similar spectrophotometric detection of Fe^{2+} ions release after the interactions of water-soluble vitamins with ferritin derivatives confirmed a more pronounced effect for magnetoferritin [41] useful in pharmaco-biochemistry or pathology.

The interaction between magnetoferritin/reconstructed ferritin and lysozyme amyloid fibrils (as pathological model systems) allowed assessment of the resulting impacts (Fig. 11.4 a, b). Laboratory conditions of these well-studied fibrils preparation include acidic pH, constant stirring and heating of native lysozyme. The final stable insoluble β-sheet-rich protein structure is the main challenge for drug design in patients with difficult-to-treat neurodegenerative disorders. Observed fibrils amount

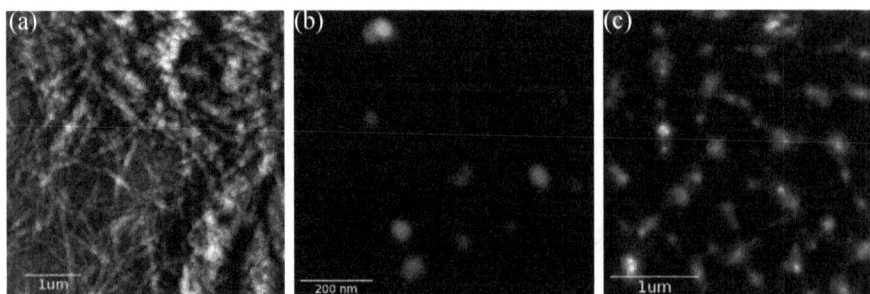

Figure 11.4. Atomic force miscroscopy illustrations of (a) Lysozyme Amyloid Fibrils (LAF) (b) magnetoferritin loaded up to 731 iron atoms per the one protein biomacromolecule (MFer), and (c) mixture LAF and MFer in mass ratio 1:10. Medium: 0.05 M AMPSO solution buffered to pH 8.6.

reduction did not depend on various magnetic characters for both tested materials. These results expand the potential applicability of synthetic ferritin derivates as modern magneto-pharmaceutical therapeutics against amyloid disorders [42].

11.2.2.4 Environmental aspect of magnetoferritin

Magnetoferritin benefit for environmental application is the destruction ability of pollutants, such as polychlorinated biphenyls (PCBs), that is still a global and local problem. Since the end of the 20th century, health risks of PCBs exposure led to their industrial production ban after the harmful effect on organisms health confirmation. The main danger of these contaminants is associated: with their high resistance against degradation, long-term persistence in the ecosystem, long-distance transport and lipophilic character, responsible for bio-accumulation of PCBs in living cells, especially adipocytes.

Incubation of a representative PCB congener 28 with magnetoferritin solution resulted in a loss of a peak characteristic of C-Cl bonds (frequency 729 cm^{-1}) confirmed by infrared spectroscopy. High-resolution gas chromatography showed a decrease in the concentration of the investigated PCB congener 28 after interaction with magnetoferritin. The highest effect exhibited the highest iron amount in magnetoferritin. The reducing ability was probably related to the redox potential of the magnetic core in magnetoferritin, depending on aqueous conditions. Magnetoferritin in a neutral and alkaline environment figures as reductants (electron donors). The effect of PCBs dechlorination is applicable in biomedicine. Thus, biogenic magnetoferritin may have a similar detoxifying function (reductive dehalogenation mechanism) in living cells against other toxins. The application possibilities of magnetoferritin may be related to the absorption, (bio)remediation or nano catalytic degradation of various other organic toxic substances, heavy metals, carcinogens or radioactive contaminants. Biosensing could help detect endangered species and protect biodiversity [43].

The advantage of various magnetic nanoparticles in environmental applications is their low cost, efficiency, the possibility of magnetic separation, degradability, simplicity and rate of their chemical preparation. Laboratory MNPs research can be transferred to the hydrological field testing to verify their applicability in huge quantities over a large area.

11.3 Ferrofluid modified high aspect ratio (nano)materials

(Nano)fibers, (nano)tubes, (nano)rods or (nano)wires are typical examples of high aspect ratio (nano)materials which are both found in nature or are prepared by human activity. Both native (nano)materials and their derivatives (e.g., textile and nanotextile) can be successfully applied in many areas of science and technology.

Magnetic modification of the above mentioned diamagnetic (nano)materials enables in preparing magnetically responsive ones. Magnetic modification with ferrofluids represent a simple procedure, applicable for magnetic modification of a wide range of (bio)materials. More detailed information about ferrofluid modification procedures are available in recently published review chapters and papers [44, 45].

11.3.1 *Ferrofluid modified (nano)fibers and (nano)textile*

Electrospinning is a widely used procedure for the formation of ultrathin fibers formed by both synthetic and natural polymers. Polymeric solution properties and process variables can be used to control the behavior of electrospun fibers. The produced electrospun nanofibers and nanotextiles can be used as "smart" mats, filtration membranes, catalytic supports, biomedical scaffolds, drug carriers, wound dressings and components of photonic and energy harvesting devices. Stimuli-responsive nanofibers have been intensively studied during the last decade and have been used in many biomedical applications [46, 47].

In our experiments, both hydrophilic nanofibers prepared from polyamide 6 and polyvinyl alcohol or hydrophobic nanofibers prepared from polyurethane and polycaprolactone were magnetically modified using two types of ferrofluids, namely water based, perchloric acid stabilized FF and chloroform-based, nonanoic acid stabilized FF. A simple procedure was used for magnetic modification of nanotextiles, namely its spraying with either acid FF/methanol mixture or with native chloroform-based FF (Fig. 11.5). Approximately 17–18 mg of magnetic iron oxides from acid stabilized FF and 7–8 mg of magnetic iron oxides from chloroform-based FF were loaded on 1 cm^2 of the modified nanotextile. SEM images of native and FF modified polyamide nanotextile are presented in Fig. 11.6. Due to the fact that magnetic iron oxide nanoparticles are known nanozyme with peroxidase-like (P-L) activity [48], ferrofluid modified (nano)textile exhibited P-L activity using typical peroxidase substrates such as N,N-diethyl-p-phenylenediamine (DPD) [49].

In another approach polyvinyl butyral (PVB) nanofibers were modified by a ferrofluid added into the PVB solution. These magnetic nanofibers were evaluated for potential magnetic hyperthermia applications, either for surface heating or as implants. The ferrofluid modified nanotextiles with different magnetic particle concentrations were studied in an alternating magnetic field; an immediate temperature increase after the field application was observed [50].

Electrospun nanofibers prepared from polycaprolactone (PCL) and gelatin (Gel) blends were modified with ferrofluid treated halloysite, a naturally occurring tubular aluminosilicate clay. It was observed that the incorporation of the ferrofluid modified halloysite into PCL/Gel nanofibers led to the formation of soft magnetic biocompatible materials. In addition, application of magnetic halloysite caused the improvement of the nanofiber structures and had strong reinforcing effects. The greatest improvement was observed for nanofibers containing 9% of magnetic halloysite which led to more than a two-fold increase in tensile strength and a five-fold improvement of the elongation at break [51].

"Standard" textile materials including woven cotton textile were also magnetically modified with the treatment with perchloric-acid stabilized FF/ methanol mixture followed by drying. SEM of native and FF modified textiles clearly showed the presence of iron oxide nanoparticles on the surface of the modified cotton fibers. All FF modified textile materials had a light to dark brown color depending on the quantity of the bound iron oxide nanoparticles. Magnetic measurements showed that the saturation magnetization values reflected the number

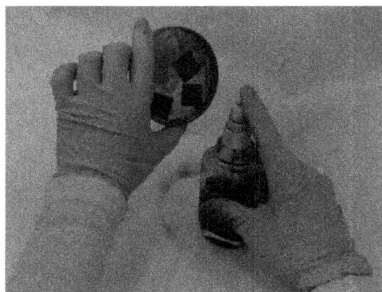

Figure 11.5. Magnetic modification of nanotextile with chloroform-based magnetic fluid using a spray technique.

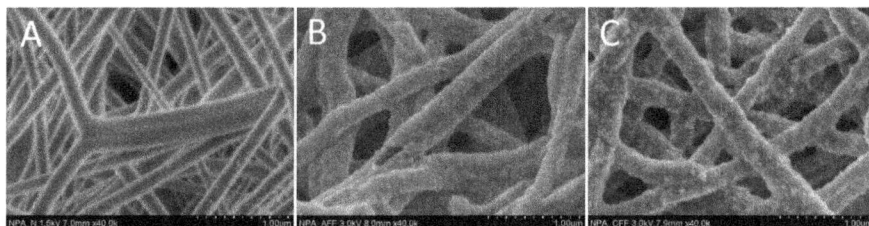

Figure 11.6. Scanning electron microscopy of native polyamide nanotextile (A) the same nanotextile modified with perchloric acid stabilized FF (B) and nanotextile modified with chloroform-based FF (C).

of magnetic nanoparticles present in the modified textiles. The detailed structural characterization at the nanoscale of both the native and magnetically modified textiles was studied using small-angle X-ray and neutron scattering measurements. The textile-bound iron oxide nanoparticles exhibited peroxidase-like activity when DPD was used as a substrate; in the presence of hydrogen peroxide this nanozyme activity enabled rapid decolorization of crystal violet. In addition, the binding of a sufficient amount of iron oxide particles on textiles enabled their simple magnetic separation from large volumes of solutions. The simplicity of the immobilized nanozyme preparation enabled its widespread application in environmental technology and biotechnology [52].

11.3.1.1 Ferrofluid modified *Leptothrix* sp.

The bacteria of the genus *Leptothrix* are characterized by the formation of a filament-looking microtubular sheath containing large amounts of oxidized Fe or Mn nanoparticles, that encases chained cells at an initial stage (see Fig. 11.7A). Partially or entirely empty sheaths, which are common in nature and culture, are formed by autolysis of bacterial cells. *Leptothrix* can be frequently found in aquatic environments in the form of ocherous suspensions. The microtubules diameter is usually 1–1.5 μm; their length can vary from a few to several hundred micrometers [53].

Mixing of water suspension of both native and autoclaved *Leptothrix* sheaths with perchloric acid stabilized FF resulted in the deposition of magnetic iron oxide nanoparticles and their small aggregates on the sheath surface, as shown by SEM (Fig. 11.7B); relatively homogeneous coverage of the *Leptothrix* sheath can be

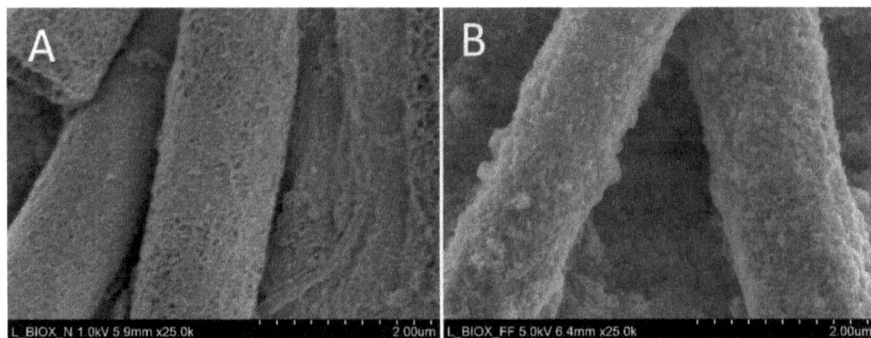

Figure 11.7. SEM images of native *Leptothrix* sheaths (A) and sheaths modified by perchloric acid stabilized magnetic fluid (B).

observed. The binding of magnetic iron oxide nanoparticles was firm and stable for at least 2 months and the same results were observed for both native and autoclaved sheaths. Both the magnetically modified sheaths and native sheaths exhibited similar energy-dispersive X-ray spectra. FF modified *Leptothrix* sheaths were easily separated using standard magnetic separators [54].

Magnetically modified *Leptothrix* sheaths exhibited adsorption of a water-soluble organic dye of crystal violet. The dye adsorption at pH 10 at room temperature reached the equilibrium within 15–30 min. The dye adsorption was reported by the Langmuir model; the value of maximum adsorption capacity was 166.6 mg of dye per 1 g of dry adsorbent [54]. Ferrofluid modified *Leptothrix* sheaths were also employed for biosorption of amino black 10B dye from aqueous solutions in a batch system. The maximum adsorption capacity was 339.2 mg of dye per 1 g of adsorbent [55]. Magnetically modified *Leptothrix* sheaths could be used for biotechnology and environmental technology applications, as a carrier for immobilization of various biologically active compounds or as high aspect ratio micro-containers [54].

11.3.1.2 Ferrofluid modified Posidonia oceanica

Posidonia oceanica is an endemic seagrass forming wide and dense meadows across the Mediterranean Sea. The dead seagrass parts form fibrous balls known as "Neptune balls", can also provide the life basis for micro-invertebrates. The fibrous balls material exhibit slow decomposition and low flammability and can be used as an insulating material [56]. In addition, *P. oceanica* fibers have been used as efficient adsorbents for removal of organic dyes and heavy metal ions [57].

P. oceanica seagrass dead biomass fibers were modified with perchloric acid stabilized FF; the modified material could be easily separated using strong permanent magnets. The modified biosorbent was employed for methylene blue removal; the calculated maximum adsorption capacity using Langmuir isotherm model was 133.3 mg g^{-1}. Alternative modification procedures were also employed successfully, namely modification with microwave synthesized magnetic iron oxide nano- and microparticles and mechanochemical approach [58].

11.3.1.3 Ferrofluid modified bacterial cellulose

Bacterial cellulose (BC) is a biocompatible biopolymer produced during cultivation of specific bacteria. BC is composed of well-arranged three-dimensional nanofibers (Fig. 11.8A) forming a hydrogel membrane with both a high porosity and high surface area. The unique mechanical properties of bacterial cellulose predetermine its wide potential application in biotechnology, microbiology, medicine and materials science.

Stimuli responsive derivatives of bacterial cellulose are of interest. Bacterial cellulose produced by *Komagataeibacter sucrofermentans* was magnetically modified using perchloric acid stabilized FF (Fig. 11.8B). Ferrofluid modified Bacterial cellulose (FBC) was used as a carrier for the immobilization of bovine pancreas trypsin and *Saccharomyces cerevisiae* cells; three strategies including adsorption with subsequent cross-linking with glutaraldehyde and covalent binding on previously activated FBC using sodium periodate or 1,4-butanediol diglycidyl ether were employed. Immobilized yeast cells (Fig. 11.8C) retained approximately 90% of their initial activity after six repeated cycles of sucrose solution hydrolysis, while immobilized trypsin could be repeatedly used for 10 cycles of low molecular weight substrate hydrolysis without substantial loss of its initial activity. FBC with immobilized reactive copper phthalocyanine dye efficiently adsorbed crystal violet from a water solution; the maximum adsorption capacity was 388 mg/g [59].

In another approach FBC nanofibers were designed as a novel selective adsorbent for efficient separation and recognition of thymidine nucleoside using the molecular imprinting method. FBC nanofibers were silanized with 3-(trimethoxysilyl) propyl methacrylate and further polymerized with a hydrophilic monomer for templating thymidine via metal chelate coordination. The developed material was optimized for thymidine adsorption and recognition. Maximum adsorption capacity was 431.3 mg/g. High selective thymidine adsorption in the presence of competitor nucleosides was observed. FBC nanofibers could be successfully used for multiple adsorption–desorption experiments. FBC can be efficiently integrated with molecular imprinting to form a new group of selective adsorbents [60].

As shown, ferrofluids can be efficiently used for magnetic modification of diamagnetic materials which leads to the formation of magnetically responsive materials. FF modified materials can be used for separation and immobilization of

Figure 11.8. SEM image of the surface structure of native bacterial cellulose (A) and magnetically modified cellulose covered by nanoparticles of iron oxides and their aggregates due to the ferrofluid treatment (B). Appearance of *Saccharomyces cerevisiae* cells entrapped and cross-linked in FF modified bacterial cellulose (C; the bar corresponds to 100 μm).

biologically active compounds, xenobiotics and cells. FF modified materials also exhibit peroxidase-like catalytic activity. The discussed materials and processes can be employed in various scientific and technological applications.

11.4 Non-polar ferrofluids in electric fields

Ferrofluids are often called magneto-polarizable or magnetically controllable liquids. Since the early research and development of ferrofluids, enormous efforts have been put into investigation of microscopic and macroscopic structural changes of ferrofluids in magnetic fields. Normal field instability [61] and labyrinthine instability [61, 62] belong to the most fascinating and popular phenomenon. These and many other ways of ferrofluid control stem from the interaction between magnetic nanoparticles and a magnetic field. Subsequently, the nanoparticle reorientation and assembly formation have a notable effect on physical properties of ferrofluids. In non-polar ferrofluids, such a magnetic field-driven nanoparticle assembly results in a significant change of dielectric permittivity. In addition, the well-known effect of magnetodielectric anisotropy may be observed in dependence on the relative orientations of magnetic and electric field intensities [63–65]. Besides the intuitive magnetic field-driven ferrofluid structures, the specific pattern formation can be induced even by an external electric field. Such a reversible patterning was observed in ferrofluids based on transformer oil when exposed to a static electric field [66]. The key requirement for the appearance of the electric-field driven formation of clusters in ferrofluids is a remarkable dielectric contrast between the base liquid and dispersed nanoparticles, allowing the induction of effective dipoles in the nanoparticles. The evidence of the electric field-driven reconfiguration of magnetic nanoparticles in non-polar ferrofluids raises further research issues related to a potential electro-magnetic coupling. One can therefore ask if electric fields can affect magnetic properties of non-polar ferrofluids in a similar way as magnetic fields affect the ferrofluids' dielectric properties. Answering such questions is of importance especially in the field of ferrofluids for electrical engineering applications. Thus, dielectric permittivity and magnetic susceptibility of non-polar ferrofluids exposed to the action of an external electric field were recently investigated. In what follows, we report on selected recent findings from this area of ferrofluid research.

The experimental verification of a static electric field influence on dielectric permittivity of a transformer oil-based ferrofluid was performed in the following manner [67]. A thin ferrofluid layer (5 μm) containing Mn-Zn ferrite nanoparticles (8.2 nm mean diameter) modified with oleic acid surfactant was confined between two electrodes of a parallel plate capacitor. The capacitor consists of two glass plates with deposited Indium Tin Oxide (ITO) layers acting as electrodes (Fig. 11.9a). The permittivity of the ferrofluid is obtained from the measured capacity of the capacitor, when an alternating voltage (measuring voltage of 100 mV) is applied to the ITO electrodes. In addition, the ITO electrodes are powered by a static voltage (DC bias voltage), thus generating the static component of the electric field acting on the ferrofluid. The DC bias voltage applied in this way is called internal, because the electrodes are in touch with the ferrofluid sample. As shown in Fig. 11.9a, there is another pair of copper electrodes attached to the glass capacitor from the outside.

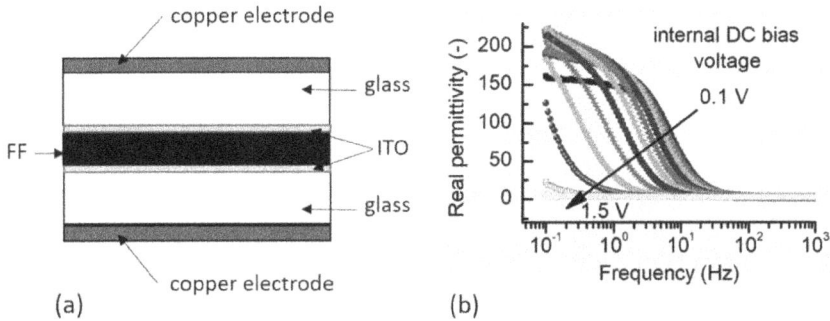

Figure 11.9. (a) Schematic illustration of the parallel plate capacitor employed in the verification of the static electric field influence on the spectrum of dielectric permittivity of the ferrofluid. (b) The spectra of the real permittivity of the ferrofluid exposed to the internal DC bias voltage from 0.1 V to 1.5 V.

The DC bias voltage applied to that pair of electrodes can be called an external one. Herein, the effect of internal DC bias voltage on the real permittivity spectrum of the ferrofluid with 2.6% solid volume fraction is presented in Fig. 11.9b.

The presented spectrum and the low-frequency dielectric dispersion can be ascribed to a nanoparticle interfacial polarization [67]. Then, under the influence of static voltage, one can observe two remarkable trends in the measured spectrum. The first is the increasing trend of the low-frequency permittivity with escalating DC bias voltage. The other is represented by a remarkable shift of the dielectric dispersion towards lower frequencies. In this way, the ferrofluid dielectric spectrum is reversibly controllable by the DC bias voltage and this effect resembles the well-known magnetodielectric response. The presented controllability of the ferrofluid dielectric spectrum by application of the internal DC bias voltage is associated with the nanoparticle cluster formation driven by the internal DC bias voltage [66]. The rising permittivity at low frequencies with increasing DC bias voltage is understood as a result of the strong DC polarizing force acting on the ferrofluid. The DC electric field polarizes the nanoparticles and gives rise to electrical dipoles in the nanoparticles. Then, the polarized state supports the nanoparticle interaction and cluster formation. However, the DC electric field also polarizes the charges on the interface and maintains them on the opposite sides of the cluster, so that their response to the alternating electric field attenuates remarkably. In other words, there will be an energy barrier along the cluster polarization axis, resulting in restricted mobility of the interfacial charges. Subsequently, the position of the low-frequency dielectric dispersion shifts to lower frequencies. It was found, that the application of the external DC bias voltage, yielding the external electric field intensity equivalent to the internal one, does not result in such a significant shift of the permittivity spectrum. The reason comes from the electric field distribution. The ferrofluid in the external electric field experiences homogenous field distribution. Thus, the nanoparticles and space charges do not undergo a drag force and trapping as they do in the internal electric field near the electrodes. The homogenous electric field distribution enables the homogenous nanoparticle arrangement. The absence of the dielectrophoretic forces is the reason of the limited nanoparticle aggregation. As a result, the interfacial charge does not experience polarization restrictions and the dielectric spectrum does not shift markedly with the increasing external DC bias voltage.

To know if the DC electric field can also influence the magnetic response of the ferrofluid, the AC magnetic susceptibility of a transformer oil-based ferrofluid containing iron oxide nanoparticles (5% solid volume fraction) was studied in a similar manner [68]. Figure 11.10a presents the problem schematically. The ferrofluid is held in a glass vial between two copper electrodes connected to a power supply. The vial is placed inside a detection coil system (excitation and pick up coils) of an AC susceptometer. The complex AC susceptibility is measured under various voltage levels applied to the electrodes.

Without the inserted electrodes, the ferrofluid containing magnetic nanoparticles (11 nm in average diameter) exhibits nearly constant behavior in the available frequency range. The constant behavior of the real susceptibility is admittedly accompanied with negligible magnetic losses (nearly zero imaginary susceptibility). The nanoparticles apparently follow the magnetic field changes via the Néel relaxation mechanism, that is the reversal of the magnetization from one to another direction of easy axis of magnetization by overcoming the energy barrier. The real susceptibility spectrum of the ferrofluid with the inserted electrodes shows a dramatic drop at frequencies above 1 kHz. This decrease is caused by the significant diamagnetic contribution of the electrodes. For this reason, attention is paid to the susceptibility spectrum behavior below 1 kHz in further analysis. From Fig. 11.10b one can observe that the ferrofluid susceptibility spectrum decreases when electric voltage is applied. In the considered frequency range, the susceptibility is again nearly constant without any remarkable dispersions, and its profile does not change with the applied voltage. It was also found that the ferrofluid temperature noticeable increases with the escalating voltage acting on the electrodes [68]. The ferrofluid heating due to magnetic losses is negligible, as the imaginary susceptibility is practically zero at these frequencies. However, the Joule heating, which stems from the electric current in the ferrofluid, is considered as a relevant reason of the temperature rise. When measuring the ferrofluid magnetic susceptibility and temperature under various

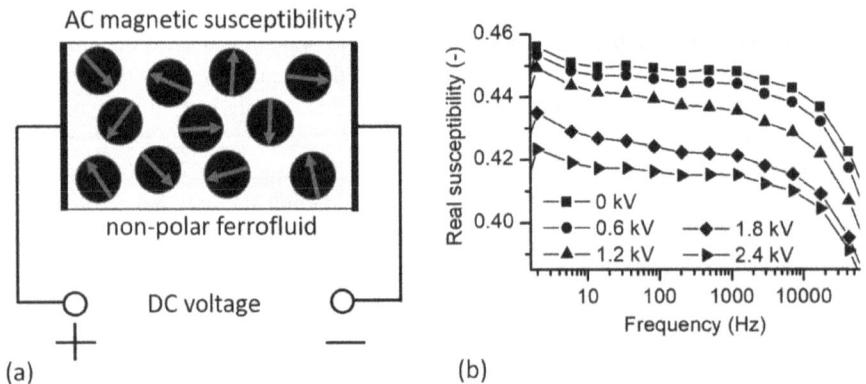

Figure 11.10. (a) The issue of AC magnetic susceptibility of a ferrofluid system exposed to a DC voltage illustrated schematically. (b) The real part of AC magnetic susceptibility spectrum of the ferrofluid under the influence of the DC voltage.

voltages (rising and decreasing voltage), it was found that the behavior of ferrofluid temperature is associated with the obtained susceptibility changes. The ferrofluid temperature increase due to the Joule heating was estimated from the measured electric current. The current of 4 µA in the ferrofluid flowing for 10 s generates the heat of 0.16 J. Taking into account the sample mass of 0.236 g and the specific heat of the ferrofluid 2.273 kJ/kg·K, one can prove that the calculated Joule heat can cause the measured temperature increase on application of the voltage step change. Furthermore, one can theoretically verify if the increased temperature is the only reason of the real susceptibility decrease. Following the simple Debye's theory, the frequency-dependent magnetic susceptibility $\chi(\omega)$ is given by

$$\chi(\omega) - \chi_\infty = (\chi_0 - \chi_\infty)/(1 + i\omega\tau) \tag{11.1}$$

with χ_∞ and χ_0 denoting the real susceptibilities at high and low frequency limit, and τ denotes the relaxation time. As Fig. 11.10b shows, the low-frequency susceptibility values decrease with the applied voltage and therefore the parameter χ_0 is mostly influenced by the voltage. This parameter is directly proportional to the nanoparticle number density n, the square of the magnetic moment m^2 and the reciprocal temperature $1/T$ as follows:

$$\chi_0 = nm^2/3kT\mu_0 \tag{11.2}$$

From Equation (11.2), where μ_0 is the vacuum permeability, it was calculated that the measured temperature increase cannot result in such a dramatic susceptibility decrease [68]. The calculated temperature dependence of the magnetic susceptibility is significantly greater than the one obtained experimentally. Thus, it is evident that the Joule heating cannot be the only reason of the ferrofluid susceptibility decrease. Another physical mechanism should contribute to the observed behavior. It was therefore considered that a fraction of greater nanoparticles exhibiting extrinsic superparamagnetism can be trapped by the gradient electric field near the electrodes. Then, the nanoparticles are restricted in responding to the AC magnetic field and the real part of the magnetic susceptibility decreases. The nanoparticle trapping near the electrodes is driven by electrophoresis or dielectrophoresis. The effect of nanoparticle trapping practically means the decreasing nanoparticle number density that results in the magnetic susceptibility lessen. However, it should be also stressed that the short distances between the nanoparticles in the electric field-driven clusters can give rise to magnetic interactions. Then, the magnetic interactions yielding reduced magnetic moments of the clusters (antiferromagnetic) can be a complementary reason of the decreasing ferrofluid susceptibility under the increasing voltage.

Based on the above-described experimental results, it is evident that both, dielectric and magnetic sensitivity of non-polar ferrofluids is controllable by the external DC electric field. This controllability is associated with the magnetic nanoparticle assembly formation in electric fields due to the dielectric contrast and electrohydrodynamics in the ferrofluids [64–66]. Clearly, the obtained results reveal the need of other experimental methods, like *in situ* polarized neutron scattering or magnetic resonance to explore the problem in depth. Then, the thorough exploration of the dielectric and magnetic properties of non-polar ferrofluids in electric fields may open new potential applications, e.g., in the field of switching or sensors.

11.5 Liquid crystals doped with nanoparticles

Liquid crystals are unique materials that combine fluidity of liquids with arrangement of molecules. The arrangement of molecules is responsible for anisotropic properties, which allows controlling them by external stimuli such as an electric or magnetic field, light or pressure. Their response to the electric field is widely used in liquid crystal displays as they respond to small electric fields by reorientation of molecules, which is accompanied with a change in optical properties. The effect of liquid crystal molecules reorientation by electric or magnetic field was named after its discoverer - Fréedericksz transition [69]. Contrary to the electric field, reorientation by the magnetic field occurs at a large magnetic field (~Tesla), which is impractical for technical applications. A method to reduce the threshold magnetic field was proposed by Brochard and de Gennes [70]. They suggested adding a small amount of magnetic particles to the liquid crystal. The new materials composed of nematic liquid crystals, i.e., liquid crystals with their molecules pointing to one direction denoted by director $\vec{n}$, and magnetic particles are called ferronematics. They are responsive to either smaller or larger magnetic fields than pure liquid crystal depending on the mutual orientation of nanoparticles magnetic moments $\vec{m}$ and director $\vec{n}$. If $\vec{m} \parallel \vec{n}$ required the magnetic field is smaller as magnetic nanoparticles encourage the reorientation and if $\vec{m} \perp \vec{n}$ is in the higher magnetic field is needed because coupling between nanoparticles and molecules of liquid crystal has to be overcome. Although nanoparticles in liquid crystals are able to decrease the threshold of the magnetic field, composites have a problem with stability, when nanoparticles are not covered with a surfactant. A surfactant, such as oleic acid, prevents aggregation of nanoparticles and significantly improve stability of ferronematics. On the other hand, a surfactant affects properties of liquid crystals, such as transition temperature from isotropic to nematic phase (T_{IN}). In composites, T_{IN} is affected by properties of particles—their size and shape. Spherical nanoparticles shift T_{IN} to lower values, while anisotropic nanoparticles shift T_{IN} to higher values [71, 72]. A surfactant can cause an additional decrease in T_{IN} as it effectively dilutes nematic matrix.

11.5.1 Fréedericksz transitions

The experiments were performed on composites of 4-cyano-4-hexylbiphenyl (6CB) liquid crystal and spherical iron oxide nanoparticles covered with oleic acid. The composites with nanoparticles sizes 10 nm, 20 nm and 30 nm and three volume concentrations for each size (10^{-4}, 5×10^{-4} and 10^{-3}) were examined. Concentration of nanoparticles in composite, size of nanoparticles and their saturation magnetization are important parameters influencing the behaviour of composites. Transmission Electron Microscopy (TEM) and Scanning Electron Microscopy (SEM) were used to determine the diameter of nanoparticles without an oleic acid layer (d) and diameter of nanoparticles with oleic acid layer (D). TEM and SEM images of nanoparticles are shown in Fig. 11.11. Values acquired from TEM and SEM allowed one to calculate the oleic acid layer thickness (D-d)/2. Notice that thickness of the layer is not identical for all three nanoparticle sizes. It decreases with increasing nanoparticle size. The obtained results are listed in Table 11.1 along with their saturation magnetization.

T_{IN} of the composites was investigated by Polarized Optical Microscopy (POM), Differential Scanning Calorimeter (DSC) and capacitance measurements. POM is a commonly used method for distinguishing the phases between which phase transition occurs. DSC allows more precise determining of T_{IN} and capacitance measurements provide information about the alignment of liquid crystal molecules. All three methods showed decreasing of T_{IN} with increasing nanoparticles concentration in composites and non-monotonic dependence on diameter [73].

Capacitance measurements in addition revealed different behavior of composite with 30 nm particles and high volume concentrations 10^{-3} (See Fig. 11.12). For capacitance measurements, the sample is placed in the cell consisting of glass plates with indium-tin oxide electrode and polyimide layer which ensures the required alignment of liquid crystal molecules. For all experiments presented in this chapter,

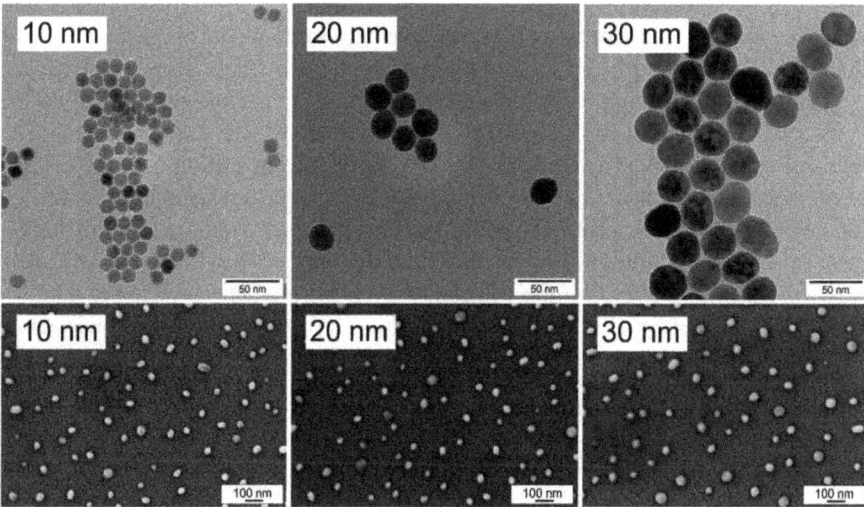

Figure 11.11. TEM (top row) and SEM (bottom row) images of iron oxide nanoparticles.

Table 11.1. Characteristics of nanoparticles – d is diameter of nanoparticles without oleic acid, D is diameter of nanoparticles with oleic acid, (D-d)/2 denotes thickness of oleic acid layer, and M_s is saturation magnetisation of nanoparticles.

	d (nm)	D (nm)	(D-d)/2 (nm)	M_s (kA/m)
10 nm	12.38 ± 0.06	36.07 ± 0.33	11.85	53
20 nm	21.81 ± 0.13	36.67 ± 0.35	7.43	173
30 nm	31.59 ± 0.12	42.74 ± 0.68	5.57	200

cells ensuring planar initial alignment were used, i.e., parallel orientation of liquid crystal molecules to a cell surface. In isotropic phase molecules are oriented randomly, but with decreasing temperature they become aligned which results in the reduction of capacity. The capacitance of composites with 30 nm particles and concentration 5×10^{-4} and 10^{-3} decline with decreasing temperature even in an isotropic phase which results in difficulty with determining T_{IN} [73]. In addition, POM showed significant widening of the temperature interval in which the composite passes from the isotropic to the nematic phase, especially for the highest concentrations.

According to Gorkunov and Osipov theoretical paper [71] T_{IN} of ferronematic T_{FN} is calculated by the following equation

$$T_{FN} = T_{LC} (1 - \phi),$$

where T_{LC} denotes T_{IN} of pure liquid crystal (in Kelvin units) and ϕ denotes volume concentration of nanoparticles dispersed in liquid crystal matrix. By taking T_{LC} from capacitance measurements 28.1°C, T_{FN} for composite with concentrations 10^{-4}, 5×10^{-4} and 10^{-3} is 28.07°C, 27.95°C and 27.80°C, respectively. Experimentally obtained temperatures of isotropic to nematic phase transition were significantly lower than temperatures calculated from theory. Therefore, the impact of oleic acid which serve as a surfactant was examined. Diameters obtained from SEM and TEM (Table 11.1) were employed to determine the volume concentration of oleic acid in composites. The volume of oleic acid coating was calculated by subtracting the volume of a sphere with diameter d from the volume of a sphere with diameter D. The total volume of the oleic acid in the sample was acquired by multiplying the volume of oleic acid layer by the number of particles present in the sample. Calculations showed that volume concentration of oleic acid in composites is considerably higher than volume concentration of nanoparticles. Mixtures of liquid crystal with oleic acid were prepared based on the calculations. T_{IN} of mixtures were lower than T_{IN} of pure liquid crystal and T_{IN} decreased with increasing concentration of oleic acid.

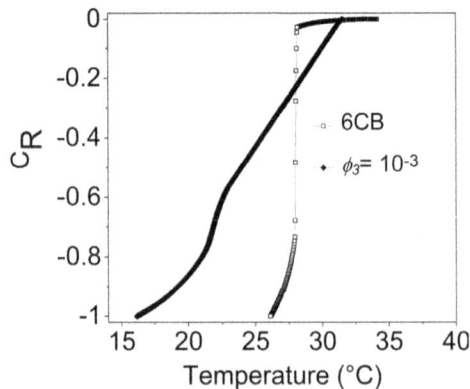

Figure 11.12. Dependence of reduced capacitance C_R on temperature for pure 6CB and 6CB with high concentration of 30 nm MNPs.

The formula established in mean-field theory needs to be modified to take into account the entire excluded volume of nanoparticles including oleic acid layer by the introduction of effective volume fraction ϕ_{EF}

$$T_{FN} = T_{LC}\,(1 - \phi_{EF}),$$

$$\phi_{EF} = \frac{V_{NP+OA}}{V_{NP}} \cdot \phi = (\frac{D}{d})^3 \cdot \phi.$$

It predicts decreasing of T_{IN} with nanoparticle concentration and non-monotonic dependence on nanoparticle size (decrease for 10 nm particles and increase for 20 nm and 30 nm particles), which is consistent with experiments. Theory and experiments lack quantitative agreement in case of 20 nm and 30 nm particles and concentrations 5×10^{-4} and 10^{-3}, which can be explained by the presence of aggregates where the excluded volume is larger than the sum of volumes of individual nanoparticles due to "free" spaces between nanoparticles in aggregates.

Fréedericksz transition in the composites were studied via capacitance measurements. Composites in nematic phase at 18°C in cells with cell gap 50 μm were exposed to the external electric and magnetic field. Liquid crystal molecules were reoriented with increasing the electric or magnetic field which caused reorientation of molecules in the cell as illustrated in Fig. 11.13a.

The reorientation of liquid crystal molecules perpendicular to the cell results in increasing capacitance. Dependence of capacitance on the voltage and magnetic field allows to determine the threshold voltage and threshold magnetic field at which the composite begins to respond as the value at which the capacitance abruptly increases [74]. High capacitance of samples with 20 nm and 30 nm particles and concentration 10^{-3} is caused by high conductivity, which was also observed in liquid crystal with Me_7GeS_5I nanoparticles [75] .

The experiments enables one to estimate the surface density of anchoring energy W and subsequently the parameter ω, which relates to the type of anchoring [76].

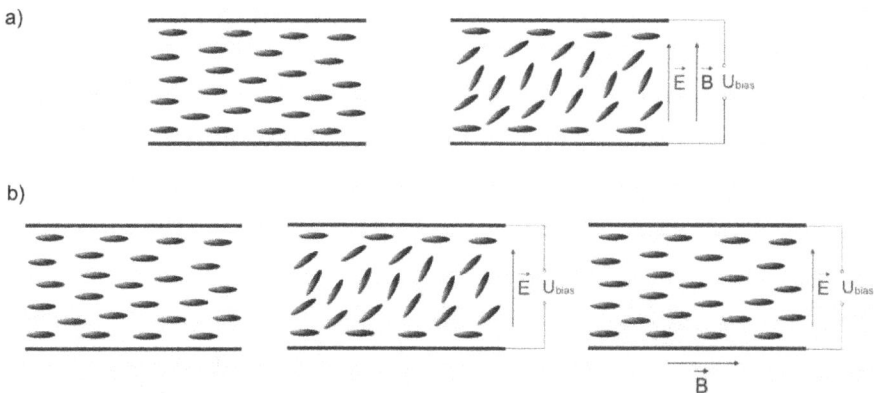

Figure 11.13. Geometry of the measurements - (a) electric and magnetic Fréedericksz transition and (b) structural transition in combination of electric and magnetic field.

The parameter $\omega = Wd/K_1$, where d denotes the diameter of nanoparticle and K_1 is the elastic constant of the liquid crystal. In the case of $\omega \lesssim 1$ soft anchoring is present and both orientations $\vec{m} \parallel \vec{n}$ and $\vec{m} \perp \vec{n}$ are allowed. In the case of $\omega \gg 1$ anchoring is rigid and only $\vec{m} \perp \vec{n}$ is allowed. The formula for threshold magnetic field of liquid crystal in studied geometry

$$B_{LC} = \frac{\pi}{L}\sqrt{\frac{\mu_0 K_1}{\chi_a}}$$

was used to calculate anisotropy of magnetic susceptibility χ_a. Here L denotes the thickness of the cell gap and μ_0 is permeability of the vacuum. The anisotropy of magnetic susceptibility was determined to $\chi_a = 1.046 \times 10^{-6}$ by taking $B_{LC} = 0.16$ T (see Table 11.2), $L = 50$ μm and $K_1 = 5.4$ pN [77] for the calculations. The relationship between the magnetic field threshold of pure liquid crystal B_{LC} and the threshold magnetic field of ferronametic B_{FN}

$$B_{FN}^2 = B_{LC}^2 \pm \frac{2W\phi\mu_0}{\chi_a d}$$

enables one to estimate the surface density of anchoring energy. For $\vec{m} \parallel \vec{n}$ a negative sign applies while for $\vec{m} \perp \vec{n}$ a positive sign is used . The surface density of anchoring energy and subsequently parameter ω for composites with 10 nm, 20 nm and 30 nm particles were determined to $W_{10} \sim 10^{-8}$ Nm^{-1}, $W_{20} \sim 10^{-7}$ Nm^{-1}, $W_{30} \sim 10^{-7}$ Nm^{-1} and $\omega_{10} \sim 10^{-5}$ $\omega_{20} \sim 10^{-4}$ and $\omega_{30} \sim 10^{-3}$, respectively. The calculated values of the parameter ω indicate soft anchoring of the nematic with the particles and shifting magnetic field threshold to lower values with addition of nanoparticles suggest orientation $\vec{m} \parallel \vec{n}$.

Additional experiments (see Fig. 11.13b) disclose how a decrease in threshold voltage affected structural transition in combined electric and magnetic field. For these experiments bias voltage U_{bias} larger than threshold voltage U_F of Fréedericksz transition was applied to the sample and held at a constant value. Voltage causes reorientation of molecules perpendicular to the cell surface. The magnetic field applied parallel to the cell surface was gradually increased while keeping the voltage at constant value. Molecules begin to align back to their initial orientation at the

Table 11.2. Threshold voltage and threshold magnetic field for liquid crystal 6CB (U_{LC}, B_{LC}) and its composites (U_{FN}, B_{FN}).

	U_{LC} (V)			B_{LC} (T)		
6CB	0.79			0.16		
	U_{FN} (V)			B_{FN} (T)		
	10 nm	20 nm	30 nm	10 nm	20 nm	30 nm
$\phi_1 = 10^{-4}$	0.79	0.79	0.79	0.16	0.16	0.15
$\phi_2 = 5 \times 10^{-4}$	0.78	0.78	0.76	0.15	0.15	0.14
$\phi_3 = 10^{-3}$	0.63	0.26	0.54	0.05	0.05	0.09

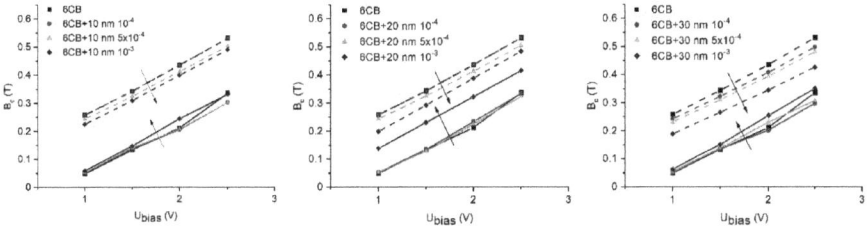

Figure 11.14. Critical magnetic field B_c vs. bias voltage U_{bias}. The solid lines represent experiments and dashed lines represents calculations.

magnetic field B_c. Experiments were repeated for various voltages and dependence of B_c on U_{bias} was plotted (Fig. 11.14 - solid lines). Increasing of concentration cause shifting B_c to higher values. However, B_c should shift to lower values according to [78].

$$\left(\frac{U_{bias}}{U_F}\right)^2 - \left(\frac{B_c}{B_F}\right)^2 = 1$$

where $U_{bias} \geq U_F$ (see Fig. 11.6 - dashed lines). For pure liquid crystal $U_F = U_{LC}$, $B_F = B_{LC}$, and for ferronematics $U_F = U_{FN}$, $B_F = B_{FN}$. Discrepancy between measured and calculated B_c can be explained by decreasing the threshold voltage of Fréedericksz transition with increasing concentration as the same bias voltage U_{bias} cause more pronounced reorientation in the sample with high concentrations, where the threshold voltage of Fréedericksz transition is lower. Reorientation back to initial orientation thus begins at a larger magnetic field B_c.

In conclusion, experimental results described in this chapter prove the importance of nanoparticles parameters on the behavior of liquid crystals. The size of nanoparticles and volume concentration of the entire impurity plays an important role. A surfactant significantly improves the stability of composites, but notably increase full volume excluded by nanoparticles, which has a large impact on the isotropic to the nematic phase transition temperature of composites. In composites with large concentration, it can result in disruption of liquid crystal properties and cause indistinct transition as we demonstrated by capacitance measurements. On the other hand, our experiments showed that the large concentration of magnetic nanoparticles causes the most pronounced shift of the threshold magnetic field to lower values. Therefore, finding optimal parameters is crucial for technical applications of ferronematics.

11.6 Study of structural changes in ferronematics using capacitance and surface acoustic waves measurements

11.6.1 Principles of SAW in liquid crystal

As we have shown, the Surface Acoustic Wave (SAW) technique can be very a useful tool for the investigation of structural changes in LCs including those doped with different kinds of nanoparticles [79–81]. Such SAW, generated on a piezoelectric substrate propagates along the surface as an elliptically polarized wave until it reaches the LC, placed on its path, when a longitudinal wave is generated into the LC due to the vertical displacement of SAW. The advantage of this SAW technique is that only a very thin layer of the LC is required. A SAW due to the strongly absorbed longitudinal wave by the LC that is subsequently indicated by the decreasing amplitude of SAW in the direction of SAW propagation can be applicably utilized for the study of LC structural changes. Next some representative results of SAW investigation of nematic LCs doped with magnetic particles (ferronematics) of various shapes and different concentrations to determine their effect on structural changes under external magnetic field are reported.

Acoustic (ultrasonic) methods have been used over several decades to study rheological, viscous and structural properties of liquid crystals. Most of the described acoustic experiments use two main types of acoustic waves, namely longitudinal acoustic and shear waves that differ in the direction of the oscillating motion of particles in respect of the direction of the wave propagation. The presented experimental arrangement using the configuration in which the generated oscillating motion of particles relative to the direction of wave propagation is normal to each other, like in the case of shear waves [82, 83], but with vertical orientation of oscillations [81].

A SAW generated on a solid substrate such as a Rayleigh wave propagates along the substrate surface as an elliptically polarized wave, with the displacement amplitude decaying into the substrate. When a LC is placed on its path, the amplitude of SAW after reaching a LC is exponentially attenuated, propagating along as a "leaky surface wave" at the LC/substrate interface due to the generation of a longitudinal wave in the LC caused by the vertical displacement of SAW on the surface. The longitudinal wave itself propagates at the Rayleigh angle, $\theta_R = \sin^{-1}(v_{lc}/v_s)$ [84], where and are propagation velocities of the longitudinal wave in the LC and the Rayleigh wave on the substrate, respectively. So that when the substrate is in contact with the LC, SAW irradiates a longitudinal wave into the liquid giving rise to propagation losses. The cumulative attenuation through that path along the fluid/substrate interface is then the sum of the attenuation of a fluid-damped Rayleigh SAW α_s given by

$$\alpha_s = \rho_{lc}\, v_{lc}/(\rho_s\, v_s\, \lambda_{SAW}) \tag{11.1}$$

and the attenuation of a plane longitudinal wave α_{lc} that the formulation in the case of LC is more complicated than in the case of isotropic fluid [85]. The flow behavior of LC differs fundamentally from that of isotropic liquids because of mutual coupling between flow direction, director orientation and viscosity. Symbols ρ_{lc} and ρ_s are the LC and substrate densities, respectively and λ_{SAW} is the SAW wavelength.

The earlier theoretical work using hydrodynamic theory of compressible nematic LCs led to the following expression for longitudinal wave attenuation [86–88]

$$\alpha = \frac{2\pi^2 f^2}{\rho_{lc}\, v_{lc}^3} \left[(2v_1 + v_2 - v_4 + 2\,v_5) + 2(v_4 - v_1 - v_5)\sin^2\theta - \frac{1}{2}(v_1 + v_2 - 2v_3)\sin^2 2\theta \right] \tag{11.2}$$

where f is the frequency, θ is the angle between $\boldsymbol{n}$ direction and wave vector $\boldsymbol{k}$ and ($i = 1$–5) are viscosity coefficients. The coefficient v_2 is shear viscosity coefficient when the molecular axes are normal to the direction of shear flow and to the velocity gradient, the coefficients and are related to the other two shear viscosity coefficients. The coefficients $v_4 - v_2$ and represent bulk viscosities. The relation for an absorption coefficient α of a longitudinal acoustic wave propagating in LC can be also expressed using Leslie coefficients α_i ($i = 1–5$) and bulk viscosity coefficients μ_i ($i = 1 – 3$) [89] the combination of which results from an account of director oscillations in an ultrasonic wave as well as have the connection with rotational viscosity, which were ignored in the work mentioned above. Later, three basic geometries for viscosimetric measurements but using shear waves were introduced [89, 90]. The anisotropic viscosity was described by three fundamental viscosity coefficients ($i = 1–3$) referring to three different orientations of director n with respect to the direction of the flow velocity v and the ∇v.

However, all the presented theoretical procedures could not be used for acknowledging the applicability of our SAW experimental technique. The character of our anisotropy measurements as well as the effect of SAW frequency [81, 91] indicated unambiguously that the bulk viscosity coefficients should dominate our SAW attenuation measurements. In view of fact that our SAW investigations are orientated to monitoring the role of magnetic dopants on structural changes under external fields, the presented SAW techniques were completely satisfying at acquiring relevant information concerning the influence of magnetic nanoparticles on nematic LCs behavior

11.6.2 *Materials and SAW experimental details*

The SAW pulses ($\sim$ 1 μs) of frequency 10, 20, 30 and 40 MHz could be generated by the interdigital transducer prepared on the $LiNbO_3$ substrate, using hf pulses and another transducer was used for receiving the SAW signal. The SAW attenuation response could be recorded then as a function of the external magnetic field, temperature or time at the magnetic field jump. The scheme of experimental arrangement configuration is illustrated in Fig. 11.15. The cell with the LC sample of thickness $D \approx 100$ μm was placed at the center of the $LiNbO_3$ substrate and installed in the sample holder that was component of the thermostatic measuring chamber. The initial intrinsic arrangement of LC molecules was supposed to have a predominate alignment in the plane of LC cell and the applied field was perpendicular to them. However, at the anisotropy measurements the sample chamber could rotate to change the angle between the magnetic field direction and wave vector k_{saw}. More detailed descriptions of experimental arrangement have been published in [79, 81].

The effect of several kinds of magnetic nanoparticles of various shapes, including nanorods, spindles and magnetosomes of different concentrations on nematic LCs behavior under external magnetic field was investigated during last period using the SAW attenuation response. The magnetic Fe_3O_4 nanorods were prepared by the processing of magnetite and goethite with an estimated particle length from 200 to 500 nm [81, 92] and added in different volume concentration (10^{-5}, 5×10^{-5}, 10^{-4} and 5.2×10^{-4}) in 6CHBT, the thermotropic nematic LC. The samples before being placed into a measuring cell that was pre-heated above the temperature of its

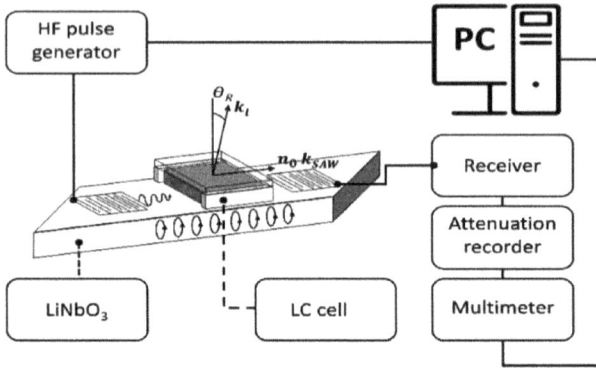

Figure 11.15. Experimental arrangement of LC cell on LiNbO₃ delay line for SAW investigation including principal geometries characterizing correlative orientation of important directions.

isotropic state (~ 50°C) and ultrasonically mixed for 2 hr to ensure homogenous distribution of nanoparticles inside the host LC.

Another type of investigated nanoparticles were magnetosomes represented by magnetic crystals of iron oxide nanoparticles (magnetite Fe_3O_4) covered and connected by a biological membrane (phospholipid bilayer) and formed chains. The chain of magnetosomes could consist of 12 – 20 crystals, which multiplies their sensitivity to the magnetic field. Magnetosomes were cultivated and extracted from the Magnetotacticum Spirillum AMB 1 bacteria in laboratory conditions using the biomineralization process [23, 91]. Using the procedure TEM analysis the mean size 43 ± 12 nm was determined. For the experiments, liquid crystals 5CB, 6CB and E7 were chosen. The composites of liquid crystal and magnetosomes with the concentration of 0.2 wt% were prepared.

The range of ferronematics was also extended by particles with non-traditional magnetic properties when the vectors of magnetic moments of the elongated particles do not coincide with the directions of their main axes [78, 93]. The size separation from the base sample was done by different times of centrifuge and the mean values of their lengths (l = 368 nm, 280 nm and 203 nm) were obtained by using the log-normal distribution function. The ferronematic system was based on liquid crystal 6CB. Magnetic nanoparticles were coated with phosphoric acid as a surfactant to prevent their aggregation. The LC samples were prepared with three different volume concentrations of the magnetic nanoparticles, $\Phi_1=10^{-3}$, $\Phi_2=10^{-4}$ and $\Phi_3=10^{-5}$.

11.6.3 SAW study of ferronematics

SAWs were used to study the effect of previously cited magnetic nanoparticles on LCs behavior in weak magnetic fields. Experimental measurements including the investigation under linearly increasing and/or pulsed magnetic field, respectively, as well as the investigation of temperature and time influences on structural changes were done. The measurement of the acoustic anisotropy gives additional useful information about the structure of nanoparticles formations in investigated LCs. In contrast to undoped LCs the different SAW attenuation responses induced by the magnetic field in studied LC samples were observed, suggesting the orientational coupling between magnetic moments of nanoparticles and the director of LCs followed by structural changes. Next some later representative results are presented.

The temperature dependencies of acoustic attenuation measured at the frequency of 10 MHz for investigated 6CBHT consisting of magnetic nanorods of different concentration are illustrated in Fig. 11.16. The SAW attenuation in the case of both pure and doped LC of all concentrations shows very similar development, characterized by unrecorded change at temperature corresponding to the crystal-nematic transition (T_{CN}) and the gradually increasing attenuation up to the temperature of $\sim 42°C$ when the structural changes to isotropic phase (T_{NI}) are represented by a strong decrease of the SAW attenuation (see inside Fig. 11.16). The presented temperature dependencies also show the role of the concentration on the shift of nematic-isotropic transition (T_{NI}) to lower temperatures, decreasing in small successive steps with increasing concentration, however, with the most expressive decrease in the case of the highest nanoparticles concentration. The shift of phase transition towards lower temperatures for doped LC can be attributed to the anchoring interactions between LC molecules and nanoparticle surfaces that disturb the LC order [94]. Another reason for this decrease could be related to the increasing volume of impurities, occurring in LC during the synthesis due to the increasing nanoparticles concentration [71].

The dependencies of SAW attenuation on linearly increasing magnetic field at a constant rate (15 mT/min) and constant temperature measured for samples of another LC (6CB) doped with the most elongate spindle-like particles ($l = 368$nm) for three concentrations, including pure LC is shown in Fig. 11.17. This figure presents the SAW attenuation changes for the magnetic field B perpendicular to the cell and wave vector k_{SAW}. According to the pure LC the role of particle concentration on the structural changes, pointing the marked decrease of the threshold field can be identified. The observed dependencies similar as in the case of other particles, can be resolved into three phases, a feeble increase in the range of 0–40 mT followed by a faster growth that gradually merges into saturation. The more rapid increase of the SAW attenuation corresponds to the LC molecules' main process reorientation running in the center of the cell. The interesting feature is that the highest concentration of particles $(\Phi_1 = 10^{-3})$ does not automatically ensure the maximal magnetic field influence on structural changes. The decrease of the magnetic field influence, even compared with pure LC, can be caused by the creation of some aggregates due to both too high concentration of magnetic particles and their length [95] .

The dependencies of SAW attenuation on the magnetic field measured for 6CB doped with shorter spindles ($l = 280$ nm) similarly, as in the case of LC samples doped with the longest spindles ($l = 368$ nm), shows the gradual decrease of structural changes with increasing concentration, from the concentration $\Phi_3 = 10^{-5}$ Over the above, the decrease of the attenuation was registered in the configuration for the magnetic field B lying in the cell plane but still perpendicular to k_{SAW}. Only negligible residual attenuations were registered for this magnetic field direction. The same character of SAW response was also obtained for the shortest spindles ($l = 203$ nm), but with a higher response for all concentrations.

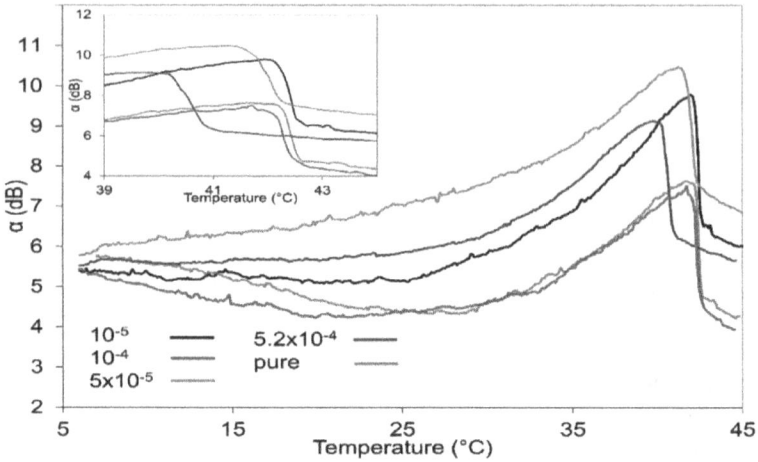

Figure 11.16. Temperature dependence of SAW attenuation of 6CHBT doped with nanorods of different volume concentrations ($10^{-5}, 5 \times 10^{-5}, 10^{-4}, 5.2 \times 10^{-4}$) including pure 6CHBT.

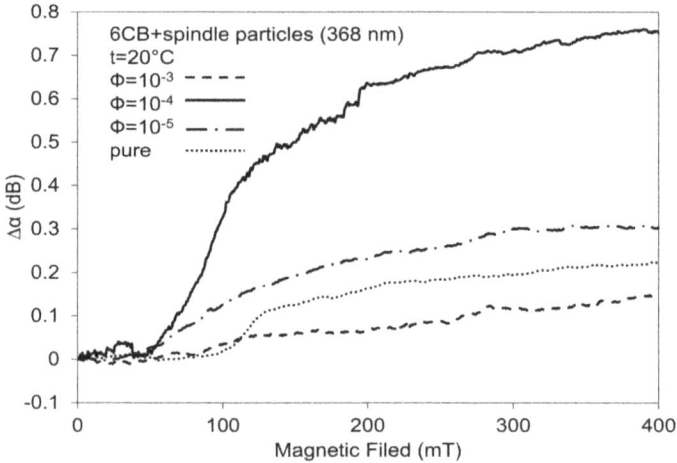

Figure 11.17. Effect of applied magnetic field on SAW attenuation for 6CB doped with spindle nanoparticles (l=368nm) of different volume concentrations ($\Phi = 10^{-3}, 10^{-4}$ and 10^{-5}) including pure 6CB.

The different SAW attenuation responses induced by the magnetic field in doped LCs compared to undoped ones were also observed due to the presence of magnetosome nanoparticles in the investigated compounds. The magnetosome nanoparticles of the same volume concentration (0.2 wt%) were added to 5CB, 6CB and E7 LCs [91]. The decrease of the threshold field for doped LCs according to pure LCs but only slight effects on structural changes were registered. Concerning the threshold magnetic fields of LCs and composites, the slight shift to lower values could be registered for 5CB and E7 LCs doped with magnetosomes. The measurement of the temperature dependencies of SAW attenuation also indicated the shift of nematic—isotropic transition to lower temperatures in the case of 5CB and 6CB, however, to a higher temperature in the case of E7. The reason for the phase

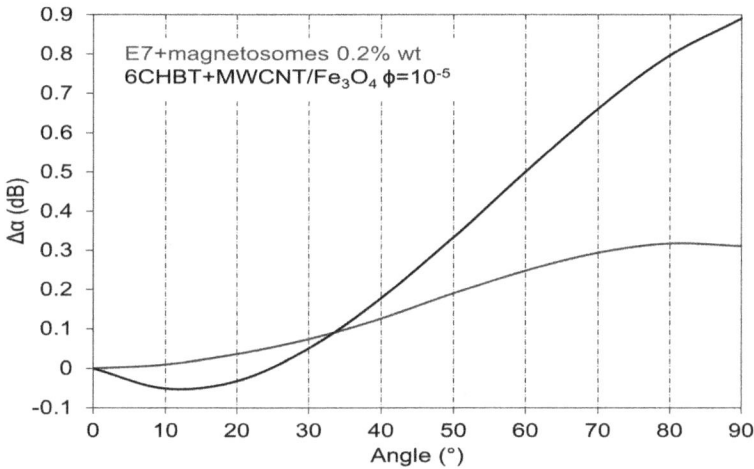

Figure 11.18. Angle dependence (anisotropy) of SAW attenuation measured at $B = 300$ mT for E7 doped with magnetosomes and 6CHBT doped with MWCNT/Fe$_3$O$_4$.

transition shift towards lower temperatures was described in the case of nanorods. However, the shift of the transition point towards higher temperature, as in E7, could be attributed to magnetosomes that induce local magnetic moments with a sufficient number of neighboring LC molecules [94]. The additional interaction can then lead to an increasing transition point. The measurement of the acoustic anisotropy (Fig. 11.18) was in agreement with our earlier assumption about the important role of bulk viscosity in present SAW experiments, typical obviously for cases when longitudinal waves were used. Capacitance and light-transmission experiments besides supported SAW results and pointed out conclusions on investigated LCs behavior in the external magnetic field.

The role of SAW frequency on temperature dependencies of SAW attenuation was investigated for 6CHBT doped with nanorods [81] or pure [91] 5CB, as representative samples, for 10, 20, 30 and 40 MHz. As it can be seen, different frequencies are differently sensitive to processes that are in progress in the LC during changing temperature. If we take into account the supposition that the shear viscosity is frequency-dependent at high frequencies and the bulk viscosity can show the frequency dependence at the region of frequencies also recognized at our SAW measurements, these results support the assumption about the bulk viscosity coefficients responsible in investigated LCs for SAW measurements [81, 96].

The progress of switching processes in a magnetic field registered by the SAW attenuation response for 6CB doped with spindle nanoparticles or magnetosomes [91, 93] and 6CHBT doped with nanorods or carbon nanotubes (CNT) [79, 81] for different magnetic fields apparently indicate that the magnetic field value is a very important parameter. For example, while in the case of functionalized CNTs for the magnetic field 175 mT the relaxation time is $\tau_{ml} = 50.8$ s, in the case 250 mT it is already shorter, $\tau_{ml} = 14.1$ s and for 400 mT it is only $\tau_{ml} = 4.5$ s. This behavior coincides well with electro-optical investigations [97]. Concerning CNTc it should be noted that although SAW responses obtained for LCs doped with SWCNT/Fe$_3$O$_4$

or MWCNT/Fe$_3$O$_4$ under the magnetic field are similar, nevertheless differences were observed in the role of their functionalization and the rise and delay of switching times was also influenced by the concentration. So that choosing their adequate concentration or functionalization the LC suspension with the required properties can be created.

The dependencies of acoustic attenuation $\Delta\alpha$ on angle θ^* (anisotropy) measured at magnetic field 300 mT for two different LCs are shown in Fig. 11.18. The θ^* is the angle between z direction which is perpendicular to the cell plane and magnetic field orientation (see Fig. 11.15). It should be noted that the acoustic attenuation is the sum of surface and volume attenuation (Equations 11.2 and 11.3). The surface acoustic attenuation is almost independent of temperature and its calculated value is $\sim$ 175 m^{-1}. However, as it can be seen the SAW attenuation grows with an increasing angle and the dependence of $\Delta\alpha$ on angle θ^* illustrated in Fig. 11.4 can be expressed as

$$\Delta\alpha = A \sin^2 (\theta^* - \theta_R), \tag{11.3}$$

that is after the application of Equation (11.2) in perfect accordance with some previous results of LC investigation using longitudinal acoustic waves [98]. Angle θ^* determined from both anisotropy measurements is $\sim$ 12–15 degrees that is, compared to calculated value ($\sim$ 22 deg.) for pure 6CHBT, somewhat of a lower value. However, the velocity of longitudinal wave decreases with increasing concentration of nanoparticles [99], that can be a reason for the difference. The comparison of these results with theoretical suppositions confirm that the bulk viscosity coefficients should dominate the SAW attenuation. In conclusion we can affirm that SAW investigations oriented above all to monitoring the role of magnetic dopants on structural changes under external fields, the present SAW technique is completely satisfying at acquiring relevant information concerning the influence of magnetic nanoparticles on LCs behavior.

11.7 Conclusion

The "Magnetic nanoparticles change the properties of traditional materials and open up new application possibilities" chapter summarizes the latest experimental findings focused on various magnetic nanoparticles physicochemical studies. The size, shape, morphology, and concentration of magnetic nanoparticles depend on the initial set-up of the synthesis in the constrained liquid environment. Magnetic nanoparticles, naked or coated with a protein (ferritin and its derivatives) and lipid (magnetosomes chains), respectively, dissolved in various polar (aqueous) or non-polar (oily) solvents, are thus useful in a wide range of applications (medicine, environmentalism). In addition, the interaction with precisely defined organic (liquid crystals) or biopolymer (lysozyme and various fibers) carbon-based materials provide a highly arranged design of various structures controllable by electric and magnetic field for displaying technology development. Experimentally proven thermal loss decreasing after magnetic fluid replacement instead of regular oil in the large transformers can save energy and contribute to the economic growth of various industrial companies.

Acknowledgements

This research was supported by VEGA 2/0043/21, 2/0011/20, 2/0044/20, the Slovak Research and Development Agency under the Contract No.APVV-19-0324, APVV 15-453, APVV-18-0160, MVTS SAV Meranet Flexible Magnetic Filaments, MVTS SKTW AMAZON, the ERDF projects No. CZ.02.1.01/0.0/0.0/16_019/0000754, CZ .02.1.01/0.0/0.0/17_048/0007399 and No. CZ.02.1.01/0.0/0.0/17_048/0007323 of the Ministry of Education, Youth and Sports of the Czech Republic, bilateral project NASU-20-04 (Academy of Sciences, Czech Republic) and by the project No. ITMS 313011T548 MODEX (Structural Funds of EU, Ministry of Education, Slovakia).

References

[1] A. P. Guimarães and A. P. Guimaraes. Principles of nanomagnetism. Vol. 7. Springer (2009).

[2] L. Balejčíková et al. Experimental assessment of interactions between liquid crystal 4-cyano-4'-hexylbiphenyl and magnetoferritin. Mendeleev Communications 30(1): 73–75 (2020).

[3] D. Weller and T. McDaniel. Media for extremely high density recording. In Advanced Magnetic Nanostructures. Springer. pp. 295–324 (2006).

[4] F. Fievet et al. Homogeneous and heterogeneous nucleations in the polyol process for the preparation of micron and submicron size metal particles. Solid State Ionics 32: 198–205 (1989).

[5] S. Mann and F. C. Meldrum. Controlled synthesis of inorganic materials using supramolecular assemblies. Wiley Online Library (1991).

[6] M. E. Evans and F. Heller. Environmental magnetism: principles and applications of enviromagnetics. Elsevier (2003).

[7] A. Bekker et al. Iron formation: the sedimentary product of a complex interplay among mantle, tectonic, oceanic, and biospheric processes. Economic Geology 105(3): 467–508 (2010).

[8] R. N. Mitchell et al. Orbital forcing of ice sheets during snowball Earth. Nature Communications 12(1): 1–9 (2021).

[9] J. L. Kirschvink and J. L. Gould. Biogenic magnetite as a basis for magnetic field sensitivity in animals. Biosystems 13: 181–201 (1981).

[10] M. R. Bellinger et al. Conservation of magnetite biomineralization genes in all domains of life and implications for magnetic sensing. Proceedings of the National Academy of Sciences 119(3): e2108655119 (2022).

[11] L. Yan et al. Bacterial magnetosome and its potential application. Microbiol. Res. 203: 19–28 (2017).

[12] F. T. De Araujo et al. Magnetite and magnetotaxis in algae. Biophysical Journal 50(2): 375 (1986).

[13] G. Fleissner et al. A novel concept of Fe-mineral-based magnetoreception: histological and physicochemical data from the upper beak of homing pigeons. Naturwissenschaften 94(8): 631–42 (2007).

[14] D. Schüler. Formation of magnetosomes in magnetotactic bacteria. J. Mol. Microbiol. Biotechnol. 1(1): 79–86 (1999).

[15] J.J. Jacob and K. Suthindhiran. Magnetotactic bacteria and magnetosomes - Scope and challenges. Mater Sci. Eng. C Mater Biol. Appl. 68: 919–928 (2016).

[16] D. A. Bazylinski and C. T. Lefèvre. Magnetotactic bacteria from extreme environments. Life (Basel) 3(2): 295–307 (2013).

[17] M. Boucher et al. Genetically tailored magnetosomes used as MRI probe for molecular imaging of brain tumor. Biomaterials 121: 167–178 (2017).

[18] C. Lyu et al. Engineering magnetosomes with chimeric membrane and hyaluronidase for efficient delivery of HIF-1 siRNA into deep hypoxic tumors. Chemical Engineering Journal 398: 125453 (2020).

[19] E. Alphandéry et al. Use of bacterial magnetosomes in the magnetic hyperthermia treatment of tumours: a review. Int. J. Hyperthermia 29(8): 801–9 (2013).

[20] N. Nakamura et al. Detection and removal of *Escherichia coli* using fluorescein isothiocyanate conjugated monoclonal antibody immobilized on bacterial magnetic particles. Anal. Chem. 65(15): 2036–9 (1993).

[21] T. Yoshino et al. Single nucleotide polymorphism genotyping of aldehyde dehydrogenase 2 gene using a single bacterial magnetic particle. Biosens. Bioelectron. 18(5-6): 661–6 (2003).

[22] M. Molcan et al. Dispersion of magnetic susceptibility in a suspension of flexible ferromagnetic rods. Journal of Molecular Liquids 305: 112823 (2020).

[23] M. Molcan et al. Energy losses in mechanically modified bacterial magnetosomes. Journal of Physics D: Applied Physics 49(36): 365002 (2016).

[24] M. Timko et al. Hyperthermic effect in suspension of magnetosomes prepared by various methods. IEEE Transactions on Magnetics 49(1): 250–254 (2012).

[25] M. Molcan et al. Structure characterization of the magnetosome solutions for hyperthermia study. Journal of Molecular Liquids 235: 11–16 (2017).

[26] M. Molcan et al. Magnetic hyperthermia study of magnetosome chain systems in tissue-mimicking phantom. Journal of Molecular Liquids 320: 114470 (2020).

[27] S. x. Chen and P. Schopfer. Hydroxyl-radical production in physiological reactions: a novel function of peroxidase. European Journal of Biochemistry 260(3): 726–735 (1999).

[28] E. C. Theil, Ferritin: the protein nanocage and iron biomineral in health and in disease. Inorg. Chem. 52(21): 12223–33 (2013).

[29] B. J. Tabner et al. Protein aggregation, metals and oxidative stress in neurodegenerative diseases. Biochem. Soc. Trans. 33(Pt 5): 1082–6 (2005).

[30] K. K. Wong et al. Biomimetic synthesis and characterization of magnetic proteins (magnetoferritin). Chemistry of Materials 10(1): 279–285 (1998).

[31] Balejčíková, L. et al. The effect of solution pH on the structural stability of magnetoferritin. Colloids Surf B Biointerfaces 156: 375–381 (2017).

[32] L. Melníková et al. Effect of iron oxide loading on magnetoferritin structure in solution as revealed by SAXS and SANS. Colloids Surf B Biointerfaces 123: 82–8 (2014).

[33] C. T. Lefèvre and D. A. Bazylinski. Ecology, diversity, and evolution of magnetotactic bacteria. Microbiol. Mol. Biol. Rev. 77(3): 497–52 (2013).

[34] L. Balejčíková et al. Influence of synthesis temperature on structural and magnetic properties of magnetoferritin. Mendeleev Communications 29(3): 279–281 (2019).

[35] L. Balejčíková et al. The impact of redox, hydrolysis and dehydration chemistry on the structural and magnetic properties of magnetoferritin prepared in variable thermal conditions. Molecules 26(22) (2021).

[36] L. Balejcikova et al. Hyperthermic effect in magnetoferritin aqueous colloidal solution. Journal of Molecular Liquids 283: 39–44 (2019).

[37] M. Koralewski et al. Morphology and magnetic structure of the ferritin core during iron loading and release by magnetooptical and NMR Methods. ACS Appl. Mater Interfaces 10(9): 7777–7787 (2018).

[38] L. Balejcikova et al. Fe(II) formation after interaction of the amyloid β-peptide with iron-storage protein ferritin. J. Biol. Phys. 44(3): 237–243 (2018).

[39] C. Quintana, J. M. Cowley and C. Marhic. Electron nanodiffraction and high-resolution electron microscopy studies of the structure and composition of physiological and pathological ferritin. J. Struct. Biol. 147(2): 166–78 (2004).

[40] L. Bossoni et al. Human-brain ferritin studied by muon spin rotation: a pilot study. J. Phys. Condens. Matter 29(41): 415801 (2017).

[41] O. Strbak et al. Quantification of iron release from native ferritin and magnetoferritin induced by vitamins B(2) and C. Int. J. Mol. Sci. 21(17) (2020).

[42] L. Balejčíková et al. Disruption of amyloid aggregates by artificial ferritins. Journal of Magnetism and Magnetic Materials 473: 215–220 (2019).

[43] L. Balejcikova et al. Dechlorination of 2,4,4'-trichlorobiphenyl by magnetoferritin with different loading factors. Chemosphere 260: 127629 (2020).

[44] I. Safarik et al. Modification of diamagnetic materials using magnetic fluids. Ukrainian Journal of Physics 65(9): 751–760 (2020).

[45] Safarik, I. and Pospiskova, K. Magnetic fluids in biosciences, biotechnology and environmental technology. pp. 343–368. *In*: L. Bulavin and N. Lebovka (eds.). Soft Matter Systems for Biomedical Applications (Springer Proceedings in Physics 266, Springer, Cham (2022).

[46] J. Xue et al. Electrospinning and electrospun nanofibers: methods, materials, and applications. Chemical Reviews 119(8): 5298–5415 (2019).

[47] R. S. Bhattarai et al. Biomedical applications of electrospun nanofibers: drug and nanoparticle delivery. Pharmaceutics 11(1): 5 (2018).

[48] L. Z. Gao, K. L. Fan and X. Y. Yan. Iron oxide nanozyme: a multifunctional enzyme mimetic for biomedical applications. Theranostics 7(13): 3207–3227 (2017).

[49] J. Prochazkova, K. Pospiskova and I. Safarik, Magnetically modified electrospun nanotextile exhibiting peroxidase-like activity. Journal of Magnetism and Magnetic Materials 473: 335–340 (2019).

[50] M. Molcan et al. Magnetically modified electrospun nanofibers for hyperthermia treatment. Ukrainian Journal of Physics 65(8): 655–661 (2020).

[51] V. Khunova et al. Multifunctional electrospun nanofibers based on biopolymer blends and magnetic tubular halloysite for medical applications. Polymers 13(22): 3870 (2021).

[52] I. Safarik et al. Cotton textile/iron oxide nanozyme composites with peroxidase-like activity: Preparation, characterization, and application. ACS Applied Materials & Interfaces 13(20): 23627–23637 (2021).

[53] T. Kunoh, H. Kunoh and J. Takada. Perspectives on the biogenesis of iron oxide complexes produced by Leptothrix, an iron-oxidizing bacterium and promising industrial applications for their functions. Journal of Microbial & Biochemical Technology 7(6): 419–426 (2015).

[54] I. Safarik et al. *Leptothrix* sp. sheaths modified with iron oxide particles: Magnetically responsive, high aspect ratio functional material. Materials Science & Engineering C-Materials for Biological Applications 71: 1342–1346 (2017).

[55] R. Angelova et al. Magnetically modified sheaths of *Leptothrix* sp. as an adsorbent for Amido black 10B removal. Journal of Magnetism and Magnetic Materials 427: 314–319 (2017).

[56] L. Pfeifer. "Neptune Balls" Polysaccharides: Disentangling the Wiry Seagrass Detritus. Polymers, 13(24): 4285 (2021).

[57] R. R. Elmorsi et al. Adsorption of Methylene Blue and Pb2+ by using acid-activated *Posidonia oceanica* waste. Scientific Reports 9: 3356 (2019).

[58] I. Safarik et al. Magnetically modified *Posidonia oceanica* biomass as an adsorbent for organic dyes removal. Mediterranean Marine Science 17(2): 351–358 (2016).

[59] E. Baldikova et al. Magnetically modified bacterial cellulose: A promising carrier for immobilization of affinity ligands, enzymes, and cells. Materials Science & Engineering C-Materials for Biological Applications 71: 214–221 (2017).

[60] Y. Saylan et al. Magnetic bacterial cellulose nanofibers for nucleoside recognition. Cellulose 27(16): 9479–9492 (2020).

[61] M. Cowley and R. E. Rosensweig. The interfacial stability of a ferromagnetic fluid. Journal of Fluid Mechanics 30(4): 671–688 (1967).

[62] R. E. Rosensweig. Magnetic fluids. Scientific American 247(4): 136–145 (1982).

[63] P. Kopcansky et al. Dielectric behaviour of mineral-oil-based magnetic fluids-the cluster model. Journal of Physics D: Applied Physics 22(9): 1410 (1989).

[64] M. Rajnak et al. Electrode polarization and unusual magnetodielectric effect in a transformer oil-based magnetic nanofluid thin layer. J. Chem. Phys. 146(1): 014704 (2017).

[65] M. Rajnak et al. Structure and viscosity of a transformer oil-based ferrofluid under an external electric field. Journal of Magnetism and Magnetic Materials 431: 99–102 (2017).

[66] M. Rajnak et al. Direct observation of electric field induced pattern formation and particle aggregation in ferrofluids. Applied Physics Letters 107(7): 073108 (2015).

[67] M. Rajnak et al. Controllability of ferrofluids' dielectric spectrum by means of external electric forces. Journal of Physics D: Applied Physics 54(3): 035303 (2020).

[68] M. Rajnak et al. Dynamic magnetic response of ferrofluids under a static electric field. Physics of Fluids 33(8): 082006 (2021).

[69] V. Fréedericksz and V. Zolina. Forces causing the orientation of an anisotropic liquid. Transactions of the Faraday Society 29(140): 919–930 (1933).

[70] F. Brochard and P. De Gennes. Theory of magnetic suspensions in liquid crystals. Journal de Physique 31(7): 691–708 (1970).

[71] M. V. Gorkunov and M. A. Osipov. Mean-field theory of a nematic liquid crystal doped with anisotropic nanoparticles. Soft Matter 7(9): 4348–4356 (2011).

[72] V. Gdovinová et al. Influence of the anisometry of magnetic particles on the isotropic–nematic phase transition. Liquid Crystals 41(12): 1773–1777 (2014).

[73] K. Zakutanska et al. Nanoparticle's size, surfactant and concentration effects on stability and isotropic-nematic transition in ferronematic liquid crystal. Journal of Molecular Liquids 289: 111125 (2019).

[74] K. Zakutanská et al. Fréedericksz Transitions in 6CB based ferronematics-effect of magnetic nanoparticles size and concentration. Materials (Basel) 14(11) (2021).

[75] I. Studenyak et al. Influence of cation substitution on dielectric properties and electric conductivity of 6CB liquid crystal with Me7GeS5I (me = Ag, Cu) superionic nanoparticles. Molecular Crystals and Liquid Crystals 702(1): 21–29 (2020).

[76] S. V. Burylov and Y. L. Raikher. Macroscopic properties of ferronematics caused by orientational interactions on the particle surfaces. II. Behavior of real ferronematics in external fields. Molecular Crystals and Liquid Crystals Science and Technology. Section A. Molecular Crystals and Liquid Crystals 258(1): 123–141 (1995).

[77] M. Bradshaw et al. The Frank constants of some nematic liquid crystals. Journal de Physique. 46(9): 1513–1520 (1985).

[78] N. Tomasovicova, P. Kopcansky and N. Éber, Anisotropy research: new developments. Hauppauge (NY): Nova Science 245–276 (2012).

[79] P. Bury et al. Structural changes in liquid crystals doped with functionalized carbon nanotubes. Physica E: Low-dimensional Systems and Nanostructures 103: 53–59 (2018).

[80] P. Bury et al. Effect of superionic nanoparticles on structural changes and electro-optical behavior in nematic liquid crystal. Journal of Molecular Liquids 288: 111042 (2019).

[81] P. Bury et al. Study of structural changes in nematic liquid crystals doped with magnetic nanoparticles using surface acoustic waves. Crystals 10(11): 1023 (2020).

[82] S. Shiokawa and T. Moriizumi. Design of SAW sensor in liquid. Japanese Journal of Applied Physics 27(S1): 142 (1988).

[83] H. Moritake et al. Properties of liquids, liquid crystals, electrolyte solutions and ionic liquids in thin cells and at interfaces studied using shear horizontal wave. In 2011 IEEE International Conference on Dielectric Liquids. IEEE (2011).

[84] B. Tiller et al. Frequency dependence of microflows upon acoustic interactions with fluids. Physics of Fluids 29(12): 122008 (2017).

[85] S. Taketomi. The anisotropy of the sound attenuation in magnetic fluid under an external magnetic field. Journal of the Physical Society of Japan 55(3): 838–844 (1986).

[86] D. Forster et al. Hydrodynamics of liquid crystals. Physical Review Letters 26(17): 1016 (1971).

[87] P. C. Martin, O. Parodi and P. S. Pershan. Unified hydrodynamic theory for crystals, liquid crystals, and normal fluids. Physical Review A 6(6): 2401 (1972).

[88] A. Sengupta, S. Herminghaus and C. Bahr. Liquid crystal microfluidics: surface, elastic and viscous interactions at microscales. Liquid Crystals Reviews 2(2): 73–110 (2014).

[89] S. Pasechnik, V. Chigrinov and D. Shmeliova. Liquid Crystals: Viscous and Elastic Properties, WILEY-VCH Verlag GmbH & Co. (2009).

[90] M. Mięsowicz. Liquid crystals in my memories and now—the role of anisotropic viscosity in liquid crystals research. Molecular Crystals and Liquid Crystals 97(1): 1–11 (1983).

[91] P. Bury et al. Effect of liquid crystalline host on structural changes in magnetosomes based ferronematics. Nanomaterials 11(10): 2643 (2021).

[92] P. Kopčanský et al. The influence of goethite nanorods on structural transitions in liquid crystal 6CHBT. Journal of Magnetism and Magnetic Materials 459: 26–32 (2018).

[93] P. Bury et al. Structural changes in liquid crystals doped with spindle magnetic particles. Physica E: Low-dimensional Systems and Nanostructures 134: 114860 (2021).

[94] Y. Lin et al. A comparative study of nematic liquid crystals doped with harvested and non-harvested ferroelectric nanoparticles: phase transitions and dielectric properties. RSC Advances 7(56): 35438–35444 (2017).

[95] A. Mertelj and D. Lisjak. Ferromagnetic nematic liquid crystals. Liquid Crystals Reviews 5(1): 1–33 (2017).

[96] B. Derjaguin et al. Shear elasticity of low-viscosity liquids at low frequencies. Progress in Surface Science 40(1-4): 462–465 (1992).

[97] H. Lee et al. Improvement of the relaxation time and the order parameter of nematic liquid crystal using a hybrid alignment mixture of carbon nanotube and polyimide. Applied Physics Letters 104: 191601 (2014).

[98] K. Kemp and S. Letcher. Ultrasonic determination of anisotropic shear and bulk viscosities in nematic liquid crystals. Physical Review Letters 27(24): 1634 (1971).

[99] A. Devi, P. Malik and H. Kumar. Thermodynamic and acoustical study of zinc oxide-nematic liquid crystals mixtures. Journal of Molecular Liquids 214: 145–148 (2016).

[1] Institute of Experimental Physics, Slovak Academy of Sciences, Watsonova 47, 040 01 Kosice, Slovakia.
Emails: molcan@saske.sk; paulovic@saske.sk; timko@saske.sk; nhudak@saske.sk; zakutanska@saske.sk; gdovinova@saske.sk

[2] Institute of Hydrology, Dúbravská cesta 9, 841 04 Bratislava, Slovakia.
Email: balejcikova@uh.savba.sk

[3] Biomedical Center Martin, Jessenius Faculty of Medicine in Martin, Comenius University in Bratislava Mala Hora 4, 036 01 Martin, Slovakia.
Email: oliver.strbak@uniba.sk

[4] Department of Nanobiotechnology, Biology Centre, ISBB, Czech Academy of Sciences, Na Sadkach 7, 370 05 Ceske Budejovice, Czech Republic; Regional Centre of Advanced Technologies and Materials, Czech Advanced Technology and Research Institute, Palacky University, Slechtitelu 27, 783 71 Olomouc, Czech Republic.
Eamil: ivosaf@yahoo.com

[5] Department of Nanobiotechnology, Biology Centre, ISBB, Czech Academy of Sciences, Na Sadkach 7, 370 05 Ceske Budejovice, Czech Republic.
Eamils: baldie@email.cz; jitka.prochazkova@bc.cas.cz

[6] Institute of Electronics, Bulgarian Academy of Sciences, Tsarigradsko haussee blvd., 1784 Sofia, Bulgaria.
Eamil: raly_angelova@abv.bg

[7] Regional Centre of Advanced Technologies and Materials, Czech Advanced Technology and Research Institute, Palacky University, Slechtitelu 27, 783 71 Olomouc, Czech Republic.
Email: kristyna.zelenapospiskova@upol.cz

[8] Institute of Experimental Physics, Slovak Academy of Sciences, Watsonova 47, 040 01 Kosice, Slovakia; Technical University of Kosice, Letná 1/9, 042 00 Kosice, Slovakia.
Emails: rajnak@saske.sk; karpets@saske.sk

[9] Department of Physics, FEIT, Žilina University, Univerzitná 8215/1, 010 26 Žilina, Slovakia.
Email: peter.bury@uniza.sk

* Corresponding author: kopcan@saske.sk

Index

For Product Safety Concerns and Information please contact our EU
representative GPSR@taylorandfrancis.com
Taylor & Francis Verlag GmbH, Kaufingerstraße 24, 80331 München, Germany

www.ingramcontent.com/pod-product-compliance
Lightning Source LLC
Chambersburg PA
CBHW060354220326
41598CB00023B/2915

9 7 8 1 0 3 2 2 2 8 0 2 0